Where Next?

REFLECTIONS ON THE HUMAN FUTURE

Where Next?

REFLECTIONS ON THE HUMAN FUTURE

A collection of essays
edited by
DUNCAN POORE

First Published in 2000 by
The Board of Trustees, Royal Botanic Gardens, Kew

ISBN 1-84246-000-5

Printed in Great Britain by Manor Group Limited, Eastbourne, East Sussex

United Nations Educational, Scientific and Cultural Organization
Organisation des Nations Unies pour l'éducation, la science et la culture
Organización de las Naciones Unidas para la Educación, la Ciencia y la Cultura

7, place de Fontenoy
75352 Paris 07 SP   **The Director-General**
Tel : +33 (0)1 45 68 10 00
Fax: +33 (0)1 45 68 55 55

It has often been said that we who live on the earth at the present time must learn to share it better. This means just as much sharing our knowledge and experience, expertise and energy with others as sharing wealth and natural resources. The interdependence of humanity has many dimensions, not least those relating to the environment. We also have to think about our interdependence in terms of sharing with future generations, taking into account an ethics of time. In short, we need a more rigorous, more conscious, more globally connected approach to actions whose impact crosses distance and time.

We face many predicaments in today's world. But we also have the means to turn the risks and challenges into opportunities, progressively transforming our current habits, customs and strategies on the national as well as the international level, so that the first page of a new culture can be written. Science and technology, together with the knowledge they bring, clearly have an important role to play in this process of establishing equitable, sustainable and peaceful development.

In order to meet the complexity and the globality of the problems besetting today's world, we need diagnoses drawing on the natural, social and human sciences, leading to action before irreversible situations are reached. A reduction, on a world-wide scale, of the inequalities and injustice that endanger the whole of humanity, has to be at the heart of this strategy. Nothing is more critical to the success of this task than the production — and sharing — of knowledge. In an information age, knowledge becomes the real wealth, the primary resource.

The establishment of the New Renaissance Group and the publication of this collection of essays, discussing some of the most pressing contemporary issues, constitute an important initiative. The authors bring together considerable expertise and experience in their respective fields. They have produced a series of thought-provoking texts that will surely contribute to the process of holistic reflection and debate so necessary if humanity is to attain and maintain a dynamic equilibrium with the rest of nature in the decades to come.

Federico Mayor

Foreword

At the beginning of this new Millennium it is a great pleasure for the Royal Botanic Gardens, Kew to be able to publish this series of thoughtful and provocative essays on the future, which has been commissioned by The New Renaissance Group. The unifying theme of this book is that the future of humanity and the future of our environment are inextricably linked through a series of interactions that play themselves out at many different levels, and in numerous, and often subtle, ways. As a result, it should be clear to everyone that the decisions we make now, and in the coming years, about many aspects of our environment and our lives, will have a lasting impact on the prospects of future generations.

In the history of the Earth more than 46 million millennia have come and gone, but never before has a single species had the capacity to so dramatically and so quickly modify the future of the planet. The nineteen chapters of this book draw together the views of leading thinkers on some of the unprecedented challenges that will need to be faced as humanity seeks to chart a wise course for its common future. The conclusion that emerges is that the key decisions can only be made and implemented in the context of a New Renaissance in which the biological, cultural and scientific components of human existence are harmonised with the realities of our dependence on Nature. This is the central challenge for our future.

Peter R. Crane FRS

Director

Royal Botanic Gardens, Kew

Editor's Preface

This book is produced under the auspices of the New Renaissance Group, a Group set up as the brain-child of Max Nicholson CB, CVO. The New Renaissance Group aims to stimulate and support the achievement of a new Renaissance by promoting a 21st Century agenda for humanity and nature, and by acting as a catalyst for public debate and action.

In the early days of the New Renaissance Group, Max drew my attention to *The Humanist Frame*, published by George Allen and Unwin in 1961. This book was the creation of Sir Julian Huxley, who later became the first Director General of UNESCO. In it he collected a number of essays by distinguished people of his day who addressed contemporary issues from their many different points of view — all within the framework of evolutionary humanism. And this was, in fact, the philosophy which he carried with him into UNESCO.

I found Huxley's approach interesting and became intrigued by the contrast between the perceptions of then and of now. Some were still valid; others had proved illusory. At the same time, my own experience in land use, environment and education in many countries had made me dissatisfied with the many books in which the authors addressed the problems of humanity from single points of view. It seemed to me that it would be valuable to examine the present human dilemma from many different perspectives and thus be able to study the connections and interdependencies.

Such a comprehensive approach would be daunting for a single author; but I thought that, perhaps, it could be accomplished by following Huxley's example and recruiting a wide variety of talent as collaborators and allies. And this was the course that I followed. Contributors were asked to 'explore the intellectual basis of the present human predicament'. This was to 'include considerations of the potential of humanity to evolve within the constraints of its natural environment, the possibilities of its attaining and maintaining a dynamic equilibrium with the rest of nature and the directions in which progress might lie'. This book is the result. There was no collusion between authors, only between the editor and authors. Any faults, therefore, are my responsibility alone.

Although I have received much encouragement and support from members of the New Renaissance Group, the views expressed do not necessarily represent those of the Group.

In a perfect world such a book should have contained contributions from even more disciplines and the authors should have been drawn from the rich world-wide storehouse of cultures and beliefs. This one, sadly, falls short of such an ideal — which leaves room for another book!

Duncan Poore,
Glenmoriston
November 15, 1999

Acknowledgements

I am most grateful to the Trustees of New Renaissance Group for their support in assembling these essays and, in particular, to Max Nicholson — for stimulus and encouragement over nearly 50 years; for persuading me to become a founder member of the New Renaissance Group; and for drawing my attention to *The Humanist Frame*. The Director-General of UNESCO has authorised the name of UNESCO to be associated with the book and has written a foreword to it; I am most grateful to him for his interest and encouragement. Thanks for their support are due also to the past and the present Directors of the Royal Botanic Gardens, Kew. I am specially grateful to Martin Sands for his criticism and sharp editorial eye and to Bob Boote for keeping me up to date with his press-cutting service. The contributors — all very busy people — deserve praise for their patience in submitting to my badgering and endless queries. I am also grateful to the following for their help and criticism: Michel Batisse, Michael Billig, Andrew Brooks, Frances Cairncross, John Celecia, Jon Cohen, Matthew Dagg, J.P. Duguid, Neil Duxbury, Charles Gunawardena, David Hargreaves, Ivan Hattingh, Sir Martin Holdgate, Ulrich Loening, Charles Oppenheim, Jane Robertson, Paul Rogers, Richard Sandbrook, Teresa Sexton, Robin Sharp, Ian Swingland and Brian Walker. Kenneth E.F. Watt has kindly given his permission to reproduce two of his figures.

Finally, very great thanks to my wife, Judy, for her support, criticism and patience.

Contents

Chapter 1

Setting the scene

Duncan Poore

The Editor

Duncan Poore is a classicist turned scientist. After a period during the war working with Japanese naval codes and ciphers, he took a degree in natural sciences and a doctorate in plant ecology at Cambridge. Since then he has worked at the interface between environment and economic land use: with Hunting Technical Services in Cyprus and the Middle East; as Professor of Botany in the University of Malaya; as Director of the Nature Conservancy (Great Britain); with IUCN (the World Conservation Union) as Scientific Director and Acting Director-General; and as Professor of Forest Science in Oxford. For the past 15 years he has been closely associated with the International Institute of Environment and Development and with international work in environmental and forest policies, mainly in the tropics.

Chapter 1

Setting the scene

Duncan Poore

Introduction

The Renaissance and the ensuing Enlightenment released the intellectual capacities of Western Europe by removing the shackles of the Middle Ages. This had far reaching and unforeseen effects wherever western influences could reach. By the end of the 20th century, everywhere on earth had been affected — and part of outer space. Nothing was ever the same again. It was a defining moment in world history.

We may be passing through another such moment today; but the causes and symptoms are different. It is no longer a matter of releasing the energies of the past; intellectual ferment is now nearly universal. The world is like a gigantic seething cauldron; bubbles rising to the surface, bursting or subsiding into the matrix; new added ingredients swirling through the liquid and constantly generating new substances, mixtures, textures and flavours. A process apparently without end and the sequence of resulting products unknowable. Are there too many cooks? Do they know what they are doing? Is the process under control or indeed controllable? Will the resulting brew be palatable?

Another important turning point was the end of the Second World War. This was, on the whole, a time for optimism. The United Nations and the UN agencies that are still with us today were set up in a climate of hope. The constitution of the United Nations Educational, Scientific and Cultural Organisation (Unesco), for example, is a ringing affirmation of confidence. 'The wide diffusion of culture, and the education of humanity for justice and liberty and peace, are indispensable to the dignity of man[1] and constitute a sacred duty which all the nations must fulfil in a spirit of mutual assistance and concern'. Peace 'based exclusively upon the political and economic arrangements of governments' would be inadequate, since it could not 'secure the unanimous, lasting and sincere support of the peoples of the world'. 'The peace must therefore be founded, if it is not to fail, upon the intellectual and moral solidarity of mankind.' It then goes on to define the purposes of Unesco as those 'of advancing, through the educational, scientific and cultural relations of the world, the objectives of international peace and of the common welfare of mankind, for which the United Nations Organisation was established and which its Charter proclaims.'

In 1946, Julian Huxley wrote an 'Essay' for the Preparatory Commission of Unesco, of which he was then Executive Secretary, *Unesco: Its Purpose and Philosophy*.[2] In it, after discussing the merits and disadvantages of adopting any particular religious or philosophical approach, he concluded that the 'general philosophy of Unesco should . . . be a scientific, world humanism, global in extent and evolutionary in background'. Huxley later became the first Director General of Unesco.

It is because of the seminal role that Huxley played in the post-war years, that

I have chosen to use his views, and those of a number of his contemporaries, as a starting point for examining changes in issues and attitudes in the last half century. To some extent this choice is arbitrary; but it is clear that these views do represent a significant stream of intellectual opinion current at the time. They were presented in a collection of essays, collated and edited by Huxley and published in 1961 under the title *The Humanist Frame*, a title which indicates the context in which all the essays were set.[3]

Huxley was a firm believer in the primacy of *Homo sapiens* as the culmination of biological evolution, the last of a long series of evolutionary experiments in which one successful type had followed another. He was convinced that, in mankind, evolution had taken a fundamentally new course. Previous advances had been largely due to the appearance and expansion of new physical adaptations, but the evolution of man had entered into a new phase which, he contended, arose from the ability of man to think, and to communicate and transmit his experiences from individual to individual, from social group to social group, and from generation to generation.[4] Among such attributes were speech, conceptual thought, consciousness of self and comprehension, all making possible increasingly complex forms of social organisation. These evolutionary developments had given man unprecedented power to influence his own future and the future of all other living things; he had become, in fact, the main determinant of the whole course of future evolution, with corresponding opportunities and responsibilities. But this evolution might fail unless he faced it consciously and used all his mental resources — 'knowledge and reason, imagination and sensitivity, capacities for wonder and love, for comprehension and compassion, for spiritual aspiration and moral effort'.

Accordingly, there was a responsibility to define desirable goals for human society. He selected: the fulfilment of the individual personality and of individuals in society; the enhancement of human significance and dignity; the greater and better realisation of possibilities, individual, social and planetary, both in the short and the long term; fuller achievement and the continued enrichment or improvement in the quality of life accompanied by the removal or diminution of suffering, misery, frustration, cruelty, injustice and inadequacy.

He believed that responsibility for the future lay squarely with man, based on his enlarged and expanding scientific understanding. There was no longer any room for dependence on the supernatural.[5] But this heavy responsibility for directing the course of future evolution was not to be faced by the individual alone; he could rely on the shared experience of past and present thought (the noösphere of Teihard de Chardin).[6]

Huxley recognised that his vision could not be realised unless major challenges were faced and that there was a real risk of failure. He compared the present moment of emergence from biological to 'psycho-social' evolution, from the biosphere to the noösphere, to the moment of evolution when the amphibians first emerged from water to land — the need to face unprecedented challenges. In the context of 1961, he identified the following challenges: the threat of super-scientific war; the threat of over-population; the rise and appeal of communist ideology; the failure to bring China into the United Nations; the over-exploitation of natural resources; the erosion of the world's cultural variety; our general preoccupation with means rather than ends, with technology and quantity rather than creativity and quality; and the revolution of expectations 'caused by the widening gap between the haves and have-nots, between the rich and the poor nations'.

Huxley then expressed his conviction that contemporary problems could only be successfully addressed with a new organisation of thought and unitary pattern of ideas — a single inter-thinking group. He conceived Humanism as fulfilling this role, as the new single religious system of the future. It should be global but concerned with the individual; it should be dominated by the concept of quality;[7] and both consumption and the economic system as a whole should be directed towards fulfilment rather than mere utility.

The last half century: then and now

It is interesting to look back upon this approach with the advantage of hindsight. Huxley's view was essentially optimistic. Reason, supported by the fuller and better organised awareness of phenomena which would be provided by science, freed from superstition and enriched by the emotional and aesthetic satisfactions of the arts, should lead to progress in the directions of his ideal, towards the 'perfectibility' of man. The challenges of the time were identifiable and did not appear to be insuperable, if approached in the right way. He clearly thought that a unified philosophy — his 'single inter-thinking group'— was achievable and need not necessarily lead to an unacceptable loss of human diversity.

How far are these views confirmed by our experiences in the last 40 years? First, I do not believe that many would quarrel with his definition of the ideal; but we might wish to elaborate and expand. His goals are perhaps too overtly anthropocentric for the taste of some, but the phrase 'the greater and better realisation of possibilities, individual, social and planetary, both in the short and the long term' covers comprehensively, among other things, the contemporary concepts of environmental responsibility and of inter-generational equity.

It is in the later stages of the analysis that we may have more difficulty. We would now, I believe, be much more cautious in our optimism, if indeed we were not downright pessimistic. We are still aeons away from a single unified system of thought, perhaps even further than we were in 1961.[8] Indeed, I doubt whether we would now consider that a 'single inter-thinking group', in Huxley's sense, is possible in the foreseeable future, or even desirable. The most significant movement in this direction has, perhaps, been the widespread and alarming acceptance of the norms of 'western consumerism'. At the same time, the gradual acceptance of English as the world language may provide a basis for future convergence.

There are, I think, a number of reasons why the spread of scientific rationalism has been much slower than anticipated. First, a high proportion of the world's population is too concerned with the problems of keeping alive and many have, in any case, never been exposed to western scientific thinking. Secondly, large numbers of people, even in the west, are deluged in a flood of 'information' without being equipped with the educational basis in critical thought and in the findings of science to filter fact from fiction and sense from nonsense. Thirdly, there is widespread disillusion with the delivery of technology; in particular it seems to be impossible to derive its benefits while limiting the associated damage. And, lastly, there is the immense power behind the propagation of consumerism; the wares it purveys are enormously attractive to most people, especially when they are only the first or at most the second generation to benefit from them.

If we were now to enumerate the challenges facing us, we would, I believe, pro-

5

duce a rather different and longer list. Some have disappeared; others have been partially solved. But many have become more intractable and a whole range of new unforeseen problems have arisen.

The threat of a nuclear war between the USSR and the USA disappeared with the collapse of the Soviet Union, but there still remain great dangers posed by the proliferation of weapons of mass destruction among nations that may be less responsible than the two great powers proved to be, and that are able to exercise less control on those who would actually throw the switch.

Communist ideology has collapsed in the USSR and its hold has weakened in China but, far from this leading to a great expansion of scientific rationalism, it seems to have been replaced by the pursuit of a blatant capitalism and, as reaction, to a rise in the appeal of Islam, and to the birth of new forms of fundamentalism in many of the established religions. The nearest, perhaps, to the Humanism foreseen by Huxley are certain forms of environmentalism.

The majority government in China does now occupy that country's seat at the United Nations, but that body is going through a crisis of confidence. The structure which was suited to the conditions immediately after the Second World War is creaking under the strain of attempting to react to those of the present-day world.

Of the other problems listed by Huxley, that of population and resources is still with us in an aggravated form; cultural variety is fast disappearing in spite of valiant efforts to preserve it; we are still preoccupied with 'means rather than ends, with technology and quantity rather than creativity and quality'; and the gap continues to widen between rich and poor nations and between the rich and the poor within nations.

Moreover, we are also beset by many new issues which were hardly apparent 40 years ago, and by some old ones in a new guise. Among these are the great and unpredictable effects of globalisation — in influences on the environment; in ease of travel; in currency dealings, financial transfers, business arrangements and trade; and in the near-instantaneous transmission of information and opinion. There are, too, mass movements of people, many of them refugees from political persecution or environmental disaster; the effects for good or ill of genetic engineering and the many ethical problems raised by the understanding of the human genome; changes in the pattern of health and disease — ageing populations, AIDS, and new vulnerability to old diseases such as malaria and tuberculosis; the many local tensions and wars associated with crises of identity, race, nationality and religion; international terrorism and trafficking in drugs; organised crime supported by all the paraphernalia of advanced technology; the prevalence of violence and the growth of crimes against the person. The numbers suffering from poverty and destitution are much higher than before and the disparity with the affluent has increased; moreover, because of near-universal access to forms of communication especially television, these differences can no longer be hidden and are often politically exploited with explosive effect. The coming of age of the computer, automation and the falling of global barriers to trade and industrial investment are leading to endemic uncertainties in employment and any prospect that there might once have been of universal full employment is fast fading from view.

At the human scale too, there are many changes. In western society, there is increased tolerance of differences of race and gender, and of sexual proclivities; but this is accompanied by a creeping intolerance in the guise of 'political correctness' and 'big brotherhood'. Insistence on rights has become a way of life with little sign of a corresponding growth in a sense of responsibilities. Any disaster or even minor mistake is always somebody else's fault; and this attitude is accompanied by a crippling growth in litigation. The nuclear or extended family is threatened without the prospect of any alternative that would be equally satisfactory in social or evolutionary terms. An increasingly competitive society has become one fraught with new uncertainties which bring stress and psychological disorders in their train. The media, too, do not help. They have developed a pervasive power that is more concerned with titillation than truth; they appear highly politicised and their anti-authoritarian stance often undermines society. Furthermore, much of the 'education' on offer no longer provides the equipment for critical judgement, social usefulness or individual fulfilment.

At the same time, more and more is expected of governments and of international organisations, but individuals are not prepared to pay the full price of those services that are the proper function of governments; nor are nations prepared to pay for those things that can only be provided internationally. There is, too, an almost universal disillusion with politicians; a tendency to populism in many countries; an uncertainty about the validity of the nation state, manifested in the rival movements towards devolution or separatism on the one hand and, on the other, in the direction of larger economic or political groupings; and a widespread feeling that the United Nations should be reformed in line with the great multiplication of its members and the fact that it now exists in a world totally transformed from that of the late 1940s.

In fact, change is so rapid and comprehensive that humanity is suffering from a crisis of confidence and identity. Possibilities for the future are immense and unprecedented; but so are the pitfalls.

These issues, and many others, are discussed in the following chapters. The authors were asked, independently, to 'explore the intellectual basis of the present human predicament; this [to] include consideration of the potential of humanity to evolve within the constraints of its natural environment, the possibilities of its attaining and maintaining a dynamic equilibrium with the rest of nature and the directions in which progress might lie'. They range widely — through:
- the relation between population and natural resources;
- the potential imbalance between supply and demand;
- the increasing urbanisation of the world and its implications;
- the various concepts embraced by 'sustainability'; and the ambiguity of 'sustainable development';
- 'sustainable livelihoods' as a universal goal;
- the impending shortage of energy;
- the blindness of economic theory;
- 'sustainable development' as an unattainable ideal;
- the problems and opportunities offered by the revolution in communications;
- the destructive elements in human behaviour;
- the need for reformed institutions of international governance;
- the new role of law as an agent of social policy;
- the absence of a background of shared moral values;
- the importance of beliefs as the mainspring of action;

- the need to realise that the findings of science do provide a genuine, though perhaps only partial, basis for action;
- what might or should be the purpose and content of education; and
- the dangers inherent in the separation of 'matter' from 'mind' and in the segmentation of knowledge;

The conspectus which follows is meant as a guide to the contents of the later chapters and to the connections between them.

Max Nicholson's theme is all-embracing (Chapter 2). He believes that *Homo sapiens* is a threatened species; but he is, in spite of all, an optimist. He makes a clarion call for a New Renaissance in which people reach a full understanding of the natural world and the place of the human species in it. He believes that new generations, inspired by this understanding and learning from the mistakes of their predecessors, should be able to attain harmony between man and nature, and among people and peoples. Only this can provide a sound basis for the future evolution of mankind.

We next move on to the relationship between population and resources. Charles Pereira (3) brings his immense experience in the world of agriculture to consider whether it is possible to balance future world food supply with future world population. Is present population growth sustainable? What are the possibilities of further increasing crop productivity? Can the advances of the 'Green Revolution' be sustained? It is crucial to make the best use of the world's limited resources of soil and water. There are links with other chapters: especially those on population (4), biological diversity (5), and sustainability (6, 9 and 11).

Virginia Abernethy (4) first discusses population projections, giving fascinating, and perhaps controversial, insights into the influences that determine the changing rates of population growth — particularly trends in fertility and mortality. She goes on to examine the influence of population on environment and *vice versa* (especially in terms of energy, food, employment and conflict). She concludes that world population will certainly exceed and perhaps has already exceeded the 'carrying capacity' of the Earth — defined as one that provides a reasonable quality of life for all.

One very important aspect of environmental sustainability is the maintenance of biological capital, represented by the variety inherent in living organisms. Ghillean Prance (5) stresses the crucial significance of biodiversity, both as the basis of life and as an insurance for the future, linking with the arguments of Charles Pereira in relation to crop plants. He expresses deep concern about the present man-induced wave of extinction, but believes that this can be stemmed by a combination of preservation, sustainable use and the equitable sharing of benefits.

We move next into the borderland between ecology and economics. Robert Goodland (6) analyses and distinguishes between the various aspects of sustainability — environmental, economic, social and human; of these, he picks out the attainment of 'environmental sustainability' (the 'maintenance of natural capital') as essential and urgent. He argues that man-made capital is rarely a satisfactory substitute for natural capital; that very high human numbers, rapid economic growth and high levels of consumption are all environmentally unsustainable; and that 'sustainable development' is a dubious and opaque concept, the popularity of which is linked to its imprecision. He maintains that many, perhaps most, economists either do not understand sustainability or brush it aside as of minor importance.

David Burns' chapter (7) is a satirical essay, modelled on 'Erewhon' by Samuel Butler, about an imaginary country, 'Stekram', in which all citizens pay the true cost of all that they consume, compete freely, receive in proportion to the value added by their activities, and pay no taxes. This country bears a quaint resemblance to certain economic theories and yet is made up of real people who think and feel, organise their finances, enjoy themselves and worry from time to time about the future of the world. This essay is a chilling reflection of the world as it might become if the theory of market economics becomes, unalloyed, the universal practice.

It has been said that the information revolution is comparable in scale to the industrial revolution; it may, perhaps, have even greater repercussions. Justin Arundale (8) argues that information, as well as becoming a valuable traded commodity, is the lifeblood of civilisation and democratic activity. Present and probable future advances in information technology have important implications for economic and intellectual development, as well as for issues such as equality, world citizenship, and cultural autonomy and integrity.

The next two chapters turn to the ways in which societies might progress to a better way of life which lies within environmental limits — the only way in which progress can be durable. David Goode (9) considers cities. He argues that it is absurd to consider a sustainable future for the biosphere without paying great attention to the functioning of cities; for, by the middle of the next century, seventy percent of the total world population will be city-dwellers. He discusses the 'ecological footprints' of cities — the vast and pervasive influence they have on the world outside; the functional elements of city metabolism; the quality of life for urban dwellers; and the tentative moves that are being made towards sustainability. Cities depend upon the countryside; but without sustainable cities the countryside is doomed.

Ashok Khosla (10) describes the immense problem of finding employment for the new millions that enter the job market each year — in the modern world self-respect depends upon the autonomy and security that comes from a regular wage. He argues that the only path for planetary survival (and to alleviate the lot of the very poor) is through the pursuit of 'sustainable livelihoods' — a concept which he believes is applicable to societies in all stages of development. He outlines and gives examples of the measures necessary to make it come about — sustainable elements of technology, enterprises, economies and institutions of governance.

David Fleming (11) takes up the theme of sustainable development, though in a rather different and less optimistic vein. He considers that the sustainable development principle has made a useful contribution in the past, but believes that the proposition that economic growth can be maintained without decisive environmental damage is no longer relevant. The critical problem that faces the market economy is that growth — which is needed to prevent unemployment rising out of control — is now hard to achieve, and will shortly be frankly impossible. He emphasises this with reference to the future behaviour of energy prices. He maintains that the focus of attention in the future will therefore soon shift from the environment to the task of sustaining social stability under conditions of extreme and unprecedented difficulty.

James O'Connell (12) discusses peace and conflict in the context of the social implications of technology — the most dynamic socio-economic factor in the contemporary world, and a force both for good and for ill. In this context, he surveys actual and potential conflicts between countries and groups, the pattern of which has changed since the break up of the Soviet Union from bi-polar to multi-polar. Weapons have proliferated; the number of intra-state conflicts has greatly increased; and countries can no longer rely on defence as adequate protection. Situations constantly arise that are apparently intractable. O'Connell is confident, however, that, by drawing on the benign features of our new technological capacities, it may be possible to underpin new forms of global and regional unity.

Shridath Ramphal (13) deals with the crucial issue of international governance in a unified if not united world. He summarises present world problems — security, refugees and displaced persons, conflict and arms, drugs, poverty and the side-effects of globalisation — and argues that, in order to ensure the security of people and the reduction of inequity world-wide, the international community should enhance its capacity to deal with emerging world problems. He outlines how institutional arrangements for global governance should be changed by reform of the Security Council, a UN rapid response force and an Economic Security Council. And he calls for a new civil ethic of shared values.

Nigel Simmonds (14) echoes the last point made by Ramphal. He reviews the position of law in the present world and argues that the role of law is changing: rather than providing a stable framework demarcating the bounds of otherwise conflicting interests, law has come to be a resource that is deployed manipulatively in the struggle between those interests. This is evidenced both in the increasing insistence on 'rights' and in the many issues now being addressed by international law. He maintains that, in the absence of a common ethical understanding, the foundations of law lie on shifting sands.

Martin Palmer (15) describes the attitudes of the various faiths to nature and the environment and explores how what we believe fundamentally shapes how we understand, interpret and, thus, use or abuse nature. He draws upon religious and ideological insights, beliefs and practices, to argue that the future will depend upon what we believe.

Rom Harré (16) gives the view of a philosopher of science. He makes a critical examination of the claims of post-modernism, which he considers 'a newly hatched threat to the celebration of human reason', and develops a case against moral relativism. He goes on to consider the ways that scientists, by adopting outmoded and naive conceptions of science, have made themselves vulnerable to anti-science criticism. The post-modernist reaction — that science is just another story — is just as damaging. Harré argues that there is a way by which the achievements of the natural sciences can be recognised while the essentially discursive character of human patterns of thought and action is preserved.

Hermann Bondi (17) advances some basic principles of education, making a clear distinction between education and training, and emphasising the importance of motivation. He continues by developing a proposal for reforming science education, considering that the prevailing anti-science attitudes have their roots in the way that science is taught. The spirit of science would be conveyed in a much more exciting and accessible way by following the evolution of certain carefully selected scientific ideas, rather than by the present prevailing approach — to teach it as a corpus of knowledge. In a democratic society, too, it is important that a citizen should understand the significance — or lack of significance — of the statistics with which he or she is provided.

Philip Stewart's essay (18) comes full circle to Nicholson's arguments (2) and to the 'psycho-social' evolution of Huxley. Human action has become the dominant factor in the biosphere, and ideas play a determining role in this action — human evolution is mainly cultural. The western tradition of segregating knowledge into compartments, separating mind from matter, has proved a barrier to understanding — and deeply damaging in practice. A proper appreciation of the ecology of ideas is the key to harmony between man and the rest of the biosphere. He advocates a 'universal ecology' as the corner-stone of knowledge, education and research. This should encompass the processes linking man with the natural world and man with man; and equally should embrace all aspects of being human — the physical, the mental, the cultural and the spiritual. Such might, indeed, open the door to a new Renaissance.

In the final Chapter, I shall attempt to weave these threads into a fabric. This may not prove to be durable, but I hope that it is not totally threadbare!

Chapter 2

The New Renaissance

E. M. Nicholson

The Author

Edward Max Nicholson CB CVO was born in 1904. He has had a long and active career, including being a founder of the British Trust for Ornithology in1932 and founder editor of the Oxford University Press *Birds of the Western Palearctic* in 9 volumes; General Secretary 1933-39 of Political and Economic Planning (PEP); an international shipping administrator in World War II: post-war head of Herbert Morrison's office during his period as Deputy Prime-Minister to Attlee; Director-General of the official Nature Conservancy for Great Britain 1952-66; Chairman and President of various national bodies; recipient of honorary degrees and author of several books, as well as a contributor to Julian Huxley's *The Humanist Frame.*

Chapter 2

The New Renaissance

E. M. Nicholson

Introduction

The plight of various threatened plant and animal species has been much in the news recently. Global and national strategies for the conservation of biological diversity have been agreed, and these include ecological assessments of the conditions that must be met if threatened species are to survive. But few people appreciate that this same requirement applies now to our own species, *Homo sapiens*.

Our situation is of course quite different from that of the Giant Panda or the Blue Whale. We are threatened by our own success — or rather because the attributes that have swept the human species to global ecological pandominance are now sweeping us into a growing imbalance with nature which threatens life on earth.

This Chapter explores how new idea systems can be applied, using new resources of technology, to help the world community to understand the human situation and the human potential. It suggests changes in approach that will equip the human species with means for the successful re-direction of its ongoing evolution towards a stable, harmonious relationship with the natural environment. That must be the principal objective of a New Renaissance.

The Human Situation

If we are to shape the human future, it must be through understanding of the human past, and of the factors that have impelled our species to pandominance.

Hominids appear to have evolved in the East African tropical forests and savannas. Their upright posture — freeing dexterous forelimbs for uses other than walking — their relatively large brain, and their excellent visual ability, paved the way for tool-making skills that greatly expanded their capacity to secure food and defend themselves against hostile rivals. Social co-operation within the group — a common feature of advanced primates — and the evolution of communication by more and more sophisticated signs and language, further aided success and geographical expansion. Still in the stage of hunters and gatherers, early humans spread into desert fringes, saline wastes, deltas, and most other land areas except steep bare mountains, ice-caps and frozen tundra. They colonised a wide range of climatic and altitudinal zones and of soil types.

Like all advanced animals, our forerunners would have been closely aware of their natural environment, its resources, its opportunities and perils, and its seasonal and other changes. Environmental challenges were the spur to human social evolution, as benefits accrued to those individuals or groups who acquired particular knowledge of their surroundings, special hunting skills, or informed teamwork. The natural environment was in fact man's first book and model for learning; and we can still see this process repeated by infants today.

The use of fire, the development of agriculture, the domestication of food-yielding and draught animals such as cattle, sheep, horses, Indian elephants, llamas and reindeer, the discovery of pottery-making and the smelting of metals, and the building of increasingly effective shelters — all these brought much enhanced food security, freed the human species from many of the vagaries of its environment and paved the way for ecological dominance. Leisure from the primary activities of securing food, water and fuel could be devoted to an expanding range of skills, including food blending, cooking, preservation and storage, and finding and shaping materials for clothing and buildings. Even water barriers became insubstantial, for simple boats came into use thousands of years ago and more and more remote islands came within the range of our species. The early extraction of metals, harvesting of crop plants and timber, and harnessing of wind and water power permitted a transition from dependence on nature to a technology-based society with organised production, distribution, barter, commerce, and a spreading money and market economy. All this happened by successive piecemeal applications of human ingenuity and initiative, and, not surprisingly, different cultural variants appeared in different parts of the world, even though many discoveries were interchanged by trade. The result was a busy, enterprising human world, basically equipped, primitively managed, and socially co-operative, but amateurish, disorganised and making only haphazard use of mainly fragmentary and often fallacious knowledge.

The handing down of knowledge by transmission from generation to generation is not unique to humans. But it was enormously enhanced by the development of graphics and handwriting, which enabled knowledge to be recorded and analysed over space and time. One body of knowledge thus recorded over the years related to the environment. Eventually the urge to know where, how and why plants and animals occurred, led to biology and modern ecology. So it came about that research — describing habitats, charting biodiversity, analysing systems and monitoring change — emerged as a scientific framework for a deeper understanding of all life. Unfortunately, as human culture developed, especially in the industrialised world, it rejected learning from nature, and switched support for science to human-centred interests and activities, heavily exploiting natural resources. It took the eventually disastrous effects on human affairs of the resulting abuses and misuses of nature to re-divert effort towards ecology and conservation, leading in turn to the modern growth of an environmental movement strong enough to insist on wiser attitudes. Even now, however, those wiser attitudes have a long way to go before they break through as determinants of social policy.

Human minds and curiosity have opened out new emotive, artistic, intellectual, and social capabilities. But, despite their enormous technological advances, humans are still driven by their biological attributes. They retain a diurnal rhythm in which eating, drinking and sleep must figure. They retain powerful sexual appetites, and an urge for dominance within and between social groups. At an early stage in human social evolution, tools useful for hunting large animals doubtless also proved valuable as weapons to secure inter-group dominance. Increasing skill in the use of metals was applied to make combat more lethal. Unfortunately, unlike many animal species, humans have not managed to make non-lethal ritualised combat the norm. Today, more than one human society has a nuclear weapons arsenal with the power to render large tracts of the planet uninhabitable. Despite the enormous powers humans now wield, many have still to learn how to master or eliminate drives towards cruelty and violence, excessive sexual appetites, and an addiction to exhibitionism, absolute dogmas and behavioural obsessions. The demographic momentum, taking humanity through the present global population of 6 billion towards an ultimate levelling off, forecast as at least half as much again in a century's time, and the increasing inter-group conflicts for environmental resources, are threatening attributes now exhibited by the ecological overshoot of the human species.

Mind and the universe

The remarkable thing about humans is that these relatively short-lived organisms, whose physical dimensions are minute on the cosmic scale, have been able to deduce mathematical, physico-chemical and biological principles that provide a relatively consistent intellectual model of the workings of the universe and of organic, including human, evolution. This intellectual capacity has broken away from emotive and habitual bases to offer the potential to change human individual and social conduct, so as to guide human evolution towards a new, sustainable, harmony with the other components of the global ecosystem.

However, in seeking to understand profound reality, human beings are constrained by their biological limitations. Although the conceptual goal may be one of total objectivity, we have to accept that our thought processes are determined by the structure and capacity of our brains, and by the ways in which we can observe the universe. Scientific method has evolved as a practical discipline that imposes a consistent framework and standard of proof. It forces us to reject the sloppy practice of taking the assertions and observations of everyday life at face value without bothering to scrutinise their validity. It stresses the need for experiments to be repeatable, and for the process of inquiry to be detached to the greatest possible extent from the prejudices of the inquirer. In seeking to guide human societies forward, it is this objectivity for which we must strive.

Science has already provided remarkable insights into the origins and evolution of the physical universe. Current evidence suggests that this commenced as a 'singularity' of immense energy, at one point. Cosmologists have provided a plausible, self-consistent, description of what went on during the first few seconds, as energy was transformed into matter and the whole system commenced expanding at immense velocity. The background radiation from that original 'big bang' 18 billion years ago has been detected. We are able to place our small planet in the cosmic context, believing that it orbits a medium sized second-generation star in the outer arm of what appears to be a fairly commonplace spiral galaxy. We also believe that we understand the processes of biological evolution on Earth over the past 4 billion years, and how the genetic template of DNA has evolved and diversified, to be expressed in the 10 million or more species we estimate to share the planet with us (of which no more than 1.7 million have been described by science).

The human dilemma

Despite all this knowledge and inference, our ability to safeguard our own biological future appears highly limited. We are only beginning to consider how to detect the asteroids whose orbits could intersect with ours, and to guard against an impact which could disrupt global ecology and very possibly cause the extinction of our species. Although some people take refuge in the comforting thought that our earth may be endowed with benign self-regulating Gaian powers, at least over climate and ecology, the fact is that we are witnessing massive, and potentially dangerous, human interference with our basic life support systems.

This interference — already degrading habitats, increasing human competitive conflict over scarce environmental resources, and leading to changes which are accelerating extinctions to a rate a hundred or even a thousand times that occurring naturally — is a direct result of the forces that have driven human evolution to unparalleled success. People have altered habitats so as to produce more of the plants and animals we use. As productive methods have been adopted on an ever-widening scale, so human populations have increased. As human numbers rise, so the demand for yet more environmental resources rises also. The pattern is one familiar in nature, and in any other species would lead to population crashes in the areas where 'carrying capacity' is most severely overshot — or to migration from those areas into adjacent territories where resources remain available — with inten-sified competition between the immigrants and those already established there. That writing is on the wall for humanity, too: already conflicts over land and water are emerging, especially in parts of the tropics, and environmental refugees are becoming more numerous.

But humanity is unique in being able to use its scientific skills to avert what, for another species, would be inevitable catastrophe. The way lies open to apply a knowledge of ecology (and especially human ecology), to the conservation of the natural resources on which the human future depends. We are just coming to the exciting point of the drama: will science succeed at the eleventh hour? Whatever our personal predictions, we are bound, while options stay open, to pursue the con-structive line with all the vigour at our command. Our species has risen by energy, confidence and imagination; and these attributes should not be allowed to fail us now. We enjoy great new gains in knowledge of the earth and how to manage it, as well as in knowledge of the potential of the human species. Practical initiatives are needed, to build on that knowledge and to apply it.

The need for action

What on earth can we do next? We now have enough knowledge of the Earth's bio-mes and human habitats to make an inventory of these according to their global distribution and scale, their economic importance and accessibility, and the quan-tity of reserves for the future, region by region and country by country. This will (at least in theory) provide a reliable appraisal of the resource-base available to the human species, and the limits of its accessibility and sustainability. It will also define where human communities and the ecosystems that support them are at greatest risk. It will, finally, allow sensible action plans to be developed for each country or other appropriate geographical unit.

But, for this to happen, the science and the scientists have to be available. While the environmental movement has made remarkable world-wide progress in the past half-century, it still falls short of meeting some key needs. There have been some ambitious international initiatives such as the International Biological Programme of the International Council of Scientific Unions (ICSU), the Man and the Biosphere Programme of UNESCO and isolated national efforts such as the early research of the Great Britain Nature Conservancy; but there have been few really comprehensive research programmes yielding a broad range of ecological and envi-ronmental management knowledge.[1] Although many national parks (the first among them, Yellowstone, founded in 1872) have been claimed to be invaluable scientific laboratories, few have been used for this purpose.[2]

At the level of academic research and training centres, and of organised professions, the unmet needs are no less great. Numerous small and narrowly focused professional bodies offer, within their tight limits, various kinds of environmental services, but none commands the breadth of scope and depth of learning required to meet the broad and demanding environmental needs of modern governments and of the international world, including the global economy. International and national voluntary bodies (non-governmental organisations), which now have a vast membership, do something to fill the breach, but they cannot do nearly all that a great profession should be doing with these world-scale problems. The trouble is aggravated by the continuing failure of the educational establishment to overcome its environmental illiteracy, and by corresponding deficiencies over much of the field of culture, except where it concerns amenity.

Even more serious gaps are evident when it comes to applying ecological knowledge to the use of natural resources and the care of land, either developed or thought suitable for development. Thanks to advances in nature conservation and land use planning, fair progress has been made, in some countries at least, with public interventions for protection of wildlife and the countryside, sometimes even where strong conflicting interests have been involved. But, in many parts of the world, powerful financial and commercial interests concerned to maximize short-term returns have escaped effective administrative control or governmental regulation in the long-term public interest. Intensive agriculture (with its dependence on chemicals and pesticides), forestry, hydro-power schemes and the expansion of cities over the fertile lands around them all illustrate the consequences. The scientific and academic establishment has not been able to prevent such abuses. The underlying principles of care for the earth and the biosphere have been neither appreciated nor understood. Even many of the most obvious practical problems of land use and avoidance of misuse have had to be belatedly urged by outsiders. The deep educational and cultural gulf between science and other subjects has remained unbridged despite all warnings. It has been aggravated by the effect of two world wars, which have also given chemistry and physics a temporary dominance over the life and earth sciences, a dominance only gradually being readjusted.

The earlier twin interaction between people and environment has now given way to a graver triple interaction between each of those and the vast newly-created 'technosphere', the world of modern technology. The latter has produced many clever and useful devices, but it has been impelled by supply-side ingenuity, and the sometimes unscrupulous use of advertising to expand demand and so create markets. Increasingly expensive and sophisticated artefacts have been produced, and the technological culture promotes continuing replacement by the 'latest models' regardless of the true needs of the users. The numerous, often costly, environmental side effects of such products have been left out of the equation, 'externalising' the costs to be borne by society at large or by future generations. The obviously false premise — that all the natural resources called for by booming human demands would be forever there for the taking — seems to have gained acceptance by economists and the market economy. As corporate bodies have grown wealthier and more powerful, often operating internationally on a scale greater than the total economies of many nations, so they have passed beyond current capabilities of public control except belatedly, half-heartedly and often ineffectually. Economics as a discipline has largely exhibited environmental illiteracy, and needs transformation if it is to contribute to the new sustainable world economy, regulated by the honest and sound global house-keeping of real resources that is now urgently necessary.

Market economics and free-market technology have acquired global dominance at a time of rapid political change and a collapse of misguided initiatives. Marxism has ceased as a significant political force, coinciding with the dissolution of the USSR and the ending of the Cold War, with accompanying dilemmas over the future of NATO and other alliances. There is also disillusionment with political parties and machines, and with the dominance of the sovereign nation-State — the essential building block of the United Nations. In many parts of the world, the rapid growth of new political systems has both facilitated and been undermined by corruption and mismanagement. Ethnic conflicts and antagonism between religious factions have become rife. There has been increasing confrontation between the world's developed 'North' and undeveloped 'South'. Consumerism and conflicting approaches to economic affairs add to the long agenda of problems, which will leave burdensome legacies to be cleared up, monopolizing much time and effort that is badly needed for facing the future.

Challenges and solutions

Our species is finding it slightly less difficult to weigh up and tackle issues arising from the material problems of war and peace than it does those of an intangible nature concerning morals, ethics, religious, spiritual and transcendental matters, or values, assumptions and ideas. At the same time we are certainly witnessing a shift away from materialist obsessions, and a growing movement towards affairs of the mind, the emotions and the creative arts and sciences. A leading British astrologer, Jonathan Cainer, claims general agreement among his colleagues that 1998 marked the end of 'the darkest era humanity has ever known' and the start of 'humanity's own verdict on its past behaviour and its decision to play the game of life by a very different set of rules in future'.

Whether or not the astrologers know something that we do not, it is clear that we stand at a real turning-point in history, requiring revolutionary changes in out-look and conduct. Yet we still face ancient dilemmas. We find ourselves enjoying remarkable advances in a wide range of human capabilities and material skills, as well as some creditable progress in our non-material activities. Yet we are still dom-inated in too much of our conduct by relics of our animal origins, and by traces of deep-rooted primitive myths and superstitions, avoiding contact with scientific thought. We cannot reconcile our personal motives with our group loyalties and beliefs, nor our passing feelings with our sense of eternity. We seek order, but we live in chaos.

A vast and urgent issue, where the material and immaterial come together, is that of world population, both quantitative and qualitative. At the most basic level of personal mating and reproduction, it stems from deep-seated instincts, overlaid by powerful (and all too often dogmatic) religious and social convictions, strongly institutionalised in many countries. Recently this legacy from the past has been complemented in the West by comparably strong trends towards almost unlimited sexual freedom, accompanied by a marked decline in parental responsibility. Yet it would seem obvious that anyone procreating children either by design or accident should be willing and able to see them properly through their long dependent years, and to help to prepare them well as members of modern society.

Public attention has been mainly monopolised by statistics, because these are concrete and easy to grasp. Unfortunately, predictions have been seriously misleading. In Britain, for example, it was expertly forecast during the 1930s that the population was about to decline. This was welcomed in some influential circles, and with reason. In fact, however, partly through the dislocations of World War II and the subsequent period of reconstruction, population rapidly increased. From around 46 millions in 1931 the UK had reached 56 millions by 1996, and it is projected that there will be a need to support 60 millions by 2001. In the same period there has been a remarkable (though socially uneven) growth in affluence, technology, social aspiration and medical skills. On the negative side, this has led to a disproportionate expansion of demands on land for housing, shops, defence, road and air transport and other purposes as well as an inevitable need to invest in social infrastructure.

The UK, however, is fortunate in having the education, medical care, wealth and employment to support these growing numbers, even if at an unwelcome environmental cost. In many other countries population growth is far more rapid and, though birth rates are now falling and medicine has vastly improved the prospects for individual survival, the lack of education, family planning facilities, and employment (especially for women) are combining to impede the trend towards smaller families. Although many people and Governments now understand the need to stabilise human numbers at a level that the environmental resources at our disposal can support, the demographic momentum is so powerful that, barring catastrophe (which cannot be ruled out), the world needs to plan to support around 10 billion people a century from now. Globally, population growth has generated enormous social and environmental pressures and ill-feeling between North and South, as well as between religious leaders, politicians, economists and environmentalists. It is obvious that sustainable social and economic development will be impossible if human population growth continues to outrun the growth in a country's capacity to give its citizens a decent quality of life. Yet even in 1999 we see a faction of irresponsible national leaders — impelled by their interpretations of religion — urging inaction that would inevitably bring misery to millions.

The population issue is just one example. Much more widely, we are suffering because our mass of new knowledge is failing to transform our worn-out culture, or to equip us with a new perspective. For that we seem to need a new and greater version of the international encyclopaedic *History of Mankind* which Julian Huxley persuaded UNESCO to produce and publish almost simultaneously with his *The Humanist Frame*. Such a 'Book of Man', regularly updating human knowledge at the highest level, would make a worthy focus for fresh United Nations' efforts and for a New Renaissance. It would also bring together the required scholarly and internationally acceptable base to underpin the new global information network. UNESCO is clearly the right body to take on this task: let us hope that it will.[3] A well-planned overview of this kind, feeding into education throughout the world, provides the key to overcoming the world's sorry legacies of reductionism, short-termism, and obsession with prejudices, pettiness and trivialities rather than enjoying the wholeness of its inheritance.

Injurious fragmentation of knowledge is nowhere more conspicuous than in the now rapidly advancing human sciences, including psychology, genetics, ethology, sociology, anthropology, medicine and history. As in the environmental sphere, we badly need a well-informed appreciation of human diversity, its interactions and values, and its evolutionary potential. Our complex, sub-divided but single, human species contains a vast genetic and cultural richness which needs conservation as much as — or, taking our self-interest into account, more than — that of any other species. Yet that diversity, especially at the cultural level, is being eroded and this may become the last generation to enjoy immunity from ill-informed and irresponsible interventions to promote the prevalence of certain genes and to discriminate against others. More profoundly embracing studies and higher levels of professional wisdom and vision are needed before entering the biologically and ethically dangerous terrain of responses by genetic research to emerging health-service and commercial demands. While, in biology, research has led the way to environmental conservation, where the human species is concerned the lead has mainly come from the health and social services rather than from research — the wrong way to arrange it.

Globalisation, the impact of expanding technology on lifestyles, and the obsolescence of so many traditional attitudes and beliefs, now demand a new outlook and awareness from us all. Inspiring and developing this new outlook, and securing its general adoption will initially fall to a pioneering minority endowed with the right backgrounds and talents. In spreading the message they will find ample support among the much expanded ranks of the young with higher education, who reject selfish and materialist lifestyles and show genuine concern for their fellows and their natural environment. This new, positive trend is strongly at variance with persisting commercial and establishment attitudes, which continue to be daily advocated through the main channels of opinion. But it is fully in harmony with the spirit of a New Renaissance; and it is to stimulate this new outlook and positive trend, and to move forward these developments, that I have established the New Renaissance Group.

How can the prevailing tune be changed? A comparable need became manifest in the middle of 1939, as Hitler's war became clearly impending. The shadow Ministry of Information, then created, commissioned a group of four picked men (of whom the present writer was one) to put into a few words the essence of the public response which would be called for to meet this vast challenge. Within days of the outbreak of war coloured posters nation-wide displayed their message:

YOUR COURAGE, YOUR CHEERFULNESS,
YOUR RESOLUTION
WILL BRING US VICTORY!

Coincidentally or not, these words were widely remarked by observers at home and abroad as a key to Britain's ensuing triumphant response to the Nazi Blitz.

Ten years later, Gerald Barry's talented team, mounting the Festival of Britain, showed how a drab and dispirited period could be transformed with new impetus by a national campaign for confidence in the nation, by individual enterprise and by lively new design. Such gifted people are still to be found in many countries and given the right lead they could focus the vast, misused, resources of publicity and advertising once more to enlist people and nations in the actions that must be taken in the interests of the human future. It has been done and it can be done again. No more is needed than the realisation, the will, and the right team. Will the Millennium achieve it?

Conclusion

In conclusion, the New Renaissance argument hangs on the recognition of the paramount importance of four great arenas of action:

1. Knowing and understanding the natural world.

2. Achieving full harmony between people and nature.

3. Achieving harmony among all members of the human species, and all human communities, world-wide.

4. Achieving a transcendent harmony with the eternal realm beyond us, both philosophic and religious.

Integration within and between these paramount arenas forms the strategic agenda for the 21st century, awareness of which it will be the concern of the New Renaissance Group to promote.

Chapter 3

Soil, water and people

Sir H. Charles Pereira FRS

The Author

After a PhD in London in soil physics and six years with the Sappers, Charles Pereira joined the British Colonial Service. He set up large scale experiments in Kenya, Uganda and Tanganyika on how and why land-use changes on whole valleys alter river flow. From 1961 to 1968 he was Director of the Agricultural Research Council of Rhodesia and Nyasaland and, from 1972 to 1977 Chief Scientist/Deputy Secretary of the Ministry of Agriculture, Fisheries and Food in London. Elected FRS 1969, Knighted 1977. From 1972 to 1976 he was on the World Bank/FAO Advisory Committee to the Consultative Group of donors to International Agricultural Centres.

Chapter 3

Soil, water and people

Sir H. Charles Pereira FRS

A century of major advances in the science and practice of agriculture

Students of the world situation are faced by a flood of information on the statistics of population, agriculture, industry, economics and health. Although there are many competent observers and well produced publications, some issues, such as population growth, attract highly emotional treatment, with contrary interpretations of the same data. Since the end of the war I have had more than half a century of exceptional opportunities, as a professional consultant in watershed management, land use, soil and water resources, to see for myself. Mainly in the tropics, I have been able to study the changes that are taking place in the ability of many developing countries to supply fresh water, food, forage, fuelwood and tolerable living conditions for a relentlessly expanding number of people. This chapter is an attempt to assemble the picture as I have come to understand it.

We are now concluding a century of unprecedented progress by the human race. There are now greater numbers of people enjoying a higher standard of living than ever before. The developing countries have attained as much progress over the past 30 years as the industrial world managed over a century.[1] Infant mortality rates have halved, from 150 to 70 per thousand live births. Life expectancy has increased from 42 years to 62 years. School enrolment has doubled. But this is the bright side of the picture. The other half of the story is that progress has been grossly uneven. Over one thousand million people live in real poverty with 800 million malnourished (unable to maintain body weight under light work). This means that for every well-fed citizen of the industrialised countries there is someone near to starvation in the Third World.

This has also been a century of unprecedented population growth. It is difficult to appreciate the meaning of numbers when people are counted in thousands of millions, but the world total has tripled, in my lifetime, to six billion. This means that the human population has grown twice as much in my own brief span as it grew in the 100,000 years since *Homo sapiens* first appeared.

Over 95 percent of the annual increase now occurs in the developing countries, most of them lying in or near the 'Inter-Tropical Convergence Zone'. This inescapable feature of world geography is the tropical band into which the Trade Winds from north and south blow warm wet air towards the Equator. The rising wet air releases torrential rainfall and formidable quantities of latent heat so that the area is one of turbulent weather. The erratic distribution of droughts and floods makes agriculture hazardous. The human race has thus been concentrating into a zone of agricultural insecurity, and continues to do so.

Food shortages would long ago have checked this avalanche of people but for the contributions of science. The most spectacular development has been labelled the 'Green Revolution'. Begun in Mexico by the Rockefeller Foundation, a combination of intensive plant breeding and skilful agronomy transformed that country by 1956 from a grain importing economy to one self-sufficient in cereals. Led by Dr Norman Borlaug, many thousands of strains of wheat were tested. I had the pleasure of walking through some hundreds of wheat plots with him early in the morning. I had expected to see elaborate computer-assisted recording, but he noted a few numbers on the back of a cigarette box. He explained that he was looking for results so exceptional that he was happy to have two or three numbers at the end of his walk. The plant breeding programme, however, used large and complex computer calculations.

Borlaug's new varieties redistributed the starch into shorter straw and larger ears. They were highly responsive to fertilisers and resistant to some pests and diseases. Very importantly, they matured rapidly so that, in the all-year round growing season of irrigated tropical crops, two harvests a year were normal and three were possible. All new varieties were tested over a range of climates on substations at high and low altitudes. Only those that thrived in all these conditions were issued.

In 1962 the Ford and Rockefeller Foundations jointly opened the International Rice Research Institute (IRRI) in the Philippines to work on the same lines. Crossing dwarf Japanese varieties with tall local varieties they rapidly achieved results parallel with the progress in wheat.

In 1966 an International Agricultural Research Centre was opened in Mexico with the acronym CIMMYT (from the title in Spanish), to work on improvement of both maize and wheat. In the same year India, faced with severe grain shortages, took the bold step of importing 18,000 tonnes of the new dwarf wheats, together with the fertilisers to grow them. The fertiliser was subsidised to a price that made the new wheats profitable to grow. They were bulked up by the large irrigated farms of the Punjab and distributed widely. More than three quarters of the wheat now grown in India is of these improved varieties, usually referred to as 'High Yielding Varieties' or HYVs. Although they have been bred to respond profitably to fertilisers, they outyield local traditional varieties even without fertiliser because of their fast growth and resistance to pests and diseases. Norman Borlaug was awarded the Nobel Peace Prize in 1970.

Internationally, it was realised that the rapidly expanding populations of the developing countries could be fed only from their own resources in agriculture, and that this required very much higher yields to be obtained from the areas already under cultivation. Crops could be genetically modified by intensive plant breeding. However, the supply and distribution of improved seeds and fertilisers required credit services to enable farmers to use them. The World Bank and its regional associates supported the credit schemes. In the West, agricultural credit is a routine component of the seasonal production cycle, with the farm title deeds as security. In Africa, tribal land is held communally and so a farmer cannot pledge his holding as security. This problem was successfully overcome in Zimbabwe in the 1970s by the government of Bishop Muzerewa with the support of the World Bank. Every family with a right to cultivate in their communal land could apply for an advance, not of money but of hybrid maize seed, fertiliser and insecticide against stalk-borer. When that part of the crop surplus to family food supply was sold to the government grain depots, the debt was deducted and the loan guaranteed for next year.

The grain output from the communal lands rose in the first two years from 41,000 tonnes to 480,000 tonnes. The repayment rate was over 95 percent. Few countries have shown the organisational ability to follow this example.

The two Foundations further developed the concept of International Agricultural Research Centres by establishing the International Centre of Tropical Agriculture (CIAT) in Colombia, mainly concerned with pastures and livestock, and the International Institute for Tropical Agriculture (IITA) in Nigeria, to improve crops in the wet tropics. Concern for the rate at which world population growth was overtaking food production continued to grow. In 1971 the World Bank, jointly with FAO and UNDP, set up the Consultative Group on International Agricultural Research (CGIAR). This brought together fifteen donors and the representatives of ten developing countries to support the existing four centres and to set up others. The declared aims were: the Security of Food Supply; the Reduction of Poverty; and the Sustainability of Environmental Management.

Scientific policy for the Group is developed by a Technical Advisory Committee (TAC), with a secretariat provided by FAO in Rome. I was appointed as a founder-member of TAC in 1972 and took part both in the missions to prepare for new centres and in review missions for CIMMYT, CIAT and IRRI. Initially, the 15 donors provided about US$20 million *per annum* for the four Centres. This has now grown into a system of sixteen Centres supported by a current US$325 million *per annum*. The fields of work range from economic policy studies to aquaculture. A major role is in the training of indigenous research workers; some 50,000 have been trained over the past 25 years.

The rate of progress was exciting. For example, a review mission to IRRI in the Philippines in 1975 gave me the opportunity to see how local farmers were coping with the new varieties. The Centre was testing one of their early varieties, IR30, by helping several groups, each of ten rice farmers, to grow it on their half-hectare plots. A number of Danish volunteer workers were attached to IRRI for training. One of these supervised each group of ten. I was able to visit three of the groups. Seed and fertiliser had been issued on credit. The whole densely populated plain grew 'Peta', a tall slow-growing local variety that gave a single annual crop of 1.5 tonne/hectare. An exceptionally good crop would yield 2 tonnes. All three groups had already harvested two crops of 5 tonnes/hectare of IR30 and had planted a third. However, a festival season had intervened, and the farmers had celebrated so thoroughly that their third crop went in three weeks late. Their Danish instructors were complaining that they would get only half a crop, about 2.5 tonnes. I asked whether the rice was of the kind that they liked. They replied happily that the new rice was far too valuable to eat. Their neighbours were offering two bags of local rice for one of IR30 seeds. Their yields thus totalled 25 tonnes/hectare of local rice, equivalent to at least 15 years of traditional harvests. I had to agree that the term 'Green Revolution' was not a journalistic exaggeration.

This was progress indeed for the research workers, but by 1975 the vast growth of the world population was alarming all who studied it. The National Academy of Sciences of the USA published a study on *Population and Food: Critical Issues*. They estimated that to feed the annual increase in people an extra 25 million tonnes of grain must be grown each year. They called for an urgent increase in research effort as a 'major mission' for the USA. The CGIAR was making great efforts to improve both the crops and the research training of the agricultural services of the developing countries, but these services were the weak link between the Centres and the farmers.

Throughout the developing countries agriculture was, and still is, associated with rural poverty. The enterprising left it at the first opportunity. National research stations had indeed been built, many of them with overseas help, but they were handicapped by a very persistent difficulty. Promotion in government service went routinely by seniority rather than by research performance. This deadened the career prospects of young research staff, even after they had been trained at the International Centres. I found this policy to be universal in India, Pakistan, Sri Lanka, the Philippines, Indonesia, Malaysia and in many African countries after they became independent. Promoting the 'Green Revolution' technologies through the local agricultural services has therefore been an uphill task. Only where a government with the exceptional resources of India faced the threat of incipient famine, was there effective support for a veritable revolution in agriculture.

The research staff of the developing countries work under very unstable conditions. The nature of agricultural research implies long-term dedication. In some countries a cause for the poor output of some national agricultural research services (NARS) is the instability of their funding and, in consequence, the high rate of staff turnover. Even where budgets are approved, there are frequent delays and uncertainty in the release of funds. The CGIAR set up an International Service for National Agricultural Research (ISNAR). In a 1997 report ISNAR found that in Nigeria, in the 20 years since independence, the rate of loss of half of the research staff over five years had been characteristic.[2]

The world drought of 1972-73 reduced world stocks of grain in storage for 1974 from a normal 90 days' supply to only enough for 35 days. The former levels have not been restored, since Canada and the USA, producing some 90 percent of the surplus between them, have reduced their storage margins. The USSR made heavy purchases for stockfeed that further reduced the world reserve. The alarm in 1974 led to a World Food Conference convened by the United Nations in 1975. The Conference called for an internationally organised 'Global Food Security System, financed and operated to meet the needs of poor people in poor nations'. Two decades of discussion have not yet solved the problems of who is to pay the very high costs of storage and who is to control the distribution of stocks. The sudden interest shown by the United Nations was in contrast to their consistent neglect of the problems of Third World agriculture over their two previous 'Development Decades'. Dr. Boerma, then Director-General of FAO, reminded the UN Economic and Social Council that only 10 percent of official development funds had been allotted to agriculture. At present no storage organisation appears likely to be set up. Both high income and low income countries express lack of confidence in the co-operation of the other group. However, the inevitable incidence of regional threats of famine in developing countries is met, with low efficiency but immense goodwill, by voluntary efforts of bilateral aid and international charities, led by the Red Cross, in addition to the efforts of the UN agencies.

Sub-Saharan Africa has been more difficult to help. Population growth rates are higher and poverty more widespread than in other regions. Water resources are more scarce, so that irrigation contributes only 12 percent of food production. FAO studies[3] predict that the development of irrigation will be enough to maintain this proportion for the growing population but not to increase it. The CGIAR Centres have been successful in improving the yield and drought resistance of rainfed food crops and have released HYVs of sorghums, millets, short-season maizes, ground-nuts, pigeon peas and chickpeas. 'Triticale', the first new cereal for several thousand years, has been bred from inter-generic crosses between wheat and rye for tolerance of drought and salinity. It is now grown on 1.5 million hectares in 32 countries. For the wet tropics they have issued disease-resistant cowpeas and pota-toes that grow from true seed, avoiding the difficulty of storing tubers in hot climates.

There is no prospect of achieving the Green Revolution success with rainfed crops. The variability of local soils and climates is too great to permit the massive responses won in the standardised conditions of irrigated crops. However, the new varieties helped to increase food production in sub-Saharan Africa by 25 percent over the decade 1978-1988; but population grew faster, so that food production *per capita* fell by 8 percent over the same decade. In 1983 a study by the World Bank concluded that: 'In Africa today population is growing so fast that even a high growth rate of human and other complementary resources, comparable to that achieved by the developed countries over the past 50 years, would not be enough to sustain them'.[4]

The major problems are social and political rather than agricultural. Robert MacNamara, for ten years President of the World Bank, declared in 1986 that all of the 21 food deficient countries south of the Sahara have: (i) mounting pressures of population growth; (ii) ecological degradation of the resource base; and (iii) eco-nomic maladjustment.

'The [last] is due, not to recession in the world of trade, but to unwise trading and pric-ing policies. Government interferences have stifled economic progress by the encroach-ment of the state into almost every sphere of economic activity. There has been an accumulation of wealth by a political elite while the average African is poorer today than he was in 1970. Unless these trends are reversed the disastrous famines that now occur in only a few countries will become everyday occurrences in the majority of the sub-Saharan nations.'[5]

Ten years later the CGIAR reported that in 1996 there were 22 million Africans facing food emergencies in the Congo, Burundi, Rwanda, Angola, Ethiopia, Eritrea, Sudan, Malawi, Zambia, Liberia and Sierra Leone.

Much of Asia has made more progress than southern Africa. Taking part in a Quinquennial Review Mission to IRRI in 1982, I was glad that we were able to report that: 'In economic terms, investments in IRRI of about US$20 million annu-ally were generating an added value of about US$1500 million *per annum* from increased production in many countries. One HYV, IR36, was then grown on more than 10 million hectares in Asia. A new HGV, IR56, was resistant to a major pest, the Brown Plant Hopper, and also to the two major diseases of 'Blast' and 'Tungro'. In order to encourage national research staff, IRRI agreed not to name its new releases, but to supply the new seeds to developing countries, allowing national research staff to name and promote them. Only other specialists with careers in research will be able to appreciate the generosity of this gesture.

How far the research and development scientists will be able to continue to gain advances large enough to match the high rates of population growth is doubtful. In rice the biological advantages of altering the plant architecture appear to be near to their limits, although gains in resistance to pests and diseases are still possible. In wheat, however, CIMMYT has recently achieved a new variety that is currently out-yielding the present HYV's. But, it will take another five years to incorporate the necessary genes for disease resistance and to test it over a wide range of conditions before the new variety can be released.

Major damage to soil and water resources.

Demand for food will continue to increase. The age-structure of the present populations of the developing countries, with some 36 percent under the age of 15, ensures a prolonged period of growth. My personal observations, however, indicate two grave threats to a continued increase in food supply. These are the continued degradation both of the irrigation areas from salinity and of the stream-source areas from soil erosion. Both are due to failures by the Third World countries in the management of their natural resources. Neither are yet being effectively corrected.

These two vital assaults on the resources needed to sustain future generations are sharply evident in the poorest section of Asia, sometimes known as the 'Bengal Triangle'. These are the lands that drain into the Bay of Bengal. This vast watershed contains the world's most spectacular scenery, the Himalayas, part of one of the world's greatest potential irrigation areas, the Gangetic Plains, and, sadly, the world's poorest half-billion people. In view of the wealth of the natural resources of soils and water it is startling to learn that in this region the average income is below that of Sub-Saharan Africa. From the criterion of the management of their natural resources, Bangladesh, Bhutan, Nepal and the eastern states of India (Bihar, West Bengal and Assam) have governments that have neglected their rural areas, as I have seen in the course of field visits to their watersheds.

In 1979 the Technical Advisory Committee to the CGIAR recommended that a study be made of the research and development needs of the developing countries to improve the management of their irrigation resources. The major Canadian foundation, the International Development Research Council (IDRC) agreed to fund the study and I was appointed to lead the team.[6] Our mission visited India and Pakistan, Iran and Iraq, Syria and Egypt, Tunisia, Algeria and Morocco, Israel, the Netherlands, France and the USA.

The world's greatest opportunities for increase in food production lie in the further development of irrigation of the great arid alluvial plains served by the major rivers of the Ganges and Indus, of the Euphrates and Tigris and of the Nile. All have climates warm enough for all year round crop growth. They have already been developed to support a total of some 40 million hectares of irrigated crops. Ominously, very large additional areas have been developed, but then unskilfully irrigated until waterlogging, salinity, alkalinity and sediment deposition have caused them to be abandoned as wastelands.

The overall problem has been the lack of provision for drainage, which requires a long-term investment of heavy capital expenditure. The drainage problems are intensified by a surprising lack of irrigation technology and management. A tenet of the international profession of civil engineering has been that the subsistence farmer, after centuries of experience, knows how to manage his soil and water. The engineer has been concerned only to get the water to the outlet from the canal, leaving the rest to the farmer. This belief has proved to be quite unfounded in practice. In Pakistan we were able to learn the measured results of a thorough study. The management of the canal system distributing water from the Indus is conducted with professional competence by the Water and Power Development Authority (WAPDA). Beyond the canal outlets we saw the results of the Mona Project, a 10-year study funded by USAID through the Colorado State University. The main losses were from the communally owned water courses from the canal outlets to the small farms. No one accepted responsibility for their maintenance. We saw them choked by weeds, trampled by cattle, and forming narrow strips of marshland, interrupted at intervals by buffalo wallows for the ploughing stock. Competent measurements showed the losses from the watercourses to be 55 percent.

Annual diversions of water from the Indus River to serve the world's largest area of contiguous irrigation amount to 120 billion cubic meters. Less than half of this water reaches the crops: some 25 percent leaks from the unlined canals and over half of the water delivered to the outlets is lost before reaching the crops. The volume lost annually is equal to the full annual flow of the Nile. A little of the lost water is recovered by pumping from tube wells, but over parts of this vast irrigation area the subsoil is saline and the groundwater cannot be used. The main danger of these massive seepage losses is that the level of saline groundwater may reach the surface, stunting or killing the crops.

The Mona Project tested various forms of lining for the watercourses and farm distribution ditches, including brickwork, concrete and plastic sheeting. All were too expensive. The best solution was to reshape the channels with hand tools and to compact the soil. Outlets to farmers' fields were protected by concrete, and concrete platforms were made for village laundries. Bathing tanks were made for water buffalo. The reduction in water-logging made the use of fertilisers profitable. The 'Green Revolution' could not flourish in these areas until water management was thus improved.

The scale of the farm distribution watercourse problem is formidable. The USAID, the Asian Development Bank and the International Fund for Agricultural Development (IFAD) have financed improvements to more than six thousand Pakistani villages, but over 80,000 villages share the problems. Some three million hectares of irrigation in Pakistan are affected by salinity and the problems are growing at about 40,000 hectares *per annum* in spite of an internationally supported campaign of reclamation by pumping from tubewells.

The problems of salinity in irrigation are easier to describe then to remedy. All irrigation water contains some dissolved salts from the upper catchment areas. When applied in the field the water acquires more salts from the subsoil. In some areas subsoils are saline from their geological histories and the concentration in the groundwater may then become toxic to crops. Where drainage is good enough to keep the groundwater level two to three metres below the surface this may not affect the crop yields. If overwatering brings the water level to a depth of about 1.5 metres, then water rises by capillarity to the surface. There it evaporates and deposits the salt, stunting or killing the crops. Where drainage is good, extra water is applied to flush away the salt.

In the plains of the Indus and the Ganges great engineering feats of dam and canal building have been accomplished over the past two centuries. The main canals were built under British rule and were designed to alternate with the drainage canals that are essential for continuous irrigation. The main canals were not, however, built for such use, but rather as famine-prevention measures for the years in which the monsoons failed. The drainage canals were less urgent and were left for a later programme: many have never been built, although irrigation has become continuous. However, some main drainage canals have been built, including one parallel to each bank of the lower reaches of the Indus River, in a major development programme supported by international finance from the World Bank and its associates.

The Nile valley and delta have been under irrigation for about 5000 years: a dam of that age was about 100 metres long and 12 metres high. A crude system of basin flooding from the annual flood has been in operation since about 3300 BC. The Greek historian Herodotus described drainage works in the Nile valley over 2000 years ago. The construction of the Aswan Dam in 1965 established over-year storage and provided irrigation water all year round, but there was no adequate preparation of the farmers for the necessary changes in the irrigation system. When our mission visited them 14 years later the farmers were still continuing their routine as for the annual flood and were applying all the water they could get. The result has been a massive overwatering and rise in the watertables of the Delta, where 800,000 hectares have developed problems of waterlogging and salinity.

With international funding, a very large drainage programme has laid drainage pipes under the fields at a rate of 75,000 hectares *per annum*. We saw contractors from many countries using the latest laser-controlled levelling methods. They were working all year round. I was intrigued to see that they drove straight through standing crops, with no signs of protest from the farmers. Agents were on the spot to pay compensation at a little above market prices, so that the farmers were very co-operative.

In addition, a large-scale UN project developed coastal desert areas for irrigation, but the essential soil and water research had not been done and soil salinity has destroyed the whole enterprise. Moreover, rapid urban growth has taken up substantial areas of first class cropland. Egypt's population growth has continued unabated and the country can no longer support the present 66 million people from the 3 million hectares of irrigated land. At the same time, the Nile water supply is under threat from upstream irrigation developments by the Sudan on the White Nile and by Ethiopia on the Blue Nile. Egypt has thus become heavily dependent on American crop surpluses, supplied through USAID.

The third of the world's great areas for irrigation development is in the most serious state of degradation from salinity as the result of lack of management. Successive regimes occupying the flood plains between the Euphrates and Tigris rivers developed large scale irrigation schemes from about 2400 BC. Ironically this is the area from which our earliest written records of irrigation management have been found. The practice of flooding heavily to flush away salt accumulations is recorded on Sumerian clay tablets. They described leaching with water from the Tigris and drainage into the Euphrates. Well organised irrigation continued for some 4000 years before it lapsed through warfare about 1600 AD. A large construction programme in the 1950s commanded about 4 million hectares but management was not well enough established. Waterlogging and salinity have caused abandonment of half the restored area. Yields on the rest are low and the area is one of general poverty.

The Government of Iraq has scheduled three and a half of the four million hectares for reconstruction in a vast new drainage scheme with World Bank support. The sparse population of the salt-damaged lands and the very large areas to be redeveloped need heavy engineering solutions. The ongoing construction work that we saw on the 'Third River' was impressive. This is a major drainage canal from north of Baghdad to the sea. The canal is fed by open collecting ditches two to three metres deep, spaced at intervals of one kilometre. These were being opened up at the rate of 2000 kilometres each year. Sadly the war interrupted this excellent work. The subsequent sanctions imposed by the UN to end Iraq's defiance of the peace-keeping precautions have further delayed progress.

Irrigation investment in developing countries will continue to be governed by popular politics in which investment in drainage schemes will continue to get lower priority than will access to new land. The relentless degradation of much of the present irrigation by salinity seems set to continue. *International funding should therefore give strong priority to drainage and reclamation.* Our study led to the setting up of the International Irrigation Management Institute (IIMI), but on a much smaller scale than we had recommended. IIMI rapidly established its value and has since expanded to the International Water Management Institute (IWMI).

The Green Revolution was well established by the end of the 1970s, but the food situation remained precarious in many developing countries. Table 1, issued by FAO in 1979, illustrates the appallingly low yields for which millions were toiling. Too many people were attempting subsistence agriculture on limited areas of land, so that farm sizes were too small to generate the crop surpluses above family food supply that are essential to pay for improved seed and fertilisers. Population growth has therefore continued to limit the increase of food production by decreasing farm sizes in both rainfed and irrigated cropland. This has become acute in the Bengal Basin. Table 2 illustrates the scale of the problem today.

Only farms large enough to rank as commercial enterprises achieve the full yield potential of tropical climate, soils and water resources. This was well demonstrated in the development of Rhodesia/Zimbabwe, where a World Bank study reported that the large commercial farms were employing, feeding and supporting

Table 1. Yields attained country wide

Field conditions	Yields (tonnes of paddy rice per hectare)	
Experimental treatments	Research farms	10.0
Optimum inputs and cultural practices with advanced water management	Japan	6.0
Fertiliser, improved seed and pest control. Full control of water supply and drainage	Republic of Korea	5.0
Water table control, drought elimination, low rates of fertiliser	West Malaysia	3.0
	Sri Lanka	2.9
	Rep. Vietnam	2.5
	Pakistan	2.5
	Thailand	2.4
	India	2.0
	Burma	1.7
Flood prevention, no fertiliser	Kampuchea	1.5
	Laos	1.5

Source: FAO On-Farm Use of Water. Committee on Agriculture. 1979.

35

Table 2. Expanding populations and shrinking farms

Country	Population (millions)		Average farm size (hectares)	
INDIA	*1972*	*1997*	*1970*	*1986*
Assam	15	28	1.47	1.31
Bihar	58	98	1.50	0.87
West Bengal	45	72	1.20	0.92
NEPAL	*1985*	*1997*		*1991*
	15	27	–	1.12 (hills)
				2.13 (terai)
BANGLADESH	*1971*	*1991*		*1984*
	76	111	–	0.86

Source: India, Ministry of Agriculture, New Delhi, 1997; Bangladesh. Bureau of Statistics, Dhaka, 1994; Nepal, Ministry of Agriculture, Kathmandu, 1991.

more people per hectare than were the densely populated areas of subsistence cultivation.

The World Bank prepared an ambitious programme of investment in new irrigation development, rehabilitation of deteriorated irrigation areas and minor refurbishments in the world's 36 poorest countries for the period 1982 to 1990. Most of the investment was to be in India, which had 43 million hectares under irrigation and was allotted US$ 57 billion out of a total of US$71 billion. A striking part of the study was the degraded state of much of the Third World's irrigation development as a result of inadequate maintenance. Rehabilitation was estimated to cost less than one third as much per hectare as the construction of new projects but new schemes have far more political appeal. This support has subsequently enabled India to carry through an ambitious programme of commanding and developing four million hectares of new irrigation; yet new irrigation areas involve increased demands on limited freshwater resources. The care of the stream-source areas becomes more critically important.

Land use in the upper watersheds.

The second of my major concerns for the future is the severe degradation of soil and water resources that I have seen to be increasing in the stream-source areas of the upper watersheds. While irrigation is usually developed in readily accessible areas with high political priority, the water flows down from remote areas of mountain ranges or high hills. In developing countries these are characteristically difficult of access. In studying the upper watersheds of the river systems of the Himalayas, I have found them accessible only on foot, often needing two days of strenuous walking above the end of the jeep tracks, or occasionally rather exciting flying in light aircraft. Such remote areas have little political weight and therefore get little attention from urban-based governments.

Such neglect of hill areas is a serious aspect of the great watershed that drains into the Bay of Bengal and contains the world's poorest half-billion people. Here the Himalayan foothills of India, Nepal, Bhutan and the older limestone hills surrounding Bangladesh are under attack by expanding populations struggling to live by subsistence agriculture on ever steeper slopes. These areas of heavy rainfall have been fully protected by natural forest cover until recent times. The felling of trees, shrubs and bamboos for fuelwood together with fire and severe overgrazing has turned the steep slopes above the terraced farms into eroding wastelands. In the Nepalese Terai, at the foot of the slopes, I have seen new irrigation schemes buried under a layer of mud and gravel washed down from the hills. An aerial study of Nepal's Middle Mountains by an FAO team showed that the serious erosion came from the steep overgrazed wastelands above the terraced cropland.

Attempts by the UN agencies and by bilateral donors to restore and stabilise these eroded slopes through scores of watershed rehabilitation projects over the past three decades have had remarkably little effect. This has been in part because of the scarcity of agricultural, forestry and administrative staff. Staff careers depend on staying near to the cities; posting to the hills is regarded as banishment. There are indeed severe disadvantages. Schools, shops, medical services and the cultural advantages of city life are missing. Even food may be scarce. In the mountains of the Western Province of Nepal food supplies are seasonally precarious and I have hitched a ride on a little De Havilland 'Twin Otter' carrying rice to feed the government staff stationed beyond the reach of the road system.

Solutions are possible and have been demonstrated by some of the projects, but in a 'pepper-pot' fashion over the very large areas of the problems. One which I assessed in Nepal demonstrated that these steep wastelands under generous rainfall and warm temperatures would grow vigorous fodder crops and fuelwood if planted and protected from livestock. The villagers had to make a difficult decision. In return for tethering and stall-feeding their cattle, sheep and goats they were offered paid employment to plant the erosion gullies with 'Elephant Grass' (*Pennisetum purpureum*), and the legume 'Stylo' (*Stylosanthes guianensis*) and to plant the slopes with the fast-growing nitrogen-fixing trees of Nepalese Alder *(Alnus nepalensis)*. The terraced fields were so exhausted by continuous cropping that farmers would not risk their seed unless they were able to apply manure. The livestock were thus gathering nutrients from the forest. Yet with free range grazing only 40 percent of the manure was collected in the pens at night, while the greater part was scattered on the steep mountain slopes. The collection of the dry dung from the forest was a major task allotted to the women. They found that cutting and carrying from the nearby plantings of tall grasses was less strenuous than collecting the manure. The local authority or Panchayat had ten wards, each under a headman. Only three agreed to join. The Forest Department supplied seedlings and advice so that the project flourished and within three years all of the other wards had applied to join.[7] At this point the clumsy and competitive system of international aid funds chose another target and funds were cut off.

Planting of trees and shrubs is a constructive use of land abandoned as a result of exhaustive cropping and soil erosion. Deep-rooted species can grow in deep soils even after the topsoil has been lost and the first metre depth has been depleted. They should not, however, be planted where the roots can reach the saturated zone above the water table unless the level of the water table needs to be lowered. A large-scale successful example that I was able to study in southern India was in Karnataka and supported by funds provided by the UK Overseas Development Administration (ODA).

Here major investment in production of paper pulp used tall grasses for the bulk of the raw material but needed a proportion of long-fibre input. Bamboo was first cut from the surviving areas of natural forest; but further damage to the forest was prevented by the substitution of *Eucalyptus* from plantations. These grew well on a large scale on land that had been worked out by traditional farming methods and abandoned. The planting, tending and harvesting provided much needed employment. Small-scale farmers whose crops had declined were offered tree seedlings to grow eucalypts for purchase and collection by the paper company. Only the trunks went to the mills, leaving one third of the trees for fuelwood. A political anti-government campaign was conducted in the Indian press with astonishing virulence, alleging that the eucalypts were drying up the country. The eucalypts were also accused of stunting other crops. A detailed study funded by ODA was carried out by the Institute of Hydrology and the Oxford Forestry Institute, in co-operation with the Karnataka Forest Department. Using advanced research methods, measuring soil moisture by neutron probes and transpiration by radio-active tracers, the research showed clearly that tall eucalyptus plantations used water at the same rate as the natural forest. When drought killed other crops these drought-tolerant Australian species survived. Part of their means of survival was their ability to use water stored deep in the soil that other crops could not reach.

Many tropical trees and shrubs are salt-tolerant. More than 20 species of eucalypts have some useful degree of salt tolerance. Areas of Pakistan's irrigation system that have been lost to salinity are growing useful crops of eucalypts for poles and fuelwood. China has planted *Casuarina equisetifolia* in coastal deserts to provide tall windbreaks on a very large scale. Known as the 'Great Green Wall' this is a belt 3 kilometres wide and some 3000 kilometres long. The plantations provide fuelwood and construction timber in addition to their main role, which is the fixation of sand dunes and the protection of agriculture from sandstorms.

Such plantings on saline land need strong encouragement, but within cropland irrigation systems they are salvage methods only and do not substitute for efficient irrigation management.

Has population growth overtaken the Green Revolution?

As populations have expanded, grain imports into the developing countries have increased. Food supplies *per capita* increased until 1995.[8] Since grain supplies more than half of humanity's calories, it provides a useful index of world supplies. World grain production per person rose rapidly from less than 250 kg in 1950 to nearly 350 kg in 1984. Subsequently production has fallen sharply to about 290 kg[9] (Fig 1). World grain in store fell to the lowest total on record in 1995. Grain production and storage are important since grain provides the only effective medium for rapid food assistance for the Third World disasters that

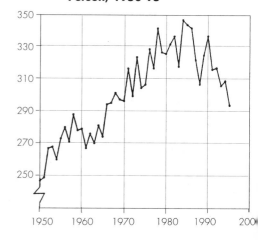

Figure 1. World Grain Production Per Person, 1950-95

are becoming increasingly a concern of the more developed countries. The steep decline in grain production per person has little impact on the industrially developed countries since there is still a margin of grain in store and the 40 percent of their grain consumption that goes to animal feedstuffs for meat supply gives substantial room for adjustment in emergency. In the developing countries a fall in grain production has an immediate effect on the food intake of the lower income groups.

The most comprehensive study yet made of the degradation of soils resources in the developing countries was by a Netherlands team in support of the UN Environment Programme in 1991.[10] Of the world total of about 1.5 billion hectares of croplands one third have been damaged enough to reduce or to eliminate yields. This includes both damage by soil erosion and by salinization. Since then an additional five to six million hectares have been damaged each year.[11] This rate of loss continues today, since, in spite of much debate and concern in the western world and very substantial aid funds, there is little sign in the developing countries of the reorganisation needed to administer and improve their rural areas and to care for the remote sources of their water supplies.

The problems of population and of poverty have continued to increase, with poverty growing twice as fast as population. The World Bank Development Report of 1995 showed that in the ten years 1984 to 1994 world population increased by 17 percent, while poverty, defined as living on less than a dollar a day, increased by 38 percent. In 1997 the UNDP reported 840 million as undernourished, including 200 million children.

Economic and social organisation to overcome poverty is needed to make use of the technology now available. A recent FAO report described 'share-cropping' at 50 percent of yield as a normal arrangement of tenancy. This eliminates the margins for the purchase of essential inputs, debars the peasant farmer the route out of poverty offered by science, and perpetuates poverty. Only Asian politicians can overcome this very severe problem. I remember visiting the Tata Energy Research Institute in New Delhi, where high level scientific work had attracted a Nobel Prize. I then drove southwards through a landscape of very small farms and thin crops to Hyderabad, to visit a group of Canadian agricultural-aid workers. They were trying to teach peasant farmers the advantages of sowing their seed in drills rather than by broadcasting. Jethro Tull published his pioneer work *Horse-Hoeing Husbandry* in 1753. The technology of weeding between the rows by a multi-tined implement was adopted world-wide, and by many advanced farms in India. I had to ask myself why the Canadians should be thus concerned to help Indian subsistence farmers to catch up with two centuries of agricultural history while India has both the skills and the means to do so.

The difficulties are not due to lack of information. In nearly all of the developing countries that I have visited I have seen agricultural solutions successfully demonstrated on the larger holdings of their more advanced farmers. These are often members of their legislatures. Some countries have evolved excellent stable patterns of land use, such as the terraced slopes of Nepal, Java and Luzon. These are now threatened by population increases beyond their capacity. The missing element is the political determination and the administrative skill to use the aid funds and technical assistance effectively. Here the UN and the bilateral aid agencies of the developed countries must take responsibility for the lost opportunities to insist that their funds go to support land improvement and rural services rather than to urban development. Within my personal experience international agencies have, over the past thirty years, been turning a blind eye to the diversion of funds to urban rather than rural areas. Although the UNDP has long held the responsibility for co-ordinating aid programmes within countries, they have not been able to prevent competition between donors. When one donor insists that it will support only rural development, others will eagerly offer aid that better suits the programmes of leading politicians. The result has been the alarming persistence of rural deterioration and the continuing losses of soil and water resources.

Part of the lack of effect of programmes to restore damaged watersheds is that the need to 'Monitor and Evaluate' rural projects attracts only lip service. It is left to local government departments to do this work, but they need to provide staff, vehicles, equipment and running costs. These are always in short supply and, since the donors do not provide the means, the work is not done. In the aid machinery, the pressures to keep the funds flowing are far greater than those to ensure their effective use.

How aware are the citizens of the western world of the living conditions of the majority of the human race?

We learn about the foregoing problems mainly through the disasters that television brings into our comfortable homes. Generous public support is repeatedly forthcoming for disaster relief, but the underlying social and economic causes are only little understood and the grim pictures are rapidly replaced by more domestic issues. As yet there is no public conception in the western world of the threat to world stability offered by the drift of hundreds of millions of people into acute poverty. The problems of stability in the world with which our grandchildren will soon be coping are very real.

In the twenty-five member countries of the Organisation for Economic Co-operation and Development (OECD), some 45 percent of present population growth is by immigration rather than birth rate. In Europe the proportion is an astonishing 88 percent. The number of refugees registered by the UN has been reduced, by resettling 4 million, to about 14 million in 1996. Population pressure has driven large numbers of illegal immigrants across their country's borders. More than 10 million Bengalis from Bangladesh have moved into surrounding territories. This caused violent clashes in the 1980s with an estimated 4000 dead. Wars and natural disasters have displaced even larger numbers within the boundaries of their own countries. The UN Population Division estimated the total as about 30 million in 35 countries.[12] Sir Crispin Tickell has described the refugee populations as the 'loose cannon' of the modern world.[13]

Clearly, immigration offers no solution to the world problem. An alternative, by which the nations of the temperate zone grow ever more food and ship it to the tropics, has no economic or logistic feasibility. Already the international aid shipments have damaged tropical agriculture by destroying the urban markets on which their farmers rely. In most of the sub-Saharan countries their climates dictate that their main crops should be maize, sorghums and millets, for which their urban areas provide the main markets. Aid shipments of wheat are, however, shifting the urban market demand towards bread, which cannot be made from the local grains. Instead of waiting for threats of famine and then improvising very expensive and agriculturally disruptive food aid, the well-intentioned western world should help to prevent the occurrence of famine by organising supplies of improved seed and phosphatic fertilisers and the essential provision of credit funds to enable the farmers to buy them. Sadly, the use of fertilisers is decried by enthusiasts for 'organic farming'. Although they claim that they wish to protect the environment, they ignore the implications of the low yields involved. To grow the world's supply of crops without fertiliser would require the ploughing of about $2\frac{1}{2}$ hectares for every one hectare now in production. This would involve the clearing of almost all natural forest, national parks and great areas of marginal land. A major change to organic farming would be environmentally devastating.

There are other important environmental issues, such as degradation of drylands through overgrazing, with subsequent erosion by wind, or the massive losses of natural forests and their irreplaceable biodiversity, but a full discussion would require a book rather than an essay. I shall only discuss two of these issues here, biodiversity and the genetic modification of crops.

The diversity of living species, the genes that they carry and the ecologies in which they live, are to most non-scientists a remote area vaguely associated with rainforests and tropical wetlands. It is difficult to arouse the interest that is urgently needed in an issue that will become serious for our grandchildren. The need for forests to remove and store the carbon dioxide that our industrialised societies are pumping into our atmosphere is slowly being recognised, yet the rate of forest destruction is accelerating. Both the forests and the plant life in the oceans are needed to maintain the precarious balance of gases in our atmosphere that rules our climate.

Our food supply depends upon an astonishingly small selection of about twenty plant species from the millions available. About 60 percent of our food comes from only three species, wheat, maize and rice. By collecting and testing their wild relatives, and by crossing those with desirable genes, plant breeders have achieved high-yielding, fast-growing varieties with critically important resistance to pests and diseases and to drought. However, the base is extremely narrow. Should a major new disease arise in any of these three cereals, famine would result.

The wild relatives are themselves under threat. Those of wheat are wild grasses indigenous to the Middle East and are threatened, as I have seen in Eastern Turkey, by persistent destructive overgrazing. The major research centres of the CGIAR have devoted much effort to the collection, storing, regrowth and testing of traditional cereal varieties and of related grasses. These are now used in elaborate patterns of computer-designed genetic studies.

The frequent references to rainforests arise because it is in them that the vast core of plant diversity is to be found. Rainforests harbour some 70 percent of all terrestrial species on only 7 percent of the land surface. Global extinction rates are now estimated at between 4000 and 6000 plant species every year. Sir Ghillean Prance, FRS, Director of the Royal Botanical Gardens at Kew, warned in a recent lecture[4] that: 'The actions of the next forty years regarding biodiversity will determine whether or not human life will survive on earth'.

Recent advances in biotechnology have achieved a new form of 'revolution' in plant breeding. This has come in timely succession to the plant-breeders' 'green revolution' that has fed the formidable surge in population numbers of the developing countries over the past three decades. The redesign of plant structures to increase grain yield is nearing its limits. The incorporation of resistance to pests and diseases has become ever more important as the chemical control afforded by spraying of pesticides is diminished by the adaptation of the pathogens. Perhaps the most economically important example is the world-wide increase in the resistance of the Late Blight of potatoes (*Phytophthora infestans*) to the routine fungicides. This blight, which caused the historical Irish famine, receives more fungicide than any other crop disease. Frequent use of toxic sprays damages wildlife, especially birds, so that the substitution of inherent disease resistance is a valuable ecological gain.

Plant breeding is inevitably a slow process, as each crossing of varieties introduces groups of genes, needing further generations for re-selection. The recent achievements in biotechnology enable selected genes to be spliced into crop DNA (the reproductive tissue in all living cells) so that only the desired properties are transferred. This offers a vitally important gain in the speed and accuracy of crop development. The non-scientific world has, however, been startled to learn that the course of evolution has resulted in all living cells having a similar basic pattern of DNA. This enables transfers of genes to be made across the familiar biological boundaries. The resulting 'transgenic' crops include potatoes with genes from chickens or from silk moths, and tomatoes with genes from fish. Unfortunately the media have seized on these strange developments and have aroused public concern about 'interfering with nature'. This has been taken up as a political issue. In a wave of emotion and scientific ignorance, parties of activists have been destroying field trials that were providing the scientific evidence that the politicians were demanding.

The Royal Society has published a thorough expert study of these issues.[15] They warned that new approaches are needed to feed a world population of 8 billions, expected by 2020 AD, just as today's food requirements of a population of nearly 6 billion could not be met by the technologies of the 1940's. The advent of genetic modification (GM) has 'huge potential for further beneficial developments'. The report considered the scientific evidence on three major issues: (i) the risk of transfer of genes from GM plants to wild species and to non-GM crops; (ii) the uptake of genes from GM foods by the digestive system; and (iii) the current state of the regulatory system. They concluded that the chances of accidental gene transfers happening are slight provided that the regulatory processes are followed, but that this must be kept under consideration. They were, however, concerned that, although many aspects of GM technology are already closely regulated, there is not as yet any means for looking at these problems as a whole. 'The Royal Society therefore urges the Government to establish an independent over-arching regulatory body'. They concluded: 'Consumer confidence, based on an appreciation of the scientific evidence and the regulatory checks and balances, is central to whether GM crops will contribute to feeding the world's rapidly expanding population'.

A practical study of the agricultural issues has been published by the Royal Agricultural Society of England in a very readable booklet entitled *Old Crops in New Bottles*.[16] By the end of 1997 some 25,000 detailed trials of GM crops had been studied in the UK. Although an estimated world-wide total of 40 million hectares of GM crops were planted in 1998, not one hectare has yet been grown commercially in these islands. There is some concern that our agriculture could suffer from such restrictions.

A development that is raising some serious ethical issues is the introduction by

Monsanto, one of the world's largest seed companies, of GM seeds containing a 'Terminator Gene' that will make cereal seeds infertile, thus requiring the purchase of new seed every year. In India, Dr. M.S.Swaminathan, FRS, a leading international scientist, has warned that such technology is unsuitable for the subsistence agriculture of developing countries. The Consultative Group on International Agricultural Research (CGIAR), the group of donors that fund the International Centres driving the Green Revolution, have agreed to ban the terminator gene.

There is thus urgent need for the close regulation of the use of genetically modified seed and also the need to convince the public that the new technology will be necessary to feed the expanding world population

World scientific leadership is emphatic that the present growth rates of population are unsustainable

After attending the Second World Population Conference in Belgrade in 1965, when the world population was 3 billion, Dr. C. J. Martin, a senior economist at the World Bank wrote: 'The world of AD 2000 looks like a townplanner's nightmare. But the harsh reality remains: it is almost certain that where every one person now stands there will then be two competing for the space.'[17] We now know, in 1998, that the world population in AD 2000 will indeed be more than 6 billion.

The decade of the 1990s has produced a solid core of agreement among the world's most senior and responsible scientific bodies that present rates of growth of population are unsustainable. In 1992, the 1600 members of the world-wide Union of Concerned Scientists signed a 'World Scientists' Warning to Humanity'. The signatories included 105 Nobel Prize Winners and other leading scientists from 70 countries. They warned that the area of cultivated land already degraded by exhaustive cropping and soil erosion was equal to the combined areas of India and China; that poverty and hunger already affected one fifth of the world's population and that the present rate of population increase was unsustainable. Also in 1992, a joint statement by the officers of the Royal Society of London and the United States National Academy of Sciences was entitled *Population Growth, Resource Consumption and a Sustainable World*. This outlined the state of the vast majority — of some 80 percent of the world's people — who live in the developing countries.

> 'The rapidly growing populations are causing substantial and increasing damage to their environment by pollution, by the clearing of forests and by their destructive methods of agriculture. With less than 7 percent of the world's scientists and engineers they have urgent need of help to solve their environmental problems. The future depends on early stabilisation of fertility rates to the replacement level of 2.1 children per woman.'

In 1994, sixty-six of the world's leading academies including those from the larger developing countries, met in New Delhi at the 'Population Summit', hosted by the Indian National Science Academy. They debated 25 studies and prepared a statement that summarised the 'unprecedented population expansion', and noted that in the last decade food production from both land and sea declined relative to population growth. They concluded: '*To deal with the social and environmental problems we must achieve zero population growth within the lifetime of our children.*'

The agreed recommendations included: the incorporation by governments of environmental goals into legislation, economic planning and the setting of priorities and incentives; universal access to convenient family planning and health services and a wide variety of safe and affordable contraceptive options; the industrialised world to assist the developing world in combating both global and local environmental problems.

The statement and the 25 studies were published by the Royal Society in 1994 under the title *Population, the Complex Reality*. The Summit Statement achieved a remarkable consensus. The National Academies from 60 countries signed it; four declared themselves not competent to sign and only two refused. These were, unsurprisingly, the Pontifical Academy of Sciences and the African Academy of Sciences. It is ironic that, in spite of the opposition of the Roman Catholic Church to contraception, Italy has the lowest total fertility rate of any country listed in the World Bank Development Reports, with an average of only 1.3 children per woman. The African Academy demonstrated that the traditional tribal thinking remains dominant in the politics of the African continent.

In Britain the President of the Royal Society, Sir Michael Atiyah OM, declared that, in the scientific world, there is: 'The strongly held view that the continuing growth of the world's population, together with the related economic and environmental questions, is the single most important problem facing mankind'.

In 1995, Ismael Serageldin, World Bank Vice-President for Environmental and Sustainable Development, convened a panel of eight leading authorities, three of them with Nobel Prizes, to make a study of 'Meeting the Challenge of Population, Environment and Resources'. Norman Borlaug, the initiator of the Green Revolution in irrigated cereals, summarised the issues:

> 'Twenty-five years ago, in my acceptance speech for the 1970 Peace Prize, I said that the green revolution had won a temporary success in humanity's war against hunger; but I warned that, unless the frightening power of human reproduction was curbed, the success would be only ephemeral. The demographic changes of this century have already exacerbated human misery and the degradation of the environment. Indeed, I would submit that human demographic increases are the greatest environmental threat to the earth in the years ahead. World Peace will not be built on empty stomachs and human misery. Agricultural experts have a moral obligation to warn political, educational and religious leaders about the magnitude and seriousness of the arable land, food and population problems that lie ahead.'[18]

The naive belief by some opponents of family planning that scientists will be able to invent new solutions to world food supply is clearly not shared by the scientists.

Information and debate

There is an abundant literature on the increasing hazards to our human life-support system. The threats to soils and water resources and the reduction in the carrying capacity of the land to support growing populations have been much discussed. The UN agencies publish data annually. The complex picture is focused by two outstanding annual publications: *World Resources*, produced jointly by the World Resources Institute, the UNEP, UNDP and the World Bank, and the *State of the World*, published by Lester Brown and the World Watch Institute. Regrettably, in the highly emotional controversies over family planning, such analyses of facts are dismissed as 'prophecies of doom'. My own observations have convinced me that the forecasts published in the 1950's have already been realised in the 800 million people now hungry. At present, neither in the industrialised nor in the developing countries do the general public and their political leaders recognise the urgency of the arithmetic of rising numbers and degrading resources of soil and water.

Priorities

The most obvious priority is to reduce birth-rates, but the effects will be slow. The high proportion of children in the populations of the developing countries will prevent any early stabilisation. If all new couples were to have two children only, the world population would continue to rise for half a century. Since delay intensifies the problem and makes the solutions more difficult, the time lag increases the priority. A woman who has three children instead of six will have 27 great-grand-children rather than 216. Currently less than 5 percent of the UK funding for overseas aid goes to support family planning, while the USA, for political reasons, refuses to contribute to the UN Population Fund (UNFPA).

The next immediate priority is to offer aid on generous terms for the improvement of land use, especially in the upper watersheds, with additional funding for measuring the effects of the improvements. An essential aspect of this form of aid is the persuasion of developing countries that their administrative and technical staff in their remote water source areas should be better rewarded and that service in such areas should be better recognised.

The third major priority is to improve irrigation management in order to avoid waterlogging and control salinity. The scale on which faulty management of irrigation is destroying areas of productive cropland must be of urgent international concern.

Admittedly, the urban problems are also acute, and by 2015 AD the UN Population Fund estimates that the developing countries will have half of their people in urban concentrations. Political power will be even more urban, but their dependence on the rural areas to supply water, fuel and most of their food will remain. Although formal priority for support for environmentally sustainable land use has become the declared policy of most donors, expediency rather than policy appears to govern the administration of aid funds.

In almost all the countries that I have been able to study, the technologies for soil conservation are either traditional, as with terracing, or have been demonstrated and, indeed, adopted by leading farmers. The need today is less for the teaching of better farming methods and much more for better organisation of marketing, of supply of improved seed and fertilisers and of credit services that will enable farmers to use them.

A note of optimism

To end on a more optimistic note, there is good evidence that programmes of birth control have achieved encouraging initial progress in many developing countries. Family planning and family health care services are proving to be effective in the Third World as in our own. The graph for the 52 largest developing countries, plotting 'Total Fertility' (number of children per mother) against 'Percentage of Couples Using Contraception' is linear (Fig. 2)

This holds over the range from 7 children in some African countries to under 3 children in China, Thailand and Mauritius, where over 70 percent use family planning. Thus many of the developing countries have made important progress in slowing their birth rates. International charities, such as the Marie Stopes

Figure 2. Decline of fertility rate with family planning

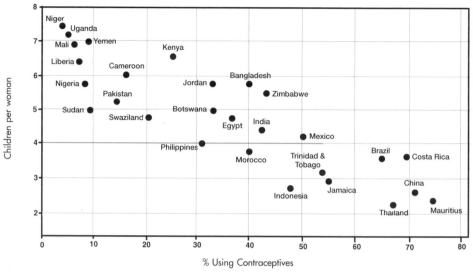

Source: John Hopkins University, Maryland, USA (1994)

International, are playing a leading role in this vital task. Family planning is not yet effectively developed in Africa; Nigeria, the largest country, is near the high end of the fertility ranking. The trends shown in Figure 2 are encouraging but the demographic change is as yet too slow to prevent further massive increases in poverty and consequent damage to the vital life support systems of soils and water resources.

When world population was less than one billion Oliver Goldsmith wrote:

'Ill fares the land, to hastening ills a prey,
Where wealth accumulates and men decay'.

A century later, with a world population of three billions, Hilaire Belloc commented:

'But how much more unfortunate are those,
Where wealth declines and population grows.'

Chapter 4

Population and environment: assumptions, interpretation and other reasons for confusion

Virginia Deane Abernethy

The Author

Virginia Abernethy is an anthropologist who has made her career studying human motivation. She is best known for work on why couples want, then have, more or fewer children. The interface of population and environment is her particular area of specialisation. Between 1988 and 1999 she was the editor of *Population and Environment* (Plenum Press, New York). She is now Professor Emeritus of Vanderbilt University, Nashville, Tennessee, USA.

Chapter 4

Population and environment: assumptions, interpretation and other reasons for confusion

Virginia Deane Abernethy

Introduction

Women and couples in virtually every country in the world are decreasing their desired and actual family size. In contrast, John Bongaarts of the Population Council, in the October 1998 issue of *Science* (page 420), describes an increase in the U.S. fertility rate, 'from 1.77 to 2.08 births per woman between 1975 and 1990.' In the 1970s, American women reduced their fertility rate to below replacement level, so that their child-bearing at young ages was low by historic standards. Extrapolation of that fertility behaviour to the remainder of their reproductive careers gave the appearance, states Bongaarts, of very low completed fertility. He concludes that the low rate in 1970 was more apparent than real because he assumes that these deferred births were made up at later ages and account for the 2.08 births per woman reported in 1990.[1]

This is a nice try, very politically correct, at obscuring the reality that the United States was the recipient of heavy immigration during the 1970-1990 period, and that this flow of newcomers raised the fertility rate. In fact, the foreign-born are approximately 10 percent of the current U.S. population (and rising) and account for a disproportionate share of births. In 1994, for example, 18.3 percent of births, that is, 731,262 out of a total 3,995,767 births, were to the foreign-born.[2]

The Hispanic sector in the United States averages approximately 3.5 births per woman and the tempo of births is rapid; women bear their first child at an earlier age than the U.S. average. Further, the teenage illegitimacy rate is higher in the second and third generations than in first-generation Hispanic immigrants, and is higher than among native-born Americans, either white or black. Mexico is the single largest source of the total immigration flow and the principal reason that the U.S. fertility rate – a blend of the high fertility rate of the foreign-born sector and the significantly lower rate of the native-born (still well below the replacement rate) – is rising.

The Bongaart example demonstrates the power of interpretation or, as christened in political circles, 'spin'. Numerous questions about spin arise, in fact, as one examines past population data and projections for future world population.

A standard demographic practice is to change the reported number of the total population in year 'x' long after the year in question has passed. The revised number is then used as the base for recalculating the rate of growth through some further year. If the revised total is upward, it absorbs part of actual growth, so the rate of growth appears lower than it actually has been. For example, the UN estimated in 1973 that the 1970 world population was 3.610 billion; the 1982, 1884, and 1988 UN estimates raised the total to 3.683, then 3.693, then 3.698 billion, respectively.[3]

Which of the 4 totals for the 1970 population should be the base for calculating the growth rate between 1970 and, say, 1990? Demographers usually, but not always, use the most recently reported number, a practice that is defended on grounds that it uses the most up-to-date numbers available. But this *ex-post facto* technique regularly astonishes mathematicians.

Components of world population change are fertility and mortality. Various projections of world population, with all their warts, are explored in this paper.

Projections of world population

The 1.77 and 3.5 in the lead example and similar numbers are a useful measure of fertility that is called, technically, the total fertility rate (TFR). This is an estimate of the 'average number of births a woman would have if she were to live through her reproductive years (ages 15-49) and bear children at each age at the rates observed in a particular year or period'.[4]

The concept of fertility rate came into broad use during the 1960s to replace birth rate. The birth rate measures the number of births per 1000 in the population at mid-year and is sensitive to the composition of the population. For example, a very large proportion of young women tends to result in a high birth rate for the population even if the average number of births per woman is small. Therefore, birth rate misses the component of individual behaviour that, over the long run, is most determinative of the population growth rate.

Some analysts prefer to measure the reproduction rate, which is a refinement on fertility rate. It refers to the number of daughters born, since females – because of the limited pregnancies any woman can successfully bring to term – place an upper bound on a generation's population growth. The fertility rate statistic, however, is that most commonly used.

Mortality rates have been less studied and are more straightforward. The mortality rate captures both infant and child survival and survival probabilities among the elderly and is usually expressed as the number of deaths per 1000 in the population at mid-year. It is, of course, sensitive to the composition of the population because, all else equal, a young population has a lower mortality rate.

The size, composition (age and sex), and fertility and mortality rates of national and world population are central to demographic forecasts of rates of growth and future population size. The population estimates impinge, in turn, on social, political, and economic scenarios. Population size and growth rates are relevant to almost all domains of personal and public life because they affect resource use, generation of waste products, public health, the need for public expenditures on education and infrastructure, the size of government, taxes, economic opportunity, crime rates and numerous other quality of life indicators.

The variety of effects linking population and environment is reflected in the term 'carrying capacity'. Carrying capacity refers to the number of individuals who can be supported without degrading the natural, cultural and social environment, that is, without reducing the ability of the environment to sustain the desired quality of life over the long term.

Principal sources of world population forecasts are the United Nations' population projections and the Statistical Abstract of the United States published by the U.S. Census Bureau. Projections are not predictions but are, rather, an array of possibilities, each with its specific probability of actually coming to pass. The demographic method of forecasting the future entails assumptions about the key parameters of fertility and mortality.

The 1980s were a transition period characterised by considerable uncertainty, but since about 1990, the United Nations has reported progressively lower growth rates and lower totals in its array of projections of future world population size. The same, generally, is true of the U.S. Statistical Abstract, although the 1998 Abstract avoids projecting a further reduction in the most probable size of world population in 2050.

The apparent stability of the 1998 Census Bureau's forecast results from offsetting the current, higher-than-expected decline in the growth rate with a higher baseline, that is, calculations begin from a higher estimate of total population in earlier years. The interested layman has difficulty in making independent calculations of past growth rates, because UN and United States publications and editions of the U.S. Statistical Abstract use different benchmark years for reporting total population size.

It is conventionally agreed, nevertheless, that the world population passed 6 billion around October, 1999. This means that it took approximately 12 years – since 1987 – to add the latest billion. It also means that, because the population in 1960 was about 3 billion, the latest doubling of population took approximately 39 years.

Some years ago it was thought that the ultimate world population size, to be attained in the twenty-first century, could be as high as 12 or even 14 billion persons. Indeed, at the growth rates of over 2 percent annually observed during the 1970s, this number could have been quickly upon us. (Doubling time is calculated by dividing the number 70 by the percentage increase; thus, $70/2 = 35$ years to double.)

By the late 1970s it began to seem that the rate of growth might be declining. This was confirmed with the following ten years' data. The UN's 'medium variant' assumption published in their 1994 revision of World Population Prospects projected a total of 9.8 billion people by year 2050. Two years later the projection was revised downward to 9.4 billion. The late autumn 1998 projection was 8.9 billion. From over 2 percent annual growth, the rate as estimated in an early 1998 publication had slowed to less than 1.38 percent annually. At that rate of growth, the time for population to double increases from 35 to over 50 years ($70/1.38 = 50.7$).

The UN middle range estimate assumes that a fertility rate of 2.1 (approximately the replacement rate for the existing population) will be achieved within the next decade or so. This would result in population reaching 10.8 billion by 2150. But with a fertility rate just one-half child less, that is 1.6 births per woman, the population in 2150 would be 3.6 billion, i.e. smaller than today. And with one-half child more (i.e. 2.6 birth per woman), the 2150 population would be 27.0 billion.[5] Thus, small differences in assumptions about fertility and mortality rates make enormous differences as one goes forward in time.

Both fertility and population growth rates are declining in virtually every country in the world, with the exception of the United States. Many populations in Europe may soon cease growing. Population growth in Japan will also soon cease. Furthermore, fertility rates below the replacement level mean that many additional national populations in East Asia also will stabilise.

In Russia, rising mortality among middle-aged men, combined with very low fertility, caused a decline in actual numbers, though not in the rate of growth, during the political tumult of 1995. A survey carried out in Pyongyang concludes that the North Korean population shrank by 3 million between 1994 and 1998, pointing probably to a combination of foregone births and premature deaths in response to floods, crop failure, and famine.[6] The UN and the U.S. Centers for Disease Control suggested in 1998 that political conflict and disease might throw sub-Saharan Africa into a negative growth mode within 20 years, in sharp contrast to its experience of extremely rapid population growth over at least 40 years.

In contrast, the United States is continuing a rate of growth slightly over one percent, and it is far from clear that this rate is declining. Indeed, Census Bureau forecasts of U.S. population size carried out after 1989 have had repeatedly to be raised. The 1989 middle projection estimated a population stabilising in 2020 at less than 300 million. Subsequently, the Census Bureau has several times raised all forecasts in their array of projections. Between 1989 and 1992, for example, it raised the middle projection for 2050 by 28 percent, to 383.2 million.[7] A revision, made in February, 1996, raised that middle scenario to 393.9 million in 2050, with no end to growth in sight.

The estimate for the year 2050 would be higher had the Census Bureau not changed a key assumption: the estimate of the number emigrating annually was raised, so net immigration does not fully reflect the increasingly large number of arrivals. Analysis of the raw data by a private think-tank, the Center for Immigration Studies, suggests that immigration and U.S. births to immigrants accounted for 70 percent of U.S. population growth during the 1990s.[8] The numbers and the rate of population growth seem unlikely to decrease without federal legislation to halt mass immigration.

The demographers Dennis Alhburg and J.W. Vaupel suggested, soon after publication of the 1989 Census Bureau projections, that the Census Bureau estimates were too low. Their calculations projected a most likely mid-century population of approximately one-half billion. 'We conclude that their [the Census Bureau's] high projection might be treated as a reasonable middle forecast'.[9] Peter Pflaumer also projects a larger population – 488.8 million in his year 2050 middle scenario – than foreseen by the Census Bureau.[10]

Looking further forward, Ahlburg and Vaupel state that, 'By 2080 our high projection is more than 300 million people bigger than their [the 1989 Census Bureau's] high projection. A U.S. population of 800 million may seem incredible, but the annual average growth rate that produces it runs at only 1.3 percent per year'.[11]

With remarkable disregard for net legal and illegal immigration numbers that are already substantially higher than their estimate, Table 4 of the Statistical Abstract, 1997, assumes that net immigration will be 820,000 from now until the year 2050. Projections based on this number inevitably under-estimate the size of the future U.S. population, if current trends persist.

U.S. demography does not, however, dominate world statistics. Most population growth is in the Third World. Country by country population surveys and censuses – however flawed – are establishing beyond doubt that the rate of world population growth is declining. This is so after three centuries of increase at an increasing rate

– becoming especially rapid after World War II.

Thus, although the exact rate may be in doubt, it appears that, if present trends continue, the population in 2050 will not exceed 9 billion. Moreover, the ultimate population size, to be reached no later than one century later, would be no greater than 11 billion. This long-range projection is nearly 7 percent lower that the 11.6 billion projected by the United Nations in 1992. The 1998 UN low fertility rate scenario suggests, indeed, a world population of no more than 8 billion, to be reached in less than two score years.

Error is likely to be on the side of a faster decline than demographers report, because of understandable reluctance to acknowledge the various Malthusian checks that appear to be operating. Moreover, UN projections are not without critics among model-builders. Mathematical curve-fitting is an alternate method for forecasting future population size, and the results are substantially different.

Although both mathematical and demographic methods begin with historic population data, this is their only feature in common. The demographic method, as described, depends upon combining the age and gender distributions of current populations with assumptions about future fertility and mortality rates (plus migration flows for country estimates). Assorted values of these parameters are combined to produce an array of projections.

The mathematical method derives the equation – through trial and error – that best fits the historic data. No use is made of assumptions about future fertility and mortality, because the equation alone is used for forward projection. The trajectory of future population growth is simply an extension of the historic curve.

The mathematical method is an adaptation of that developed by the American geologist, M. King Hubbert, who used past oil production figures to estimate all ultimately recoverable oil (and, by subtraction, remaining recoverable oil). Contrary to all accepted wisdom of the era, Hubbert predicted in 1956 that U.S. oil production in the lower 48 states would peak in 1969, plus or minus some months. The production peak came in 1970, well within Hubbert's interval.[12]

Application of mathematical curve-fitting to forecasting future population size is an innovation developed by the ecologist Kenneth F. Watt.[13] Using U.S. Census Bureau figures for global population in years 1950-1996, Watt finds that the equation that best describes the historic curve of modern population growth is:

$$P = 1849 + \exp(-1121 + \cdot8387Y - 6\cdot853E - 8Y^3),$$

where P denotes world population size in millions, and Y represents the actual year corresponding to this population number.

Watt's curve for actual and projected human population size from approximately 1925 to 2070 is shown in Figure 1 (overleaf). A solid line that represents actual population numbers extends to 1996. The points after 1996 are an extrapolation of the historic curve, based on the equation that was derived from fitting to historic data. The model suggests that total world population will peak at less than 7 billion around the year 2019, that is, in less than two decades. The 2019 A.D. population is projected to halve by 2070.

Figure 2 (overleaf) plots the historic data from the beginning of the Industrial Revolution to 1996, and uses the equation to project forward to the period beyond the estimated end of the Oil Interval. The fundamental idea of the Oil Interval is that the world economy depends upon an inheritance of petrochemicals that was

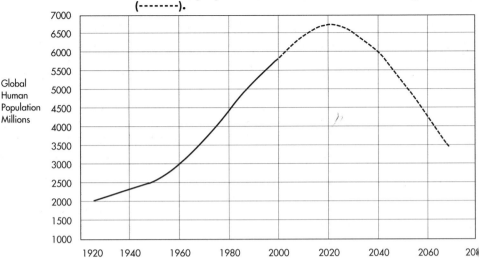

**Figure 1. Total Human Population: Actual data until 1996 (———);
thereafter, projected from mathematical forecasting model
(- - - - - - -).**

Global
Human
Population
Millions

Source: Kenneth F. Watt, 2000. [see Note 13]

stored up approximately half a billion years ago. This dependence on oil as a principal energy source gradually increased from its beginning with Col. E.L. Drake's
first oil well in Titusville, Pennsylvania, which began production in 1859. U.S. oil
production peaked in 1970, and world oil production is expected to peak sometime
around 2012, plus or minus ten years. The exact year of the peak will depend upon
production, which is to say, world demand for energy, in the next decade. Economic
recession reduces demand.

**Figure 2. Total Human Population: projected numbers from mathematical
forecasting model (———) compared with UN medium variant
projection (········).**

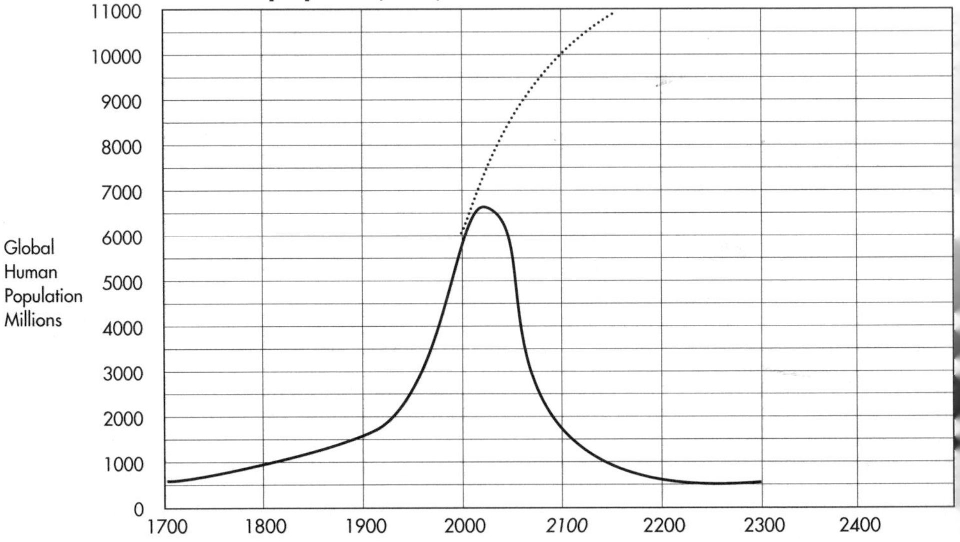

Global
Human
Population
Millions

Source: KennethF. Watt, 2000 and UN medium variant projection.

The largest remaining fields are thought to be in the mid-East and the Caucasus. The transition from cheap and plentiful energy to more expensive and specialised sources is likely to be extremely difficult and may affect both fertility and mortality rates.[14]

The twentieth century availability of cheap and plentiful oil, as well as technology that turns oil into an exponentially growing list of useful applications, may explain much of the recent population growth. Note, however, that projections from the mathematically fitted equation do not depend upon any assumptions about the adequacy of future energy resources to support the current population. The equation derives solely from historic population data.

In summary, the demographic and curve-fitting models both show that the rate of growth of world population is declining. Their projections of the timing and ultimate population maximum are vastly different, however, as shown in Figure 3, the work of Kenneth Watt.[15]

Figure 3. Past Human Population Growth Rates until 1996 and projected rates thereafter: U.S. Bureau of the Census projection (——); mathematical forecasting model (········).

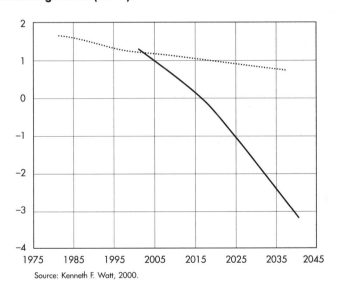

Source: Kenneth F. Watt, 2000.

The thin line in Figure 3 shows growth rates of the actual world population, by year, up to and including 1996, and projections by the U.S. Census Bureau Statistical Abstract – 1997 thereafter. The thick line is a projection of the curve based on the mathematical equation shown above, as in Figures 1 and 2.

Note that the U.S. Census Bureau projection (the thin line after 1996), which is similar to the UN middle projection, indicates a rate of descent that is slower than the computed (thick) trend line. The diverging forecasts mean either that the population trend line revealed by the curve-fitting procedure is wrong or, if it is correct, demographers are underestimating future mortality or overestimating future fertility, or both.

Perhaps official projections are about to repeat the pattern of the 1960s and 70s, lagging behind the rate of change, except that now they will underestimate decline, instead of acceleration, in the growth rate. Such systematic error could reflect a desire to avoid forecasting rapid change, which might be expected to alarm the public.

On the other hand, official sources may be correct; their assumptions about future fertility and mortality rates may justify deviation from a mathematically established trajectory. But demographers have often been wrong in forecasts of fertility, most notably during the 1930s when they confidently predicted continuing low fertility in industrialised countries – they missed the 'baby booms' – and then suggested that 'development' would automatically reduce fertility in the Third World. In my view, the burden of proof is on those who use no theory, or demonstrably wrong theory, to make assumptions about future patterns of behaviour.

Evaluating scenarios

The corollaries of rapid decline in the rate of population growth are that growth will stop sooner than expected and proceed into decline in actual population numbers, that is, negative growth. The two processes that could lead to rapid decline in the growth rate and then to negative growth are: (i) fertility that is far below the replacement rate; and (ii) an increase in mortality.

Fertility

In fact, periods of rapid fertility declines are already well documented. For a six-month period after the collapse of Communism, the East German fertility rate declined to below one child per woman. Similarly, the Russian fertility rate fell to approximately one child per woman in the aftermath of Communism's demise. Very rapid declines, but from a high level, have been observed in Moslem countries in North Africa, Peru, other South American countries, and East Asian countries among others.

A sense of scarcity and insecurity, such as is sweeping many nations and regions, appears to promote caution in entering upon marriage or increasing one's number of dependents through childbearing. Historically and cross-culturally, indeed, delayed marriage, long spacing between births within marriage, and early termination of child-bearing are very frequently observed patterns of behaviour.

The sense of scarcity is a subjective phenomenon. Perceived scarcity does not depend on either absolute poverty or threats to subsistence, so extremity is not a necessary condition of fertility decline. Afflictions no greater than frustrated aspirations or marginal deterioration in economic conditions stir a sense of scarcity in most people. The native-born sector in most industrialised countries has low fertility – not because they live in absolute want, but because aspirations for themselves and their children exceed their grasp.

Stable societies in all times and places have probably had an understanding of scarcity, that is, more attractive uses for resources than could be satisfied. Development economist Georg Borgstrum was cited in 1971, to the effect that,

> A number of civilisations, including India and Indonesia, 'had a clear idea of the limitations of their villages and communities' before foreign intervention disrupted the traditional patterns. 'Technical aid programs . . . made them believe that the adoption of certain technical advances was going to free them of this bondage and of dependence on such restrictions'.[16]

Data from many societies show that understandings of scarcity were the principal source of the motivation to limit family size.[17] Conversely, the perception that resources and opportunity are expanding usually leads to accelerating marriage plans, raising family size targets and, almost inevitably, the number of children born per woman. So great can become the tone of optimism that many people believe that economic security and even windfalls of opportunity will continue indefinitely, and traditional values and practices that once limited family size are cast off.

These responses are summed up by the 'fertility opportunity' hypothesis, an explanation which suggests that most humans have a biological urge toward reproduction and are only deterred by the perception that early or frequent child-bearing is detrimental to the family or individual. In sociobiological terms, selection for sensitivity and responsiveness to environmental cues would be expected to enhance individual 'inclusive fitness'.[18]

Anecdotal data, suggestive rather than conclusive, offer an intuitive introduction to the fertility opportunity hypothesis. Note the regret, even wistfulness, with which people describe their decisions to avoid having children: a Russian mother of two who relied on abortion to limit family size muses, 'I would have had more children if life were better'.[19] In Thailand, a labourer and father of two 'would like to have one more child, but he understands that is beyond his means'.[20] Near Mexico City, a mother of two defends her use of contraception with, 'Things are difficult here . . . Jobs are hard to come by'.[21] Pondering what she would do differently if she were rich, an Ethiopian mother of five says, 'If I were wealthy, say if I had horses and a better house, I'd have more children'.[22]

A few examples of the associations of growing optimism with high (even rising) fertility – and of the sense of contracting opportunity and scarcity with declining fertility – are summarised below. I have also predicted fertility declines (from already low levels) in the former 'tigers' of East Asia that experienced shattering economic collapses after summer 1997,[23] and I confidently await these data.

In fact, fertility declined in many developing countries after collapses in world commodity prices, a reduction in international aid, disillusionment with experiments in democracy as well as with populist revolutions, or increasing competitive stresses within the lower and middle working class. For example, phosphates were Morocco's key export when a price collapse in 1974-75 forced withdrawal of government subsidies on food, education, and health care. This reduction in families' real income, coupled with higher taxes, added insult to injury, and was immediately followed by a decline in the fertility rate – 20 percent over 4 years.[24]

Similarly, after two decades when Peru's fish exports and fertility rate had risen in tandem, the anchoveta industry reached a plateau in the 1960s. This decline in economic expectations was followed by a fertility decline that accelerated as the economy collapsed in near free-fall and political institutions became chaotic.[25]

Further, the growing wealth disparity between the upper-income 20 percent and almost all others in certain industrialised countries appears to be depressing the fertility rate of the native-born population.[26] The consumption ethos that overtakes the middle class in many societies also promotes caution before undertaking to establish a family. In a consumer society, whatever one has could be more – and caring for children is usually in competition with the acquisition of goods.

World-wide, in fact, people with enlarged aspirations are being forced to become more self-reliant, and most no longer have unquestioning faith that their government or international aid will make the future brighter. This shift in expectations is associated with declining fertility. Even in sub-Saharan Africa and countries such as Pakistan and most oil-rich – but temporarily cash-poor – Arabian countries, fertility rates that were persistently high for decades are now falling rapidly.[27]

The experience in Nigeria combines many of the elements cited above. During the 1970s to mid-1980s, oil wealth and international assistance rolled into Nigeria, the infant mortality rate declined, health-care and education programs became widely available, and many more women had access to education. Consistent with prevailing theory associated with the demographic transition-development literature, these elements of modernisation were expected to reduce fertility rates from the very high level where they had remained for over two decades. As the 1980s rolled on, however, early marriage, closely spaced births, and disinterest in contraception showed no sign of abating. While disappointing, the lag was dismissed as inconsequential when, finally in 1990-91, the Nigerian fertility rate began to fall. The operative factors, later marriage and a new level of receptivity to modern contraception, were readily attributed to development and modernisation.[28]

A different interpretation is possible. The fertility opportunity hypothesis finds support in both the initial persistence of high fertility, and then the timing of the eventual fertility decline. The unprecedented advantages enjoyed by the average person and the pervasive optimism of the earlier period predicts no fertility decline. In contrast, the economic downturn, diminishing international aid, reduced subsidies for education and health-care, more difficulty in finding remunerative employment, and widespread disappointment in the late 1980s and early 1990s, is consistent with individual adaptive efforts that include delayed marriage and avoidance of child-bearing. Indeed, Yoruba-speakers interviewed in two Nigerian villages said that an important factor in their decision to use contraception was the 'hard economic times'.[29] Specifically, 'Most of the respondents believed that child mortality had risen over the previous five years of economic difficulty',[30] and 'Two thirds of all respondents claimed that the major force behind marriage postponement and the use of contraception to achieve it was the present hard economic conditions'.[31]

A search of the literature turns up many more parallel instances, of which only a small fraction is mentioned here. The conclusion I draw is that fertility is falling at an accelerating rate, nearly world-wide, because individuals and families are aspiring to consumption opportunities they cannot afford. Alternately, they are trying to adapt to, and avoid, the worst effects and insecurities of the deteriorating environmental, social, and economic conditions around them.

The cause of such deterioration is, in some cases, population growth itself. Under these circumstances, population overshoot – that is, growth in excess of the long-term carrying capacity – is self-correcting. What is less easy to contemplate is that destruction of the carrying capacity also leads to loss of species diversity, the impoverishment of human populations and, indeed, much suffering.

Mortality

Mortality is the further dimension that may be changing very fast. In fact, the UN acknowledged in 1998 that AIDS-related deaths are changing population projections for sub-Saharan Africa and parts of south-eastern Asia. In addition to the rising burden of disease, various political disruptions may be associated with loss of life and a greater toll than is reported internationally – or, at least, reported in time to be integrated into the most recently available demographic projections.

Disease as the source of rising mortality is likely to involve pathogens as yet unknown, as well as tuberculosis, HIV/AIDS, and malaria.[32] The likelihood that new diseases will arise relates to the increasing probability that infectious pathogens will jump the species barrier from other animals to humans, as the human population grows and people increasingly impinge on the natural habitat of other species. In fact, it is now thought that the HIV virus jumped from chimpanzees, where it is harmless, to humans about thirty years ago.

Mostly, but not wholly, in the Third World, tuberculosis, HIV/AIDS, forms of hepatitis, and malaria have been becoming more prevalent, as more people crowd together. Travel between continents speeds transmission. Cholera, leprosy, tuberculosis, malaria, and hepatitis B and C have all been detected in the United States within the last dozen years as the result of contagion from overseas. Tuberculosis, for example, was nearly eliminated in the United States by the 1980s but has made a strong comeback, particularly in cities with large foreign populations and among vulnerable patient groups such as those infected with HIV.

New treatments and preventive measures have led to decline in the annual number of new HIV infections in the United States and certain other countries including Thailand, but HIV/AIDS is nonetheless reaching unforeseen proportions in some parts of the Third World. In Thailand, the prevalence of HIV infection in the general population peaked at 3.7 percent in mid-1993 but levelled off at 1.9 percent by 1997;[33] in contrast, prevalence probably remains high in Myanmar (Burma) and southern China, the other points of the Golden Triangle. Most Asian and African governments did not follow Thailand's lead in taking an extremely proactive public health approach toward the epidemic.

HIV/AIDS is rising in India, particularly within the lower classes. Truck drivers spread AIDS between communities; prostitution transfers it into the resident heterosexual population.[34] This transmission vector also accounts for the early spread of the HIV virus in sub-Saharan Africa during the 1970s. In 1993, the U.S. Centers for Disease Control modelled HIV/AIDS transmission in East Africa, by which time its prevalence was highest among the upper class. At that time, the AIDS epidemic, alone, was expected to reduce annual population growth in sub-Saharan African by about one percentage point by the year 2015, that is from nearly 3 percent to less than 2 percent.[35]

Subsequent reports suggest that the epidemic in Africa is more widespread and more serious than previously thought. The prevalence of HIV infection among adults, estimated up to 1997, includes a stunning 12.8 percent in South Africa with rates of 'at least 25 percent' in Botswana and Zimbabwe and 4.1 percent in Nigeria. Outside Africa, countries where prevalence is high include Cambodia (2.4 percent) and Honduras (1.46 percent). World-wide, nearly twice as many people as previously thought are being infected annually, and the incidence of new cases has doubled.[36] 'AIDS is now one of the leading infectious causes of death in the world and has brought down life expectancy at birth in the worst-affected African countries to levels of the 1960s'.[37]

High young-adult mortality from AIDS results in many children being orphaned. Moreover, between 17 and 25 percent of babies born to HIV-infected mothers are also infected with HIV if the mother and infant are not treated perinatally, and the virus is further transmitted through breast-feeding. The result is a 50 percent increase in infant and child mortality.[38]

Another contributor to rising mortality is famine. Death is seldom attributed to starvation, however, because malnutrition creates vulnerability to opportunistic disease, and death often intervenes at an early stage. Such vulnerability particularly affects the young. Mortality and morbidity among pre-adolescents due to contagious disease and infection – rather than by the chronic diseases of the elderly – is reminiscent of the mortality regime through the nineteenth century in many parts of the world including Europe.

Mortality rates have fallen steadily through the twentieth century. In consequence, perhaps, the magnitude of premature mortality sometimes goes unrecognised. An early-1960s famine in China in the aftermath of the failed 'Great Leap Forward' contributed to a probable thirty million deaths. During the 1970s in Nigeria, the Hausa were instrumental in isolating and starving several million Ibo in reprisal over Ibo claims to control the rights and profit from oil found in their homelands. The international community intervened to avert famine in Ethiopia, the Sahel, and Somalia during the 1970s and 80s. But famine, reportedly, had caused 2 million deaths in North Korea by spring 1998, in part because North Korea refused to give guarantees of nuclear non-proliferation in exchange for aid. This said, it is unclear that the international community will be able, even if willing, to forestall devastation from most future famines and outbreaks of disease.

A factor that often drives famine and outright killing is ethnic, religious, or clan conflict. In Somalia, fighting continues among ethnically-almost-identical opposing clans despite major 'peace-keeping and nation-building' efforts by the United States and UN – action which, incidentally, resulted in loss of life also among the military units detailed to help. In Rwanda, Burundi, and the Democratic Republic of Congo (Zaire), ethnically distinct Hutu and Tutsi tribes remain locked in a long-standing contention that no international peace-keeping force seems able to neutralise. Antagonists in Indonesia are the indigenous peoples and overseas Chinese. Sudan is divided by the antagonisms of Arab Muslims and Sub-Saharan African Christians. The former Yugoslavia pits Muslims, Greek Orthodox Christians, Russian Orthodox Christians and Albanians each against the other, although all but Albanians are Slavs by ancestry. Ethnic diversity in the Caucasus region, including Eastern Turkey, seems likely to be a non-ending source of dispute, feuding, or guerrilla action. This short list of conflicts accounts for violent deaths climbing into the millions.[39]

Such conflicts result in population loss at the same time that population pressure may be a factor that drives conflict.[40] Pursuit of goals by armed force often requires one to define the enemy as 'the other.' Ethnic cleansing is the whole point when one group believes that its living space and opportunity – call it carrying capacity – have been usurped.[41]

Thus, localised instances of conflict, separate from and yet contributing to malnutrition and outbreaks of disease, suggest that an increase in the mortality rate is occurring and must soon be reported, even though most developing countries have a very young population, which lowers the general mortality rate.

Thus, a mortality higher than that assumed in current projections would be the second factor, along with rapidly declining fertility, which would preclude future world population from attaining the high levels now foreseen by the UN and by the U.S. Census Bureau.

What to wish for

'Be careful what you wish for, because you may get it,' is ancient wisdom. Wishing for declining rates of population growth, population stabilisation, and eventual reduction in human numbers has come to be a habit. Retrenchment to lower levels of population in all continents could lay the basis for environmental sustainability. I have no doubt that reduction in human numbers is devoutly to be desired, but getting decently from here to there is the challenge.

The UN middle scenario for stabilising population size is probably too gradual to avert continuing catastrophic losses in species diversity and environmental quality in all parts of the world. This scenario leads in the long run to enormous human suffering – and perhaps to non-recoverable losses of human genetic pools and civilisation. On the other hand, population stabilisation and decline within the next two or three decades, as projected by the equation-fitting method, can occur only if there is a substantial short-term rise in mortality. One assumes that such population loss would not be allowed to occur unless major industrial nations were unable – because of their own travails or the scope of disasters – to intervene. How could this happen?

The population-environment connection

The controversy surrounding the upper limit to the carrying capacity of the Earth often reverts to Malthus – was he right or wrong? Will human numbers inevitably outpace food production or devastate the quality of most people's lives?

This battle is often waged on the economic rather than the environmental front. Although most experts in international development have concluded that rapid population growth is an obstacle to economic growth and human dignity, some analysts of industrialised countries assert that population growth is an essential ingredient of prosperity. A common fear is that population stabilisation or negative growth leads to economic stagnation. The recession in Japan, where population size is nearly stable and 15 percent of the population in 1997 was over age 65, is alleged to support this hypothesis. The prescribed correctives include higher birth rates and immigration.

Those who dwell on the positive aspect of a growing population tend to confound growth in GDP with economic health. However, the economic well-being of the average person is reflected in *per capita* GDP, rather than the aggregate figure. Moreover, a distinction should be made between mean *per capita* income, which is influenced by extreme values, and the median value, which is a better indicator of the distribution of wealth. Median income describes the economic condition of the middle class.

Bearing on the prosperity of the middle class is the relationship between labour supply and the availability of capital and technology. In fact, growth in the supply of labour must be matched by an appropriate increase in capital investment, or else labour productivity declines and the real compensation for work (inflation-adjusted salary and wages) declines as well.

The capital requirement for creating a job comparable to those in the U.S. varies from approximately $150,000 to $200,000 per net new job, and capital is in perennially short supply – otherwise the interest rate would be zero. When the number of workers increases in the absence of commensurate investment per job, wages and salaries are forced down by competition. This process is occurring world-wide, as globalisation, instanced by free trade and immigration, prevents a nation from protecting its own people.

Economic discussion, however valid, should never obscure the relationship between human well-being and natural resources. Nations not endowed with natural resources do everything they can to import them, and some financial and trading powers – such as Great Britain, the Netherlands, Singapore, and Japan – have been notably successful. As a matter of logic, however, not all nations can be net importers of natural resources. Economics does not make the Malthusian question disappear.

The clear fact that more people are supported today than ever before, and at a higher level of material well-being, obscures four important observations about resources. First, world population growth has been enabled by technology that uses increasingly concentrated forms of energy, progressing from wood to coal to petrochemicals. Second, fossil fuels are limited in quantity. Third, many essential renewable resources – such as groundwater, timber, and topsoil – are being depleted at rates that exceed their ability to be replenished or reconstituted. Fourth, the confident assertion that technology will somehow save us, is more often heard from politicians than from scientists and engineers.

Energy sources, particularly oil, are especially important in modern human history. A good example is nitrogen fertiliser, which is manufactured in a fossil-fuel-intensive process perfected in the 1920s. Vaclav Smil[42] shows that the introduction and more widespread use of nitrogen fertiliser was followed, in one country after another, by exponential growth in population density and argues convincingly that new applications of energy to agriculture have, in fact, played the key role in enabling population growth. In the United States, indeed, about 17 percent of petrochemical use is dedicated to the agricultural sector.[43] The present level of food exports could not be maintained if agriculture were deprived of oil. *Per capita* energy use is an accepted proxy for standard of living.[44] Each new energy source increases the material well-being of the population – until population growth diminishes the amount of energy available per person. As an illustration of this point, energy consumption *per capita* in the United States, barely increased between 1970 and 1990, although total consumption grew by approximately 25 percent; population growth accounted for 93 percent of the increase.[45]

Between 1991 and 1996, a period of substantial economic growth, total energy use in the United States grew at an annual rate of 2.14 percent. Population growth in this period was just over 1 percent annually, so it accounted for not quite half of the increase in total consumption.[46] If continued, a 2.14 percent annual rate of growth doubles energy consumption in just under 33 years. Not only the dwindling supply of domestic oil, but also the 'greenhouse gases' produced as it is burned – which may be altering climate – suggest that this rate of increase is unsustainable.

No known source of energy, even solar, fully substitutes for oil. All of the nation's plant biomass together captures less than half of the solar energy equivalent of the fossil fuel energy burned annually.[47]

Oil production is projected to peak within 20 years, and then the world will turn for its petrochemicals to natural gas, with a life-expectancy at this higher rate of use of about 40 years. Coal, if it must shoulder the burden of being the world's major energy source, is projected to be adequate for about 60 years after this. By then nuclear fusion may (or may not) replace fission in the production of electricity, and assorted localised technologies such as solar and wind power have the potential to supplement electricity supply. No known or likely technology seems likely, however, to be as cheap, plentiful, and versatile as was oil in the twentieth century.[48]

Thus, the U.S. and world agricultural sectors seem at risk from the expense of energy inputs, even if topsoil and aquifer depletion were not a concern. However, topsoil in arable regions of the United States is being depleted about 13 times faster than it is rebuilt under ordinary conditions of agricultural use, and underground water is being consumed at a rate about 25 percent greater than it is replenished. After having been farmed for about a century, Iowa – with some of the richest agricultural lands in the nation – has lost about half of its topsoil. Three million acres of U.S. farmland or forest are degraded or paved over annually.[49] Modern agriculture is sometimes described as a process for using land to turn oil into food. With less land and oil, the world must await the revolutionary technology of post-modern agriculture.

Conclusion

Environmental and conservation battles are perennially re-fought because the needs of people, reasons for intruding into the habitats of wild species, and the imperative to use now rather than to save, grow approximately in proportion to the number of people. People's wants trump protection of natural resources. Consumption habits compound the effects of large populations. Consumption and population in excess of the carrying capacity degrade the environment. Environmental degradation shrinks the carrying capacity, so fewer people can be supported on a sustainable basis in the future.

It is small wonder that numerous students of carrying capacity, working independently, conclude that the sustainable world population, one that uses much less energy *per capita* than is common in today's industrialised countries, is in the neighbourhood of 2 to 3 billion persons. Note the congruence with Watt's projection of rapidly declining population size near the end of the Oil Interval. The absence of cheap, versatile, and easily used sources of energy, and other resources, seems likely to change the quality of human life and may even change, for many, the odds of survival.

Do such scenarios suggest that certain economic and demographic policies should change? Does one deny that demographic facts and their interpretation have potentially major social and economic ramifications? The demographers Dennis Ahlburg and James Vaupel are not in doubt. They write:

> 'Population projection is not a bloodless technical task, but a politically charged craft of great interest to policy makers and the public'.[50]

Chapter 5

Diversity: insurance for life

Sir Ghillean T. Prance FRS

The Author

Professor Sir Ghillean Prance FRS, VMH is Scientific Director of the Eden Project in St. Austell, Cornwall and was Director of the Royal Botanic Gardens, Kew from 1988 to 1999. Prior to that he spent 25 years holding various positions on the staff of the New York Botanical Garden and was Senior Vice President and Director of the Institute of Economic Botany from 1980 to 1988. He spent much of his ealier career leading a series of botanical expeditions to Amazonia. He was awarded the first International Cosmos Prize in 1993 and has received 12 honorary doctorates. He is a visiting Professor at Reading University.

Chapter 5

Diversity: insurance for life

Sir Ghillean T. Prance, FRS

Introduction

Plants are the basis of life on Earth. Without their capacity, through the process of photosynthesis, to use sunlight in combining water and carbon dioxide to produce sugars, life as we know it today would not exist. Plants are green because they contain chlorophyll which is capable of absorbing the energy from sunlight and causing a series of chemical processes that lead to the synthesis of a series of organic compounds and the release of oxygen into the atmosphere. This process, which is in fact extremely complicated, can be summarised by the following empirical equation:

$$CO_2 \quad + \quad 2H_2O \quad \xrightarrow{\text{chlorophyll}} \quad [CH_2O] \; + \; H_2O \; + \; O_2$$

Carbon dioxide   water   light   carbohydrate   water   oxygen

Whether this is done in a single-celled *Euglena* or an oak tree, it is this capacity to fix energy and produce oxygen that sustains most of life on our planet. Animals cannot photosynthesise and so they are dependent on plants somewhere at the base of their food chain. This means that the future of the human race also depends upon maintaining a diversity of plant species as well as the animals with which they interact in interdependent relationships. Yet we are throwing away our capital by allowing so many biological species to become extinct. This chapter seeks to demonstrate that, if there is to be a human future, then we must take greater care of our fellow species upon which we depend. The conservation of other species is basic to the survival of the human race on Earth, and it depends upon the preservation of the habitats in which they live and enough genetic diversity for them to adapt to the future.

Biodiversity

Before looking at the decline in species or the way in which we should care for them, some facts about diversity will help to put the actual situation into context. The relatively recently created term 'biodiversity' is a useful one because it covers the three essential components of biological variation.[1]

1. The diversity of ecosystems[2] that occur in the world, whether large and extensive such as rainforests, coral reefs or boreal forests, or much more restricted, such as the complex of bacteria that inhabit a human body or the myriad interacting micro-organisms that inhabit a gram of soil.
2. The diversity of species including all the more familiar plants, animals and fungi, but also the less conspicuous bacteria, protozoans and viruses.
3. The genetic variation within the population of a single species that causes it

to vary in its appearance (or phenotype) or in its ecological responses to local environments. The occurrence of genetic variation is crucial to the process of evolution, because it enables different individuals to react differently to different external selection pressures such as climate change.

Biodiversity is an important concept because it focuses attention on the relationship between ecosystems, species and genes. It is an idea that has developed in response to a time in history when the loss of species, habitats and genetic diversity has become a major preoccupation of biologists. In order to maintain life on Earth in the future, it is essential to consider the interactions between all three components of biodiversity. The disruption of any part of biodiversity has an effect on the other components and so an adequate conservation policy for the future must take into consideration all three aspects; for they are inextricably woven together in the intricate web of life.

Numbers of species

Species are the basic units of life without which the other components of biodiversity would not exist. Each habitat supports a group of species which, in their turn, carry the genes that comprise genetic diversity. Species conservation, therefore, is vital to the conservation of all biodiversity. Such, however, are the intricate relationships surrounding biodiversity, that species cannot survive in isolation, without a habitat in which to live and other species with which to interact.

The diversity of species around us today is the result of four billion years of evolution which include many phases of natural extinctions and the further evolution of new species. Estimates of the present number of species on our planet range from five to fifty million and many people agree that ten million is a conservative estimate.[3] However, only about 1.7 million of these have so far been named and classified. We are, therefore, losing many species before they have been recorded or their role in sustaining life on Earth has been examined. The uncertainty about how many species exist is mainly because of the vast number of insects, fungi, bacteria and nematodes that remain undescribed. The reliability of the estimates of species numbers shown in Table 1 varies greatly from one group of organism to another. About 4000 species of bacteria have been described, but a study in Norway indicates that there are between 4000 and 5000 species in a single gram of soil from a beech forest! Thus, the present estimate, that there are 400,000 species of bacteria, could be a considerable underestimate. Although only 70,000 species of fungi have been named and described, the mycologist, David Hawksworth, has estimated that there may be as many as 1.5 million species.[4] But, for some groups of organisms, we can be confident about the statistics; for example, in 1990, there were 9,881 known species of birds,[5] and the estimate of 300,000 species of seed plants seems reasonably accurate. Such showy insects as butterflies are well studied but we have scarcely begun to classify other insects. We must use the statistics from these better-known groups to extrapolate numbers and to develop conservation policies for all groups of organisms.

Species do not occur evenly around the world and, in order to protect them, therefore, we need to know about the distribution and density of species in different countries and in different habitats. McNeely et al.[6] estimated that, in 1988, 70 percent of the total species diversity in the world occurred in only twelve countries: Australia, Brazil, China, Colombia, Ecuador, India, Indonesia, Madagascar,

Table 1. Numbers of species of major groups of organisms described and estimated to exist

Organism	Number of species described	Estimated number of species
Viruses	5,000	500,000
Bacteria	3,058	400,000
Fungi	70,000	1,000,000
Protozoa	40,000	200,000
Algae	40,000	350,000
Lichens	13,000	17,000
Bryophytes	14,000	18,000
Pteridophytes	10,000	12,000
Gymnosperms	600	600
Seed plants	250,000	300,000
Insects	950,000	8-50 million
Arachnids	75,000	150,000
Molluscs	70,000	200,000
Nematodes	15,000	100,000
Birds	9,881	9,900
Other vertebrates	35,000	40,000
TOTAL	1,600,539	

[Source: Data adapted from Groombridge, B. (Ed.) (1992) *Global Biodiversity: Status of the Earth's Living Resources*. Report compiled by the World Conservation Monitoring Centre. Chapman Hall, London.]

Table 2. Eleven of the most species diverse countries in the world for higher plants, and the ten countries with the highest numbers of restricted-range bird species

Country	Number of plant species	Country	Number of restricted-range bird species
Brazil	55,000	Indonesia	410
Colombia	35,000	Peru	210
China	30,000	Brazil	200
Mexico	25,000	Colombia	190
South Africa	23,000	Papua New Guinea	170
Former USSR	22,000	Ecuador	160
Indonesia	20,000	Venezuela	120
Venezuela	20,000	Philippines	110
USA	18,000	Mexico	102
Australia	15,000	Solomon Islands	98
India	15,000		

[Sources: Groombridge, B. (Ed.) (1992) *Global Biodiversity: Status of the Earth's Living Resources*. Report compiled by the World Conservation Monitoring Centre. Chapman Hall, London; and Bibby, C.J., N.J. Collar, M.J. Crosby, Ch. Imboeden, T.H. Johnson, A.J. Long, A.J. Stattersfield & S.J. Thirgood (1992) *Putting biodiversity on the map: priority areas for global conservation*. International Council for Bird Preservation, Cambridge, England.]

Malaysia, Mexico, Peru and Congo. These have been termed 'megadiversity' countries. Table 2 compares the number of plant species in the eleven most species-diverse countries (seven of which are on McNeely's list) with the ten countries containing the highest number of birds species whose range is restricted. These are obviously priority countries for species conservation.

Some types of ecosystem are also much more species diverse than others, and rainforests are in first place for number of species. The forest with the greatest known diversity of tree species is at Serra Grande in Bahia, Brazil, where the botanist William Wayte Thomas found 450 species of trees measuring five centimetres or more in diameter in a single hectare of forest. 307 species of trees with a diameter of 10 cm. or more have been reported to occur in a hectare of Ecuadorian rainforest,[7] and at Yanomono, near to Iquitos in Peru, 300 species of trees and lianas have been found in a single hectare.[8] Table 3 shows the diversity of tree species in some rainforest areas; this varies from 307 to 81 species, depending upon local conditions such as soil, topography and rainfall. Forests in more humid climates, without a pronounced dry season, tend to have a greater diversity of forest species than those where there is an extended dry period. In addition to the wealth of tree species, rainforests contain a large number of herbs, shrubs, and epiphytic and parasitic plants that usually more than double the total number of species recorded in any area. In the tiny Rio Palenque Reserve in western Ecuador, 365 species of vascular plants have been found in a tenth of a hectare.[9] The total number of species recorded for this Reserve, which covers 1.7 square kilometres of rainforest, was 1033. In the rainforest of Costa Rica, 233 species of plants were found in just 100 square metres[10] – an area half the size of a tennis court contained about a sixth as many species as the total flora of the British Isles! Ten hectares of forest in Borneo contained 700 species of tree – the same number as in the whole of North America.[11]

For most groups of terrestrial organisms, rainforest areas usually have the greatest species diversity. For example, the Amazon basin has 2000 species of fish and, second to this, is the basin of the Congo River with 790. Both of these rivers are surrounded by rainforest for most of their length and, between them, account for just over 33 percent of the world's total of 8,400 freshwater fish species. It is insects, however, that contribute the greatest species diversity to rainforest and to many other ecosystems. Insects account for 56.4 percent of all described species and, as more become described, the proportion is certain to increase eventually to at least 65 percent. A canopy tree in tropical rainforest acts as host for at least ten times as many insect species as a tree of the same size in temperate latitudes. In recent years, insect diversity in tropical forests has been studied by the technique of fogging.[28] In one instance, 90 percent of the insects sampled in this way were undescribed. It is this richness of insects in the rainforest canopy, as well as the unknown masses of species of all groups on the seabed and soil, that lead to such uncertainty about the total number of species in the world. In fact, extrapolations from various fogging experiments led the entomologist, Terry Erwin, to advance the much debated figure that there are 30 million species in the world[29] – a figure which he reached through calculations stimulated by finding 1,200 beetle species in 19 trees of the tropical linden tree (*Luehea seemannii*) in Panama.

The high biodiversity of rainforests is well known; but many people are unaware that their complexity is rivalled by that of coral reefs, the richest of all

Table 3. Tree species diversity for some tropical rainforests

Locality	Number of species per hectare	Minimum diameter (cm)
Cuyabeno, Ecuador[12]	307	10
Yanomono, Peru[13]	300	10
Mishana, Peru[14]	295	10
Chocó, Colombia[15]	255	10
Añangu, Ecuador[16]	228	10
Papua New Guinea[17]	228	10
Gunong Mulu, Sarawak[18]	214	10
Pasoh, Malaysia[19]	210	10
Cocha Cashu, Peru[20]	201	10
Manaus, Brazil[21]	179	15
Tambopata, Peru[22]	168	10
Xingu River, Brazil[23]	162, 133, 118	10
Breves, Brazil[24]	157	10
Oveng, Gabon[25]	131	10
Queensland, Australia[26]	108	10
Alto Ivon, Bolivia[27]	81	10

marine environments. The largest group of coral reefs, the Great Barrier Reef of Australia, contains over 300 species of coral and supports 1,500 species of fish. This area, which covers only 0.1 percent of the ocean surface, is the home of 8 percent of the world's total fish species. In addition there are 4,000 species of molluscs; and 252 species of birds nest on the coral cays. Other marine habitats also foster much unknown biodiversity. The most recently discovered new phylum of animals, the Loricifera, was described only in 1983, since when other new species of this benthic group have been discovered.

Many other areas of vegetation are species rich and need the special attention of conservationists; for example, the Cape Region of South Africa, especially the communities of the *fynbos*, is even richer in plant species than rainforest. Southwestern Australia is another particularly rich area and many oceanic islands that have been isolated over long periods are full of endemic species of plants and animals. Some figures for Hawaii, where many of the endemic species of plants and birds are already extinct, are given in Table 4.

The conservation of ecosystems

Early work by conservationists tended to concentrate on the preservation of species, leading to many lists of threatened and endangered species of plants and animals. But, as the practical aspects of conservation of these species were better studied, a greater emphasis has gradually been placed on ecosystem conservation. It is obviously more cost effective to operate conservation at the ecosystem level than at the level of the single species. 'Conserving biological diversity equals conserving ecosystems'.[30]

Species do not exist alone; rather, they are members of a community within which each has a distinct niche and role. Each species is dependent on others within the community in some way; for example, many flowering plants depend upon animal pollinators for cross pollination and these animals, for their part,

Table 4 – A summary of the statistics for native species of Hawaii

	Birds	Plants
Number of native species	67	980
Endemics	49	874
Extinct	18	23
Endangered	20	237
Population structure of endangered plant species of Hawaii		
1 individual left	8	
2-10 individuals left	38	
11-100 individuals left	82	
101-1000 individuals left	78	

depend upon the plants for food. A community is bound together in a loose web of mutualistic relationships on the one hand and by competition between species for survival on the other.

I will illustrate this from two Amazonian species I have studied. First, the Brazil nut (*Bertholletia excelsa*), an economically important tree of the Amazon rainforest. The flowers have a most complex structure and can only be pollinated by large bees that are able to lift up the hooded androecium to gain access to the nectar within. The males of these Euglossine bees, in turn, visit various species of orchids to gather scent which they pack in special pockets in their hind legs to attract female bees for mating. The large woody fruit of the Brazil nut falls from the tree to the forest floor, and only a rodent, the agouti (*Dasyprocta* sp.), has teeth strong enough to open it and extract the nuts or seeds, which it then buries in caches some distance from the trees. Also, the nutrient uptake of the Brazil nut is assisted by mycorrhizal fungi in the soil. Thus, we see that the Brazil nut is dependent for survival on a canopy bee, a terrestrial rodent, fungi and orchids. Such is the intricacy of the inter-relationships that make up biodiversity.

Another Amazonian plant of considerable economic importance is the calabash (*Crescentia cujete*), the fruit of which forms the most useful utensil for the local people of the Amazon region, who use it as a water pot, soup bowl, canoe bailer and even as a hat. The mature fruit can reach the size of a football and has an extremely hard outer shell which is why it is so useful. The green flowers of the calabash tree are produced on its trunk and branches and are, therefore, easily accessible to bats, which visit the flowers to sip their nectar and, in the process, cause pollination. The flowers are not the only nectar-producing organs of the calabash. The young developing fruit is studded with minute whitish glands which exude abundant nectar. Once this nectar appears, the fruit is attended by numerous aggressive ants which rigorously defend their source of food from any other animal. The developing fruit is protected by the ants; but, as it matures, the nectaries dry up and the fruit becomes hard and woody. It is now protected mechanically and no longer needs the ants; and it is the hard woody shell which makes it so useful. Thus, this multipurpose fruit abounds along Amazonian riversides because of the bats which pollinate it and the ants which protect it through adolescence. This is another example of the interdependence between the different species which make up any ecosystem. To lose one, in this web of interactions, will have a profound effect on others. Thus, in order to formulate adequate conservation strategies, it is essential to understand the way in which species interact through pollination and dispersal relationships, predator-prey relationships, defence mechanisms, decomposition of biomass and many other connections.

An ecosystem also depends upon the ecological processes that occur within it and are performed by the species that it contains. These processes, such as recycling of nutrients, the maintenance of soil fertility and water quality and even the regulation of local climate, depend upon the component species. The removal of some species may not affect these processes greatly; but the removal of certain 'keystone' species upon which the others depend will cause the breakdown or radical change of that community. For instance, there are communities and plant species in Africa whose survival is dependent upon the disturbance caused by elephants. If this keystone species, the elephant, is removed, then many other species will gradually die out. Often a predatory species that controls the population of herbivore species is the keystone. If it is removed, the herbivores will overgraze, leading to the loss of both plants and other animals.

Genetic diversity

Genetic diversity is the variation that occurs *within* a species or a group of individuals. In natural populations of a species there are small differences between individuals and larger ones between populations. For example, humans constitute one species but it is possible to recognise individuals because they have genetic differences (except identical twins who are also genetically identical) and to distinguish the more general features of certain populations such as the different races – Caucasian, Mongolian, African, Polynesian etc. The genes are borne as discrete segments of DNA molecules that are themselves made up of four different nucleotide bases; these form the genetic code that determines the characteristics of an individual. Each gene is formed of up to several thousand nucleotide pairs. Genetic diversity arises from the numerous possible variations of this apparently simple genetic code. Variation is caused by mutations, which are changes of both the genes and of their arrangement on the chromosomes; and is spread in a sexually outbreeding population by the process of recombination. Since an average organism contains about 100,000 genes, the possible combinations are infinite; and it is estimated that our planet contains some 10^9 different genes. Some genes that control basic processes, such as photosynthesis in plants, are extremely stable and show little variation, while others are extremely variable, such as the genes that control flower colours. Variation also occurs in the amount of DNA *per* cell and in the number of chromosomes within an individual. Both of these also contribute to the genetic diversity of organisms.

The greatest significance of genetic diversity is that it allows evolution to take place. In a natural population, selection occurs for the combination of characters that is best suited for the environment in which that population occurs. It thus enables an organism to adapt to environmental change. Selection acts upon the mutations occurring by changes in the genes themselves, on the recombination of parental material within a new individual and upon rearrangement of genes on the chromosomes. If the expression of a new genetic combination – for example resistance to a disease – is more favourable to an organism, then it is more likely to survive to reproduce and form part of the next generation than another individual that does not have this character. Conversely, if a change is harmful or lethal, which the majority of genetic changes are, it is less likely to survive. Many of the genetic changes that take place in organisms are neutral in their effect and are likely to survive, neither influencing evolution nor reproduction, but adding to the genetic diversity within the species. Ultimately all genetic diversity has been derived from the process of mutation.

It is vital to preserve the genetic diversity of a species in order to maintain its potential for evolution and its ability to survive. Species with little genetic variation are less able to adapt to new conditions such as changes of climate. The effects of inbreeding with similar genetic material, on the other hand, soon produces deleterious and often lethal effects. The entire world population of cheetahs has been found to be genetically almost identical. It is therefore not surprising that, in both wild and captive populations, abnormal animals are frequent, sperm fertility is reduced and they are easily susceptible to a number of diseases. At some stage in their history, cheetahs were reduced to a small number of genetically similar individuals and the species has not yet recovered from this genetic bottleneck. It could, therefore, be much more prone to extinction than other species of wild cat.[31]

Many genes exist in several forms ('allelomorphs' or 'alleles'), each expressing a particular character in a different way, for example hair or eye colour in humans. Some alleles, usually the most favourable, are dominant over others and an individual only shows the characteristic of a non-dominant or recessive allele if it has inherited it from both parents. A good example of this is haemophilia. A haemophiliac is only born if both parents carried the gene and both contributed the recessive allele to their offspring.

Humans have made much use of the genetic variation of plants and animals to change natural species into more useful ones by the gradual artificial selection of certain traits. For example, all our domestic dogs (*Canis familiaris*) are derived from a single species, the wolf (*Canis lupus*). The Saint Bernard and the Chihuahua are very different in appearance and behaviour, but breeders have been able gradually to select for the characteristics they wanted because of the genetic variation available in the original species. The vegetables – cabbage, cauliflower, broccoli, kohlrabi, kale and Brussels sprouts – all come from the same original species. Breeders have produced six different vegetables by selecting for edible leaves in cabbage and kale, edible flowers in broccoli and cauliflower, edible buds in Brussels sprouts and edible roots in kohlrabi; and, again, this is only possible because of the genetic variation within the original species, *Brassica oleracea*.

Molecular techniques enable us to fingerprint the genetic make-up of an individual or to study the genetic variation of a population. For the conservation of species these techniques have many applications. In programmes for the breeding of rare species, it is vital to know the degree of genetic similarity of individuals, so that unrelated individuals can be crossed and thus avoid the deleterious effects of close inbreeding. This is because many species of plants and animals, whose populations have become small, will require careful genetic management if they are to survive. Inbreeding tends to reduce genetic variation, and so it is to be avoided at all costs, if the purpose of the work is to rescue a species already on the brink of extinction.

The Royal Botanic Gardens, Kew is involved in the rescue of many species of island endemics that are on the verge of extinction and whose populations have been reduced below limits which will suffice to maintain their genetic diversity. Table 4 gives some data from the islands of Hawaii.

The Table shows that the populations of many Hawaian endangered species have been reduced below an acceptable limit. An example is *Alsinidendron trinerve* (Caryophyllaceae) which in 1989 had 20 plants left in Hawaii, most of which have since been destroyed by feral pigs and the invasive introduced species *Rubus fruticosus* or blackberry. There are now seven remaining individuals of Alsinidendron.[32] The seven are all from one genetically similar population. We have in cultivation at Kew, however, an accession from a different locality, Mt. Kaola, collected by Joseph Rock in 1961 from a now extinct population. We immediately propagated our accession for reintroduction to the wild to increase the genetic diversity of the species in Hawaii and give it a better chance of survival.

Another example of an island endemic whose population has been reduced below an acceptable level is *Trochetiopsis melanoxylon* from St Helena. The St Helena ebony was thought to be extinct for over 100 years, until two dwarf individuals were found growing on a cliff inaccessible to goats. The wild population is only two but it has now been successfully propagated, and about 2,000 individuals have been established around the island in an effort to increase the population and to get it through the genetic bottle-neck which it has suffered.

Genetic variation is the basis for the variety of life on Earth. Without it evolution could not have taken place. Species with little genetic variation are ultimately less competitive. It is therefore not sufficient to conserve species without studying their genetics. We must, therefore, examine them genetically and ensure that an adequate sample of their genetic variation is conserved. Ultimately survival depends upon their ability to adapt to change in their habitat, such as climate change; and to have enough variation to overcome deleterious mutations. This means that it is necessary to maintain enough individuals of any species to keep it genetically viable.

Within the genetic diversity of plants and animals there is still a great treasure of unexplored properties that could be of considerable use to humanity. Moreover, a single gene can be worth large sums of money; yet we are allowing them to be destroyed. In 1962, the botanists Donald Ugent and Hugh Iltis discovered a new weedy species of wild tomato (*Lycopersicum chmielewskii*). When this was crossed with the commercial tomato, the hybrid was found to contain a much higher sugar content.[33] The commercialisation of this new sugar-rich variety has benefited the tomato industry to the extent of US$8 million a year.[34] This valuable gene could easily have been lost through habitat destruction in Peru and elimination of a weedlike plant. Such considerations about the value of individual genes are fast becoming more pertinent as the genetic engineering of plants and animals is increasing. We are already losing much genetic diversity, including genes that might be valuable for increasing disease resistance, improving food crops or broadening the range of cultivated species. This cannot be wise, even if judged solely in financial terms. Between 1976 and 1980 alone, genetic material from wild relatives of crop species contributed US$340 million per year for US farmers in increased yield and disease resistance.[35]

The ability to isolate and identify individual genes has also enabled the possibility of genetic modification of organisms with genetic material from unrelated organisms. An animal or a bacterial gene can now be transferred to a plant. Plants have been produced that glow in the dark after receiving an animal gene for luminosity. However, the technique has generally been used for more practical purposes such as implanting resistance to herbicides into a crop plant, for example oil seed rape, or for the modification of a food product. The rapid introduction of genetic manipulation of organisms has led to considerable controversy and to much public misunderstanding, and there are certainly advantages, as well as dangers and uncertainties, about the technique.

Genetic modification has been much used for the production of medicines, such as insulin and interferon. However, this work is carried out *in vitro* in the laboratory and there is little danger of anything escaping into the wild. Many people are receiving benefits from the medicines that have been produced through the use of genetically modified organisms and there has been little complaint about this. A greater worry is about field crops such as cereals, oil seed rape and sugar beet. The sort of modifications in use are herbicide or pesticide resistance. It could be a disaster for biodiversity if these traits were to spread to wild species and produce 'super-weeds'. For this reason there are strict controls within Europe, but less so in many other parts of the world, and so the danger that implanted genes might spread is certainly real. Another technique that has been proposed, but not yet applied is terminator technology where a crop produces sterile seed that cannot germinate to produce the following year's crop. This could be disastrous for the small farmer in the developing world, but on the other hand it could stop the unwanted spread of a gene from a genetically modified plant.

The rush to use genetic modification has been spearheaded by large international corporations who also own both the chemical and seed companies that benefit from the sales of their products. This has meant that much genetic modification has been geared towards the crops of the developed world rather than solving the problems of poverty and starvation in the developing world. Genetic modification has huge potential to help the areas where famine occurs through developing crops capable of growing in arid regions, on saline soils and on other marginal lands. Unfortunately it is not being used for this purpose at present.

Many products from genetically modified plants are now available on the market, for example, from modified soy beans and tomatoes. Despite frequently expressed public concern, the use of these is unlikely to have any adverse affect on health, but a close monitoring of all products needs to be made; the introduction of alien proteins into a product could, for example, produce allergies. But many of the genetic modifications made to plants do not affect the chemical composition of the final product consumed.

Genetic modification is a technique that has great potential for the future, but the public is right to demand labelling of food products and knowledge of when it is used. The use of this technique needs the strictest possible control and much more research before it is allowed to be used widely, but it is certainly here to stay and is part of the future. It must be used to better the lot of the starving and must not be employed in a way that will endanger or pollute wild species.

Molecular studies are also enabling systematists to produce better evolutionary classifications of organisms. As an understanding of the relationships between species, genera and families is developed, we can also make more rational decisions about which species are more important to conserve. Biologists are now examining cladograms to see that representatives of each major clade are conserved.[36]

Protecting biodiversity may require giving different priorities for conservation to different taxa. If a choice has to be made, a much greater genetic diversity will be preserved by selecting taxic diversity from an evolutionary system, than if only one part of a genus is conserved.[37]

Biodiversity under threat of extinction

The term biodiversity was coined to stress the link between species, their genetic diversity and their habitats, when all of these three components began to be seriously threatened.[38] Extinction is a natural process; the fossil record shows that there have been a number of major extinction spasms in the geological past, such as the meteor impact that is thought to have eliminated the dinosaurs 65 million years ago. However, current rates of extinction are far in excess of natural ones. Because of the great uncertainties about even the number of species, it is impossible to estimate accurately the rate or number of extinctions that are occurring today.[39] But there is no doubt that we are in a period of mass extinction of similar magnitude to those in the Permian and Cretaceous eras. Evidence from well studied organisms such as birds, mammals and plants, combined with data on the loss of natural habitats, should provide enough warning signs that we are heading towards disaster that could even threaten the continued existence of the species *Homo sapiens*. There is no doubt that Man has been the cause of the current phase of extinctions, whether directly, by destroying habitats, or indirectly, by causing environmental changes that precipitate extinction.

This process began with the hunting to extinction of certain large and slow moving animals. It did not take humans long after their arrival there to eliminate the megafauna of South America. This phase was followed by the invention of agriculture in several different parts of the planet, which began the destruction of habitats in order to accommodate crops and provided the conditions for population to expand as more food became available. However, the really alarming rate of extinction seems to have begun in the 1850s and to have accelerated ever since, fuelled by the industrial revolution and European colonial expansion.[40] Farming, the industrial revolution and colonisation between them triggered the exponential growth of the human population, which, without any doubt, is the fundamental reason for the increased rate of biological extinction.[41] In this connection, considerations of carrying capacity of the planet for humans must include the maintenance of its capacity to preserve biodiversity on a sustainable basis.

The primary cause of the present high rate of species extinction is the loss of natural habitats. Rainforests harbour more than fifty percent of all terrestrial species and we have already lost just over fifty percent of the world's rainforest. The destruction of rainforests has been uneven with some places suffering more than others. It is apparent that some of the rainforest areas that are particularly noted for their diversity and numbers of endemic species (the 'hotspots' of Myers[42]) have also been the foci of destruction – for example, such well known centres of endemism as Madagascar, the Atlantic coastal forest of Brazil and the Pacific coastal forest of Colombia and Ecuador. A recent report by WWF[43] showed that the world's forest cover decreased by 13 percent between 1960 and 1990, from 37 million km^2 to 32 million km^2, an annual loss of about 160,000 km^2. Most of this has occurred in tropical regions (see Table 5 for recent forest loss in Brazil). However, we should not just point to forest loss in tropical countries. Temperate forests have historically

Table 5. Deforestation in the Brazilian Amazon

Year	Area km^2
1978-87	22,000/year
1988	19,000
1989	14,000
1990	11,000
1991	11,000
1992	13,000
1993	14,000
1995	29,000
1996	18,160
1997	13,277

fared at least as badly, and it is estimated that over sixty percent of temperate broadleaved and mixed forest has been lost. Because this loss has occurred over a much longer period of time and the forests have lower species diversity, the loss in temperate regions has been less noticed. The same WWF report shows that the populations of many species of marine and freshwater ecosystems have also dropped alarmingly.

Humankind seems to have an extraordinary capacity for overusing even the most valuable of natural resources. The population of Easter Island in the Pacific Ocean crashed in the early eighteenth century because the people exhausted their forests and extinguished various species upon which their life depended. In recent years there have been many reports of the collapse of fisheries in different parts of the world, especially around North America, Europe and Japan. Despite warnings from scientists for many years, the populations of bottom-dwelling fish such as cod, haddock and flounder are at all-time lows and the fishery industry has collapsed in many places. The average annual marine catch for 1990-1995 was 84 million tonnes, double what it was in 1960. On top of this at least 27 million tonnes of unwanted fish were caught and discarded, making the minimum estimate of fish caught 110 million tonnes a year. Thus, we are threatening species with extinction by vastly exceeding the carrying capacity of the oceans.

A recent bout of extinction, which should be a grave warning, is the alarming decline in numbers of many species of amphibians and the complete extinction of some.[44] So far the exact causes of this decline have not been determined, but the four prime suspects are: increasing ultraviolet light radiation resulting from the thinning of the protective ozone layer, global climate change, pesticides, and new diseases, including a recently discovered skin infection caused by an aquatic fungus not previously known to infect vertebrates. It was in the late 1970s that herpetologists began to see that amphibians were becoming rarer even in areas that seemed unaffected by human activity. Several extinctions have occurred, including the recently discovered beautiful golden toad (*Bufo periglenes*) of the Monteverde Cloud Forest Reserve in Costa Rica. Other declines have been reported, mainly from the mountainous regions of Australia, Central America, South America and the western United States and Canada. In Costa Rica, Alan Pounds conducted a study of the decline of amphibians in the mountain region that was formerly the range of the golden toad. This five year survey showed that 20 species of frogs and toads, 40 percent of the local total, had disappeared.[45] Necropsies performed on dead frogs and toads in Central America have shown them to be infected with a chylrid fungus which has also infected amphibians in several zoos. It is still uncertain whether the fungus is the primary cause of death or a secondary effect of some other environmental cause.

In California, Gary Fellers of the US Geological Survey has documented significant decline in several frog and toad species in pristine areas of the Sierra Nevada Mountains. The most alarming result of this work is the finding that the declines are all occurring on the western slopes of the mountains – the side facing California's heavily agricultural Central Valley. Pesticides, atmospherically transported to the mountains, may be the cause of the decline of these amphibians.

Perhaps this world-wide decline of so many different species of this environmentally sensitive group of animals, the amphibians, is a warning, like the miners' canaries, that something is wrong with our environment. David Wake, who has done much to draw attention to the decline of the amphibians, pointed out that: 'if a canary died, the miners got out of the mine. We don't have that option. We don't have any place to go'.

Coral reefs are second only to rainforest in the biodiversity which they sustain but, since the early 1980s, there has been a series of 'bleaching' episodes in places as far apart as the Caribbean, Taiwan, Australia, Hawaii and the Maldives, indeed in almost all tropical waters where coral reefs occur. Bleaching causes extensive mortality of coral and, consequently, of many other species sustained by the reef. The causes of this phenomenon are also uncertain but may be connected to global warming, or to white-band disease, or to a combination of these two factors working together.

It is abundantly clear that the expanding human population is causing extinction in many ways – through deforestation of rainforest; pollution of the atmosphere, soils and water with agrochemicals; by causing climate change, and acid rain, and by many other means. To secure our future, any new renaissance must reverse this alarming trend; for if we do not take better care of our companion species, we are at the beginning of a catastrophic phase of mass extinction which could end with the demise of *Homo sapiens*.

Conclusion: towards sustainable living

From a biologist's viewpoint the present world situation is not good. Biodiversity is fast disappearing and, with this loss, our options for the future are being reduced. It is biodiversity that supplies our food, medicines, fibres, wood and countless other products. It must be preserved, not least because of its potential to be of further use to humankind. But we should place an even greater value on biodiversity for its contribution the ecosystem services which enable life to exist – the atmosphere, the control of climate etc. Robert Constanza, of the University of Maryland and his colleagues, attempted to estimate, for the entire biosphere, the economic value of these ecosystem services. They placed it at between US$16 and 54 trillion a year, with an average of $33 trillion a year, compared to the annual global gross national product of $18 trillion. Yet such a value is not reflected in the way we treat biodiversity, which is often given little value. This trend must be reversed in the future.

There are signs of hope. Strong international action has been taken to address the problem of the thinning of the ozone layer in the stratosphere through the

Montreal Protocol and subsequent improvements of it. The 1992 Earth Summit in Rio de Janeiro brought together more than 120 heads of state to discuss environmental issues; and this is surely a sign of hope. From the Earth Summit emerged Agenda 21 and the two important international conventions on climate change and biological diversity. What is disappointing is the way in which many countries, particularly industrialised ones, have tried to dilute their obligations at the subsequent Conferences of the Parties. This is to ignore the warning signals provided for us by the dying amphibians, bleaching coral reefs and, in Britain, the declining song birds.

Biodiversity would be served well, if the world adhered to the three principal tenets of the Convention: to conserve biodiversity, to use it sustainably and to ensure equitable sharing of the gains made from exploiting it. Perhaps the key word, that is mentioned many times in the Convention document, is sustainability. This was also the theme of the earlier Bruntland report in 1987.[46] Sustainability means not fishing every last cod out of the North Sea or cutting down every mahogany tree in the rainforest. It involves controlling population at a sustainable level and realising that climate change, pollution, soil erosion and excess use of agrochemicals are not part of sustainable living. It is good that these issues are beginning to be addressed at an international level through international Conventions and World Commissions.

Biodiversity knows no frontiers. The corn crop of the United States depends upon the wild species of maize in Mexico for genetic material to improve its yield, to introduce disease resistance and to adapt it to climate changes. The coffee crop of Brazil depends upon the wild species of *Coffea* in Ethiopia and Madagascar. The world climate depends as much on the forests of the boreal zone in Canada and Siberia as upon the rainforests of Brazil and the Congo Republic. Our future needs more concerted international efforts, more globalisation, not of the economy, but of environmental action. Without this we are narrowing the options for our descendants and will become known as the generation who diminished biodiversity and altered the world environment to an unacceptable extent. It is almost too late; the environmental clock is ticking too fast. But by turning to sustainability there is still just time to reverse the trend and to slow down extinction. There is no room for complacency. If we continue the present extinction rate of between 4,000 and 6,000 species a year, by 2030 there will be between 160,000 and 250,000 less species to hold our biosphere together, or for us to use, or even just enjoy. The actions of the next two decades towards biodiversity and towards the environment will determine whether or not human life will survive on earth. Plants, that photosynthesise, are the green glue that holds life on earth together. I cannot stress too strongly the need for much greater global action to ensure that life on Earth continues.

Chapter 6

The urgency of environmental sustainability

Robert Goodland

The Author

Robert Goodland is a tropical ecologist. After a doctorate on Brazilian ecosystems, he became a free-lance consultant in Brazil, Bangladesh, Malaysia and New York's Ecosystem Center. He created the World Bank's environmental assessment unit and served as chief of its Latin America environmental division. He is now the World Bank's environmental advisor. He has published 20 books on environment and development. He was elected president of the International Impact Assessment Association and Metropolitan chair of the Ecological Society of America. His latest co-authored book is *Ecological Economics* (1998).

Chapter 6

The urgency of environmental sustainability

Robert Goodland

Introduction

This paper defines environmental sustainability (ES) partly by sharply distinguishing it from social and human sustainability and, to a lesser extent, from economic sustainability (see Table 1). While all four concepts overlap, there are especially strong linkages between economic and environmental sustainability. There is a danger that environmental sustainability may become a bandwagon on which social desiderata are dumped, such as gender balance, freedom, and the alleviation of poverty. ES is biophysically rigorous and will persist whether critics 'believe' in it or not; the universal source and sink capacities of the environment are not negotiable.[1] In contrast, human and social sustainability are not governed by immutable biophysical laws; they are far more appropriate fields in which to deal with important human and social goals, such as poverty alleviation, equity, and democracy.[2] Sustainability in general can be fostered by application of the 'constant capital rule', the meaning of which will become apparent.

If the term 'development' is introduced discussion becomes more ambiguous. There is an important distinction between 'sustainable development' and '*environmentally* sustainable development'. The latter is the subject of this chapter; but, because the term sustainable development is in such wide circulation, some mention of it must be made here.

'Sustainable development' is the concept that our descendants should inherit a world no worse than ours and that well-being *per capita* should not decline: see the definition on page 86. In this sense, sustainable development is a mode of development that is sustainable in all the four forms listed in Table 1 – socially, humanly, economically and environmentally. It is an enabling concept rather than a prescription. Its aims are those of economic development in general, and very important they are – the improvement in human well-being by the reduction of poverty, illiteracy, hunger, disease and inequity. But sustainable development is sometimes taken to imply, and to be dependent on, 'sustainable economic growth'. This is not only a contradiction in terms but unattainable in practice.

'Environmentally sustainable development' is a very different concept and is far more specific. It implies the unimpaired maintenance of human life-support systems – the environmental source and sink capacities. Sustainable communities accumulate stocks of capital assets; unsustainable communities deplete them.

There are four main forms of 'capital' that follow different laws and are amplified by different disciplines. Economics has emphasised financial and manufactured capital (money, factories, roads etc.) to the near exclusion of natural capital (envi-

Table 1. Comparison of human, social, economic and environmental sustainability (after Goodland, 1999)[3]
[See Note 4]

Human sustainability	Social sustainability	Economic sustainability	Environmental sustainability
Human sustainability means maintaining human capital. Human capital is a private good of individuals, rather than between individuals and societies. Health, education, skills, knowledge, leadership and access to services constitute human capital. Investment in health, education and nutrition of individuals has become accepted as part of economic development. As human lifespan is relatively short and finite (unlike institutions), human sustainability needs continual maintenance by investments throughout the lifetime. The start of sustainability is fostered by promoting maternal health and nutrition, safe birthing and infant and early childhood care. Human sustainability needs 2-3 decades of investment in education and apprenticeship to realise some of the potential that each individual contains. Adult education and skills acquisition, preventive and curative health care may equal or exceed formal education costs	Social sustainability means maintaining social capital. It lowers the cost of working together and facilitates cooperation. Trust lowers transaction costs, for example. This can be achieved only by systematic community participation and strong civil society, including government. Cohesion of community, connectedness between groups of people, reciprocity, tolerance, compassion, patience, forbearance, fellowship, love, commonly accepted standards of honesty, ethics, commonly shared rules, laws, discipline, etc. constitute the part of social capital least subject to rigorous measurement, but essential for social sustainability. Social (sometimes called 'moral') capital requires maintenance and replenishment by shared values and equal rights, and by community, religious and cultural interactions. Without such care it depreciates as surely as does physical capital. The creation and maintenance of social capital, as needed for social sustainability, is not yet adequately recognised. Violence is a massive social cost incurred in some societies because of inadequate investment in social capital. Violence and social breakdown can be the most severe constraints to sustainability.	Economic capital should be stable. The widely accepted definition of economic stability is the 'maintenance of capital', or keeping capital intact. Thus Hicks' definition of income – 'the amount one can consume within a period, and still be as well off at the end of the period' – can define economic sustainability, as it devolves on consuming interest, rather than capital. Historically, economics has rarely been concerned with natural capital (e.g. intact forests, healthy air). To the traditional economic criteria of allocation and efficiency must now be added a third, that of scale (Daly, 1992) [See Note 32]. The scale criterion would constrain throughput growth – the flow of material and energy (natural capital) from environmental sources to sinks. Economics values things in money terms, and is having major problems valuing natural capital, intangible, intergenerational, and especially common access resources, such as air. Because people and irreversibles are at stake, economics needs to use anticipation and the precautionary principle routinely, and should err on the side of caution in the face of uncertainty and risk.	Although environmental sustainability (ES) is needed by humans and originated because of social concerns, ES itself seeks to improve human welfare by protecting the sources of raw materials for human needs, and ensuring that the sinks for human wastes are not exceeded, in order to prevent harm to humans. Humanity must learn to live within limitations of the biophysical environment. ES means that natural capital must be maintained, both as a provider of inputs (sources) and as a sink for wastes. This means holding the scale of the human economic subsystem to within the biophysical limits of the overall ecosystem on which it depends. ES needs sustainable consumption by a stable population. On the source side, harvest rates of renewables must be kept within regeneration rates. On the sink side, this translates into holding waste emissions within the assimilative capacity of the environment without impairing it. Non-renewables cannot be made sustainable, but quasi-ES can be approached for non-renewables by holding their depletion rates equal to the rate at which renewable substitutes are developed.

ronment), and now human and social capital are being accorded increasing attention. Part of the reason for such asymmetry is that natural capital and social capital are public goods with no market value. Therefore they are undervalued in economics. But both can be regenerated or accumulated to a certain extent, and both can be depleted or run down. We must now recognise that natural capital has become scarce for the first time.[3] Natural capital could, indeed, be worth more than twice as much as global GNP (US$33 trillion compared with US$15 trillion);[4] so the increasing attention to natural capital and environmental sustainability is overdue.

Historians of future generations may well be interested in the social constructs we devised with respect to biophysical reality, but those generations will be confronted primarily with the reality and not with whatever contemporary social construct we made of it. It makes no difference to actual climate change due to greenhouse gas accumulation whether societies or governments 'believe' in it or not; it is a biophysical reality.[5]

A potted history of sustainability

If one addresses only the last century or so, some notion of economic sustainability was firmly embodied in the writings of J. S. Mill in 1848[6] and of Malthus in 1798[7]. Mill emphasised that environment ('Nature') needs to be protected from unfettered growth, if we are to preserve human welfare before diminishing returns set in; people should be content to be stationary (steady-state), for the sake of posterity, long before necessity compels them to it. Malthus emphasised the pressures of exponential population growth on the finite resource base. The modern version (Neo-Malthusianism) is exemplified by Ehrlich and Ehrlich and by Hardin.[8]

The interactions between population and resources are synthesised in Daly's *Toward a Steady-State Economy*[9] and *Steady-state Economics*.[10] The latter is the seminal work in which population and consumption pressures on environmental sources and sinks are integrated and extended into the single critical factor of scale – the throughput of matter and energy from the environment, used by the human economy, and released back into the environment as wastes. This is the basis for defining sustainability.

Neither Mill's nor Malthus' views on the relationship between the environment and the economy are held in great esteem by today's economists who follow the technological optimism of David Ricardo.[11] Ricardo rightly believed that human ingenuity and scientific progress would postpone the time when population would overtake resources or 'the niggardliness of nature'. As the absolute numbers of poor are increasing world-wide,[12] that postponement seems to have ended.

The definition of environmental sustainability hinges on distinguishing between 'throughput growth' and development. The 'growth debate' started to become mainstream two decades after World War II. The wisdom of infinite 'throughput growth' in a finite earth was seriously questioned by a number of thinkers.[13] Nevertheless, Toman claims that the World Development Report (1992) basically treats sustainability as another way of espousing economic efficiency in the management of services derived from the natural endowment.[14] 'Throughput growth' is defended by many economists, including Beckerman, who still reject the concept of sustainability.[15] The two influential books, *The Limits of Growth* and *Confronting Global Collapse*,[16] concluded that 'it is possible to alter these growth trends and establish a condition of ecological and economic stability that is sustainable into the future.' Barney's 1980 *US Global Report to the President* amplified and clarified the limits argument. Large populations, and their rapid growth and increasing affluence or consumption, are unsustainable.[17]

The optimistic, Ricardian tradition still dominates conventional economics and is exemplified by the Cornucopians, Simon and Kahn, in their 1984 response to the *Global 2000* report;[18] while a number of authors such as Panayotou, Summers and Fritsch find growth compatible with sustainability and even necessary for it.[19] On the other hand, the 'limits to growth' conclusions were reinforced by the 1980 *World Conservation Strategy*[20] and Clark and Munn's 1987 IIASA report *Sustainable Development of the Biosphere*.[21] In 1989, Daly and Cobb's prize-winning *For the Common Good*[22] estimated that GDP growth in the USA for the last two decades had become uncoupled from all measures of well-being. These seminal findings, that growth does not improve well-being, have subsequently been shown for most OECD nations, and for an increasing number of low-consuming nations too.[23]

The definition of sustainable development that gains most universal agreement is that of the UN Brundtland Commission in 1987: 'Development that meets the needs of the present without compromising the ability of future generations to meet their own needs.'[24] The Commission defined sustainability in a growth context; but some believe that part of the success of this definition stems from its opacity.[25] Yet, when the Commission reconvened five years later, calls for growth were striking by their absence. This report also elevated the population issue higher on the agenda to achieve sustainability.[26] Prince Charles commended the Commission, in its report of this meeting, for dropping its previous (1987) calls for huge '5- to 10-fold increases in economic growth'.

Few Nobel prize winners in economics write on sustainability; but, among those who do, are Haavelmo and Tinbergen who repudiate 'throughput growth' and urge the transition to sustainability.[27] Solow's earlier writings questioned the need for sustainability, but recently he is modifying that position.[28] The World Bank adopted environmental sustainability in principle rather early, in 1984, and is now promoting it actively.[29]

Of all these publications, Daly and Cobb's of 1989 is the most influential and durable, because it shows that more growth has started to do more harm than good. Moreover, it outlines pragmatic operational methods to reverse environmental damage and reduce poverty. Others, such as Goodland, Daly and El Serafy, supported by Tinbergen and Haavelmo, advanced the 'limits' arguments: that the human economy has reached its limit in many places; that it is impossible to grow into sustainability; that source and sink capacities of the environment complement human-made capital but the latter cannot substitute for their environmental services. Thus, it is highly unlikely that the South can ever catch up with the North's present lifestyle of consumerism.[30] The conclusion is that it has become more urgent to concentrate on redistribution.[31,32]

The status of sustainability

The recent history of sustainability as a new element in economic development is the exceptionally successful call by the Brundtland Commission for development to become sustainable. Soon thereafter, at least at the rhetorical level, nearly all development agencies added sustainability to their priorities. Brundtland's political feat was achieved only by being deliberately vague in defining sustainability. In the years following, the definition of sustainability became sharpened and operationalised. This dampened sustainability ardour by some agencies, which sidetracked it into 'Green Accounting' and 'genuine savings', from which meaningful guidance has yet fully to emerge. Meanwhile, elsewhere than development agencies, especially in the Netherlands, sustainability has become much clearer, sustainability indices have been agreed on, and there is increasing consensus on the way forward.[33]

Inter-generational and intra-generational sustainability

Most people in the world today are either impoverished or live barely above subsistence; the number of people living in poverty is increasing. Developing countries can never be as well off as today's OECD average. Future generations seem likely to be more numerous and poorer than today's generation. Even if the human population starts to decline after about 2050, it will inherit, and have to make do, with damaged life-support systems. How damaged they will be is up to us of today's generation. Sustainability includes an element of not harming the future (inter-generational equity), as well as not harming society today (intra-generational equity). If the world cannot move toward intra-generational sustainability during this generation, it will be that much more difficult to achieve inter-generational sustainability sometime in the future. This is because the capacity of environmental services is being impaired and their ability to deliver is, therefore, likely to be lower in the future than it is today. At the same time, the demand for such services from the world's larger population will be much greater.

What should be sustained?

Environmental sustainability seeks to maintain environmental services indefinitely, especially those supporting human life. The source capacities of the global ecosystem provide inputs of raw material – food, water, air, energy; the sink capacities assimilate outputs and wastes. These source and sink capacities are large but finite; any overuse impairs their ability to provide life-support services. For example, the accumulation of CFCs is damaging the capacity of the atmosphere to protect humans and other biota from harmful ultraviolet radiation. UN Conventions have, therefore, promoted sustainability by severely restricting the production of CFCs in order to conserve sink capacity or assimilation rates.

Being anthropocentric, the main reason that humans seek environmental sustainability is the protection of human life, which depends on other species for food, shelter, breathable air, moderate rainfall, plant pollination, waste assimilation, and other environmental life-support services. Economists grossly underestimate the huge instrumental value of non-human species to human beings; they largely ignore the vast number of species which are apparently of no 'use' to humanity, and the even bigger number of species as yet unknown to science. The intrinsic worth of species of no present value to human beings is almost entirely unaccounted for.

Although the conservation of biodiversity is becoming a general ideal for nations and development agencies, there is no agreement on what and how much should be conserved, nor at what cost. Leaving aside the important fact that we have not yet learned to distinguish more useful from less useful species, agreeing on how many species other than *Homo sapiens* to conserve is not central to the definition of environmental sustainability. Reserving habitat for other species to divide among themselves is important; let evolution select the mix of species, not us. But if the non-human habitat is to be reserved, a limitation must be placed on the scale of the human habitat. 'How much non-human habitat should be reserved?' is an important question. But the answer to it is moot – probably 'no less than today's remnants'. In the quest for sustainability, we should probably seek to reduce human-induced species extinctions to zero, to conserve as much remaining habitat as possible, and to restore those that are degraded.

This brings us to the precautionary principle: in cases of uncertainty and in view of the possibility of irreversibility, sustainability requires us to err on the side of caution. Because survival of practically all the global life-support systems is uncertain and non-linear, we should be very conservative in our estimate of the capacities of the various inputs and outputs, and particularly of the role of unstudied, apparently 'useless' species.

The dictionary distinguishes between growth and development. 'To grow' is defined as: 'to increase in size by the assimilation or accretion of materials'; while 'to develop' means 'to expand or realise the potentialities of; to bring to a fuller, greater or better state'. Growth implies quantitative physical or material increase; development implies qualitative improvement or at least change.[34] Quantitative growth and qualitative improvement follow different laws. Our planet develops over time without growing. Economies, on the other hand, usually start with quantitative 'throughput growth' during the period that infrastructure and industries are built; but eventually they mature into a pattern with less 'throughput growth' but more qualitative development. Our economy, a subsystem of the finite and non-growing earth, should follow this path towards development without 'throughput growth'. The time for such adaptation is now. To attain sustainability, development should, as far as possible, take the place of 'throughput growth'.

The definition of environmental sustainability

The definition of ES as the 'maintenance of natural capital', given in Table 1, is expanded as input/output rules in Table 2. The two fundamental environmental services – the source and sink functions – must be maintained unimpaired as a condition of sustainability. ES constitutes a set of constraints on the four major activities regulating the scale of the human economic subsystem. These are: on the source side, the use of (a) renewable and (b) non-renewable resources; and on the sink side, (c) the assimilation of pollution and (d) the assimilation of waste. This short general definition of ES is the most useful produced so far, and is gaining adherents; it seems to be robust, irrespective of country, sector, or future epoch.[35] The fundamental point is that ES is a concept derived from the natural sciences and that it obeys biophysical laws.

Causes of unsustainability

When the human economic subsystem was small, the regenerative and assimilative capacities of the environment appeared infinite; but we are painfully learning that environmental sources and sinks are indeed finite. Although, originally, their capacities were very large, they have now been exceeded by the scale of the human economy and have become limited and limiting. In the past the source and sink capacities of the environment did not have to be taken into account, economics dealing only with scarcities. Conventional economists still hope or claim that

Table 2. The Definition of Environmental Sustainability
[From: Daly 1973, 1974, 1992, 1996, Daly and Cobb 1989]
[See Notes 4, 10, 30 & 32]

1. Output Rule: Waste emissions from a project or action being considered should be kept within the assimilative capacity of the local environment without unacceptable degradation of its future waste absorptive capacity or other important services.
2. Input Rule: (a) Renewables: Harvest rates of renewable resource inputs must be kept within regenerative capacities of the natural system that generates them. (b) Non renewables: Depletion rates of nonrenewable resource inputs should be set below the historical rate at which renewable substitutes were developed by human invention and investment according to the Serafian quasi-sustainability rule (Table 3). An easily calculable portion of the proceeds from liquidating non-renewables should be allocated to the attainment of sustainable substitutes.

economic growth can be infinite or, at least, that we are not yet reaching limits to growth. This accounts for the fierce recent repudiation of *Beyond the Limits*, and the endorsement by them of Brundtland's view that '5- to 10-fold more growth will be needed'.[36]

The scale of the human economy is a function of 'throughput' – the flow of materials and energy from the sources in the environment, used by the human economy, and then returned to environmental sinks as waste. 'Throughput growth'

Table 3. Serafian Quasi-Sustainability of Non-Renewables

The Serafian rule pertains to non-renewable resources, such as fossil fuels and other minerals, but also to renewables to the extent they are being 'mined'. It states that their owners may enjoy part of the proceeds from their liquidation as income, which they can devote to consumption. The remainder, a user cost, should be reinvested to produce income that would continue after the resource has been exhausted. This method essentially estimates income from sales of an exhaustible resource. It has been used as a normative rule for quasi-sustainability, whereby the user cost should be reinvested, not in any asset that would produce future income, but specifically to produce renewable substitutes for the asset being depleted. The user cost from depletable resources has to be invested specifically in replacements for what is being depleted in order to reach sustainability, and must not be invested in any other venture – no matter how profitable. Hueting et al. (1995) goes further, especially for non-renewable energy, by basing a future acceptable rate of extraction of the non-renewable resource on the historic rate at which improved efficiency, substitution and re-use became available. These calculations show the folly of relying on technological optimism, rather than on some historic track record.

[Sources: El Serafy, S. (1989) The proper calculation of income from depletable natural resources pp. 10-18 in Ahmad, Y. et al. (Eds.) Environmental Accounting for Sustainable Development. The World Bank, Washington DC. 195pp. Hueting, R., P. B. Bosch & de Boer (1995) The calculation of sustainble national income. Netherlands Organization for International Cooperation, Den Haag & Indian Council for Social Science Research, New Delhi.]

is a function of population growth and consumption, which translates into increased rates of resource extraction and pollution, or of the use of sources and sinks. Unsustainability occurs when the scale of throughput exceeds the capacities of environmental sources and sinks. The present evidence for unsustainability is pervasive. Just a few examples: greenhouse gases are accumulating; the ozone shield is being damaged; scarcely a drop of sea water can be found free of human pollution; the rates of species extinction caused by human beings are high and rising; oceanic fish catches are declining fast; and natural forests are disappearing. The increasing number of the poor and starving suggests that low cost foods, both gathered and harvested, have become scarce or unobtainable.

Yet few people are prepared to admit that consumption above sufficiency, however defined, is not an unmitigated good. The human economy has grown to a level that is unsustainable; because we are living off inherited and finite capital such as fossil fuels, fossil water and accumulated soil fertility; because we do not account for losses of natural capital such as the extinction of species, nor do we adequately count the costs of environmental harm. The over-consumption of the developed countries and third-world over-population require most attention.

The time for environmental sustainability

It is urgent to move towards sustainability. Three examples will illustrate this. First, consider that, if the release of all substances that damage the ozone shield were halted today, the shield might need as much as one century to return to pre-CFC effectiveness. Secondly, the world has only a brief 25 years or so to make the fundamental transition from fossil fuels to hydropower and other sources of renewable energy, unless we are to accept perilous risks of climatic instability.[37] For, as of 1999, it has become clear that greenhouse climate change is exacerbating the ruptures in the ozone shield. Thirdly, the rates of topsoil erosion, of depletion of fossil aquifers and of species extinction are high and soaring. The effects of each of these three are compounded because, with every passing year, sustainability has to be achieved for a world population which has increased by an extra 80 million. Technology seems highly unlikely to produce substitutes for the ozone shield, for stable climates, for topsoil, for readily available water, and for species. Though environmental sources and sinks have been providing humanity with free services for the last million years, and until recently have seemed vast and resilient, we have now begun to exceed their capacities and to damage them world-wide. That is why environmental sustainability is so urgent.

Sustainability and substitutability

Where a substitute for any environmental service can be found, the substitution achieved has been marginal. Most forms of natural capital and environmental services recover or regenerate themselves slowly, and the process cannot be significantly hastened; nor is it possible to find substitutes for them. However, the proponents of conventional economics and technological optimists claim that substitutability is the rule, rather than the exception. Substitution is achieved by technology; and this can be beneficent or maleficent. It is true that, if the new technology is good, declining assets may turn into non-declining benefits. If the new technology is bad – CFCs damage the ozone shield; Chernobyl erupts; species extinctions increase – declining assets turn into declining benefits. There are grounds for concern if it proves impossible to find a substitute for a depleted asset. Here there is a strong case for development assistance. The limited extent to which human-made capital can substitute for natural capital, though marginal, is rarely discussed by neo-classical economists; yet it is central to the issue of sustainability. The services at issue are those which provide healthy air and potable water *et cetera*, and which absorb wastes.

Similar reasoning may be applied to changes in population. In theory, some increase in population could be beneficial if it were to stimulate technological change as the society approached some limiting factor. But mainly, population increase leads to the loss of capital; both *per capita* benefits and the capital itself are dissipated. Because increase of population dissipates capital, stability in human numbers is essential if sustainability is to be attained.

If it proves always possible to find substitutes, there can be no limits to economic expansion, because, if an environmental good is destroyed – so it is argued – it can be replaced by a technological substitute. When White Pine or Sperm Whales became scarce, there were acceptable substitutes for their products. When easily gathered surficial oil flows were exhausted, new drilling technology enabled very deep deposits to be tapped. In Europe, when the native forest was consumed, timber for houses was replaced with brick, steel and plastic. All these raise secondary issues of depletion, and hence of substitution. If bricks did not substitute for timber, then timber was imported.

The realisation, then, that the possibility of substitution is the exception rather than the rule is not yet widespread. Once imports cease to mask substitutability,[38] it becomes plain that many, if not most, forms of capital serve less as substitutesfor each other, but, rather, tend to be complementary to each other – or neutral.

Ecologists attach great importance to Baron Justus von Liebig's Law of the Minimum – the whole chain is only as strong as its weakest link. The factor in shortest supply is the limiting factor in a process, because factors are complements not substitutes. More irrigation and fertiliser cannot increase crop yield if day-length is limiting. If scarcity of phosphate limits the rate of photosynthesis, then photosynthesis would not be enhanced by increasing soil nitrogen, light, water, or CO_2. More nitrogen fertiliser cannot substitute for lack of phosphate, precisely because they are complements, not substitutes for each other. If one wants faster photosynthesis, one must ascertain which factor is limiting and then invest in that one first, until it is no longer limiting.

Environmental sustainability is based on the premise that most natural capital is a complement, not a substitute, for human-made capital. Complementarity is profoundly unsettling for conventional economics, because it means that there are limits to the capacities of environmental sources and sinks, and thus limits to growth. Human-made capital is a very poor substitute for most environmental services; and it is impossible to provide substitutes for some life-support systems. The sooner economics provides for the internalisation of these facts, the sooner the world can start to draw near to environmental sustainability.

Weak versus strong sustainability

A prerequisite for sustainability under the 'constant capital rule' is that we should at least maintain, and if possible enhance, the overall stock of capital assets – the sum total of all forms of capital. The question arises, however, whether it is sufficient to hold the total capital stock constant – weak sustainability – or whether this total stock should be disaggregated and each of the four types of capital (Table 1) – human, social, economic and environmental – be maintained separately, 'strong sustainability'. Weak sustainability is a value-free requirement that has little to do with environment.[39] While insisting that the 'constant capital rule' be obeyed, it attempts to equate natural capital with any other kind of capital under the umbrella of keeping total capital intact. Strong sustainability, in contrast, is a normative concept, relating to the maintenance of the immensely complex stock of environmental assets.

There is a sharp difference of opinion about which form of sustainability should be the goal. Traditional economists and technological optimists tend to support weak sustainability as they assume substitution is usually possible; and, where it is not, technology will eventually provide.

Ecological economists know that substitutes cannot be found for much, if not most, natural capital; hence they prefer strong sustainability. Beckerman dismisses weak sustainability, but finds strong sustainability to be morally unacceptable – placing environment before people – as well as totally impracticable.[40] Most of the world is far from achieving even weak sustainability; to achieve it would be a first, but significant, step in the right direction.

Weak sustainability means that the liquidation of one form of capital can be offset by augmenting another form of capital, as long as the aggregate total capital stock remains constant; that we can afford to deplete environmental assets as long as we build up other assets. Weak sustainability would be achieved if we liquidated all tropical rain forests and used the timber and cellulose sales to endow a thousand universities in perpetuity. It would permit us to deplete oceanic fish stocks and use the protein to improve human capital. It depends upon the extent to which substitution is possible. Sombreros and sunscreen substitute for the protective ozone shield. We can use CFCs because we can prevent cataracts and skin cancers by staying indoors during the day. We can burn the world's 300 or more years of coal stocks because higher sea walls can hold back sea level rise and stronger homes can resist hurricanes . . . . if substitution works.

Unfortunately, the proceeds from depleted assets are seldom, if ever, reinvested; so even weak sustainability as a normative guide is not achieved. Poor countries in Africa exemplify this; poor countries may liquidate their natural resources and feel they are getting rich until the resources run out; in fact, they become poorer and they don't even know it. Sustainability of whatever sort forces us to distinguish between growth and the liquidation of natural assets. Even rich countries fail to live up to the weak sustainability criterion. The UK, for example, has depleted its bonanza of North Sea oil and gas without reinvesting the proceeds, using them instead for current expenditure.

Natural capital – the environment – is exceedingly heterogeneous (as well as being very poorly known). Strong sustainability does not mean we cannot *ever* extract oil or cut timber; but we do need some kind of objective measure (yardstick) that might be applied to all kinds of natural capital and which could be used to judge whether we are, indeed, holding the stock of disparate natural capital constant. On this, too, there is little agreement. One alternative is to use money as the common measure; but this carries both advantages and disadvantages. Willingness to pay and induced preferences are fraught; contingent valuation is even more so. Neither do we want to descend to the lowest common denominator of 'an unexamined and unworthy purpose, such as unconstrained aggregate satisfaction of unstructured private tastes'.[41] Personal charity, the gigantic world-wide philanthropy business, self-interest and altruism all show that using money is unreliable. Preferences change over time, and irreversibles are omitted – the market may actually raise prices of a depleting asset during scarcity. Proper accounting means using constant prices, but accountants have forgotten this. Many valuation specialists, on the other hand, prefer relying on physical stock measurements.[42]

Ecological economists defend strong sustainability for three reasons. First, because of uncertainty. There are severe limits to the degree to which the main forms of capital can substitute for each other; this is especially so in the case of natural capital or environmental functions. There are few substitutes for agriculture, for example. We should not use the excuse of weak sustainability to deplete soil fertility, even if we compensate by augmenting food storage.

Secondly, much environmental damage is irreversible. There is a major asymmetry between the forms of capital. A worn-out road can be rebuilt; extinction is forever. Children can be educated and new factories can be built, but the recuperation of the ozone shield, after CFC release is halted, may take more than 100 years; or, indeed, it may never recover.

Thirdly, natural capital is poorly known. For example, taxonomists do not know the number of non-human species to the nearest order of magnitude; estimates range from 3 million to 100 million or even more. Although mammals are the best known and smallest phylum of organisms (about 4,000 species), new species of mammals, even new genera, have been discovered in the last few years. We know that a total 1.7 million species have been named so far, and that about 13,000 more species are named each year. Naming a species, preserving a few bits of its dead body, and recording where it was found certainly is a useful and necessary start, but it does not tell us which species are essential, which are nice but not essential, and which can be safely disregarded – even from a narrow anthropocentric utilitarian point of view.[43] Habitat conservation is justified because it conserves species – even those not yet known, as well as for the utilitarian purpose of protecting human life-support systems.

Moreover, we do not generally know how resilient natural systems may be; we suspect that stressed ecosystems have thresholds beyond which they may crash, rather than gradually dwindle. We do not know if the extinction of one species will cause the extinction of several or many others which may be biologically dependent upon it. Many species are linked through such dependencies as species-specific food sources, pollination, mycorrhiza, and the viability of propagules, symbiosis and other obligate mutualisms. Ideally, under conditions of sustainability, there should be no extinctions caused by human beings; if we are forced to make exceptions, we do not know how to do this prudently. While slow extinction rates are far better than fast rates, they are still incompatible with sustainability.

In the face of uncertainty and the limited possibility of substitution, the prudent goal seems to be *strong* sustainability. But more important than the distinction between weak and strong is that we achieve the focus of this chapter – the climb towards environmental sustainability.

Acknowledgment

I warmly appreciate the many years of help, including comments on this manuscript, generously provided by Herman Daly and Salah El Serafy.

Note

The personal opinions expressed in this chapter should in no way be construed as representing the official positon of the World Bank Group.

Chapter 7

Economic fiction

David Burns

The Author

David Burns is Treasurer of a multinational industrial group. He has worked for more than twenty years in finance and industry, with the exception of a period in the early eighties when, with the editor of this volume, he created a non-profit interdisciplinary research and consultancy programme for the protection of natural forests in developing countries.

Chapter 7

Economic fiction

David Burns

For affluent households in many affluent societies, the present appears to contradict the future. The latter is full of fears; the former, full of satisfactions. Nowhere is this paradox more apparent than in the minds of adolescents, who fear not to be able to reproduce the economic success of their parents. Failure to find satisfactory employment, on the one hand, and general environmental degradation, on the other, are two fears which have dominated our lives for a generation whilst we all consume – and many of us work – as never before. There is a sense of personal inadequacy in the child who doubts his capacity to compete for a secure livelihood in such a world. There is a sense also of the inadequacy of the context, particularly in countries where the primary school teachers sincerely and indeed appropriately educate young children about the extent and gravity of environmental problems.

Different countries reflect this pattern to different extents. American culture, with a bias towards optimism and a relatively unregulated employment market combined with high levels of employment, is relatively unaffected. Escapist media products sustain the American dream; the latent anguish is felt more through the violence expressed in much of this escapism. European societies and many Asian ones are exposed to this incoherent mixture of gratification and violence through the export of American media products without having the benefit of a European equivalent of the American dream; there is much less optimism about personal opportunity, and besides, the employment situation is much less reassuring.

Personal inadequacy and the inadequacy of the context make a depressing blend.

When we cast our minds forward to the future, it is often not clear whether we are looking at a desire, a forecast or a fear. Utopia, Sir Thomas More's sixteenth-century neologism meaning 'No-place', is a vehicle for discussing these things without limitation as to fear, forecast or desire.

The Republic of Plato, the first and most distinguished utopia, purported to be an ideal state. Implicitly located in the past, it is also presented as a blueprint for an ideal state in the future. Most modern readers agree to find it abominable. All we can say is that, if Plato was in any doubt about the desirability of the Republic, he hid it well. The features of the Republic are justified and recommended as they are developed.

Thomas More, on the other hand, although a distinguished theologian and an influential public figure, was such an inveterate farceur that it is difficult to know where he stands. The world of More's Utopia is described to the reader by a man called 'Talker of Nonsense', and yet it does appear that More seriously recommended the society of Utopia as his version of the desired republic.

Samuel Butler's Erewhon – 'Nowhere' approximately written backwards – is perhaps the most famous utopia in the English language (More having written in

Latin). It is overtly ambiguous. The reader affects disdain for many aspects of Erewhon culture, but in a way which is clearly satirical. The ridiculous nature of many Erewhon conventions reflects the shortcomings of contemporary European society, as it were in a distorted mirror. As a vehicle for satire, utopia is indeed ideal.

Science fiction is a genre which derives from the idea of utopia. The preoccupation is more obviously with the future, as reflected in the French expression 'futuriste'. Cinema and television now offer a continuous diet of such fantasies, superficially premonitory but often merely macabre and perverted.

Purely political futurisme, exemplified by Orwell's 1984, also exists – even if, with added poignancy, 1984 itself as a location in time has slipped into the past. I cannot, however, think of a good example of purely economic futurisme. Perhaps this is because economics is not the stuff of fiction. Perhaps, on the contrary, it is because, economists from the outset are purveyors of idealised pictures of modern life – the so-called 'economic models'.

In order to read Orwell's Animal Farm, it is useful to know something about Marxism as well as a little about pigs. To enjoy Samuel Butler, a prior acquaintance with the works of Rousseau is recommended. To get the most out of the economic satire which now follows, it is helpful to be familiar with such colourful notions as Indifference Curves, Liquidity Preference, Free Riders, Perfect Equilibrium, Dividend Irrelevance, Poverty Traps – the world of Keynes, Samuelson and Modigliani. For those without this ideological baggage, but nevertheless disposed to consider that the economists may have as much to say about our future as the politicians and the scientists, I can only wish you an enjoyable trip.

STEKRAM

or

A HOMAGE TO SAMUEL BUTLER

If the reader will excuse me, I will say nothing of my antecedents, nor of the circumstances which led me to leave my native country. Suffice it to say that I thought that I could better my fortunes more rapidly than at home. It will be seen that I did not succeed in my design, and that however much I may have met with what was new and strange, I have been unable to reap any pecuniary advantage.

It appears that the rapid economic development seen in Stekram in the last few years can be traced back to the Millennium Débâcle, in which whole regions of the then politically fragmented continent of Stekram were left without the Internet for several months, and indeed in midwinter, when in these boreal latitudes the reliance on the information highways for all forms of moral and physical sustenance is of course at its greatest. The reputation of Government never recovered from that chaotic experience, and important domains of the old administrative way of life succumbed to the wave of viruses, password recognition failures and subversion which continued throughout the year 2000. By the time that taxation was definitively abolished in the Happy Revolution a few years later, even the poorest Stekram families had little to lose from the final cessation of any form of either redistributive or

CommComm[1] activity. The severe revolutionary penalties attaching to the crime of ValueWaste[2] in fact remain a mere historical curiosity. In Stekram today, I cannot think of any citizen who would even know how to break this particular law.

The catalyst for these momentous transformations was of course the appearance on the market of InterPay, the network-based mutualistic clearing-house which led to the disappearance of Transaction Banks shortly before the revolution itself. For readers unfamiliar with Stekram, it is worthwhile to attain a fuller understanding of this so-called MarshmallowWare.

InterPay uses interactive psychological databases to trace each individual's indifference curves and thus his consumption profile. Total consumption forecasts are transmitted to Value Producers who are able to take steps to align supply with demand. InterPay declares prices for most commodities, whether goods or services, on the famous InterPrice screens. For each commodity, value continues to fluctuate with political, economic and fashionable sentiment – no doubt it always will; but the fluctuations can now be read from second to second, just as one has always done with financial asset prices.

By the time that 90 percent of the population had subscribed to InterPay it ceased to be economic for the food retailers and most other retail companies to operate shops. Those not subscribed to InterPay are in any case those with no financial resources. As a result, though not a perfect reflection of the underlying consumption demand of the whole of Stekram, InterPay has become an indispensable part of the social and economic fabric of the continent.

To my mind the most exciting feature of this new technology is not the competitive pricing of consumer goods, nor the seamless home delivery and automatic debit facilities, important though these features undoubtedly are for daily life in Stekram. Much more important in the long term has been the ability of InterPay's SoftSensors, like the crude 'truth drugs' of twentieth century popular mythology, to detect the true demand for CommComms, and hence the value placed on them.

It is true that Stekram has seen the progressive commercialisation of goods and services which many of my readers would consider as utterly non-viable – including such everyday amenities as churches, side-roads and municipal parks. It is true that other goods and services have been exposed by InterPrice as fundamentally too expensive and not worthwhile, and have quietly disappeared from the scene: the melting away of almost all parliamentary and representative institutions is a case in point, and opera – equally pointless though somewhat more musical – may well be following the same path. But of far greater significance is the survival of those institutions which are of real benefit to almost everybody, provided that the necessary payment is extracted in suitably homeopathic doses.

Following the chaos of the transition period, an early success was the reintroduction of Traffic Wardens; but these are of course self-financing, and the extravagant demands of the freelance wardens still give rise to some amusing but not altogether reassuring street scenes.

A better example is the survival of Stekram embassies in foreign countries – and the equally effective reception services of the embassy contractors in Stekram itself, offering cocktail parties to visiting envoys – entirely financed by microscopic con-

1. Common interest commodity
2. Giving of Money without Valuable Consideration

tributions from every man and boy in Stekram.[3] InterPrice asks people separately what value they assign: a) to a probable, but unprovable contribution to world less-war; and b) to an occasional, measurable contribution to world trade. A limited amount of low-cost advertising by ASC and JUSTASC[4] convinced Stekram citizens of a proposition which the old Government had never even sought to defend, let alone believe in: that a cocktail party was of real value even to those people not attending it.

The breakthrough came with the successful application of InterPay technology to the Police Forces. It had been feared that policing activity would not survive the withering away of the State. As with all CommComms, there was a perceived risk that individuals would adopt the old Free Rider position of pretending not to consume. The ability of InterPay's SoftSensors to pick up market sentiment at its source, deep in the human breast, had only a limited impact on the stock markets – where the heaving breast of sentiment had long ago been embraced, as it were, by market participants. But on the seemingly unfinancially-minded constabulary the effect was radical. People want policing. They will pay for it. Taxation, for all those years, was but a council of despair.

People in Stekram relate that the first days of the Happy Revolution witnessed a horrifying explosion of criminality. Since, of all the combines to survive the demise of the public sector, the Police seemed initially the least affected, there was a moment of collective uncertainty. Mind you, everyone knew what was happening. The old-style Single Police Force was bathing in the reflected glory of crime. Every new criminal act was increasing the value that citizens attributed to policing. Needless to say, the constables busied themselves with inciting criminal acts. The SPF thus benefited from a substantial and growing situational rent, and moreover drained liquidity out of the InterPay system. This led to the rather melodramatically titled Constabular Hyper-Inflation, which was in fact of fairly short duration and modest proportions, probably never posing a real threat to the Revolution.

Constable and criminal have always had a symbiotic relationship in any society, even in value-confused societies such as those familiar to my readers. One way to break the cycle of violence is of course to have a Force of supermen who are in some way free of the essential human realities of value creation and can simply chase criminals for the sake of it, as in the Wild West; or for a salary, as in the administrative society of before the Revolution. It was thus perhaps inevitable that siren calls were heard for a return to 50 percent marginal tax rates and other forms of ValueWaste. But breaking the law (in this case by Giving of Money without Valuable Consideration) never serves to defend the rule of law, and happily a better way was found. The increased demand for policing attracted a wave of new entrants to the market. Soon there was an independent Police Company in each barrio. Fantastically detailed statistical analyses sprung up, and the Law and Order lobby became adept at sorting the high-performing from the corrupt. Policing league tables now figure alongside investment performance in the Financial News. Of course, citizens have not always been able to move to the best-policed neighbour-

3. A traditional Stekram expression – boys, however, in most cases do not actually pay since they are taken care of in Family Units. Obviously those not subscribing to InterPay make no contribution, but they are of negligible importance.
4. The Ambassadorial Services Company and the Justicial and Ambassadorial Services Company respectively, Stekram's only two providers of diplomatic and cocktail services.

hood. But the detailed information about police efficiency and courtesy available on the InterPolice screens, as well as the individual personal pressure on the neighbourhood police companies to meet best-in-class performance levels – finally all this began to do the trick, and reported criminality to decline. As I travel around Stekram today, I never cease to admire the market offering of these little local police companies, which give so much employment to UFUs[5], and at the same time so much assistance to the old people taking exercise in the parks: this is Stekram organisation at its best.

The continued existence of UFUs is an unfortunate subject which I must ask my readers to consider in some depth. Technology in Stekram has developed to a stage where work is no longer necessary for the operation of most major industries. A handful of general managers and similar janitorial staff is all that is required. This even applies to many services: there are just a few, such as the police, restaurants, television and prostitution, in which employees are still required by consumers to add the human touch. As a result, Stekram is no longer fascinated by the working condition and follows the great majority of human societies since the beginning of civilisation in considering it undignified for someone to have to work. However, it is even more undignified not to eat, and so salaries have been driven down to a level which would distress my readers.[6] With salaries so low, so unpopular and so few and far between, and even freed from the burden of taxation, the vast majority of citizens would be quite unable to finance the demands of InterPay from employment income alone. A surprising result of this has been the resurgence of that old pre-industrial institution, the family. Prior to the Happy Revolution, threatened by divorce rates and undermined by mindless administrative redistribution, the family seemed to be consigned to the dustbin of history. Now, obliged to pay cash through InterPay in order to live (or at least in order to eat), and paying also for all those CommComms hitherto provided by the State, Stekram citizens have become intensely attached to the one institution where irrational cross-subsidisation and other ValueWaste is allowed: the family.

In Stekram so-called Paradoxical Inter-Generational Mortgages ('paradigms' for short) have become a commonplace, with conventional loan terms now stretching to five generations and backed up by the obligatory paradises (Paradoxical Inter-

---

5. Unfinanced Family Units

6. A most surprising finding of SoftSensor research has been that Stekram citizens, when the question is put to them in the right way, do not attribute exorbitant value to media stars and sports players. This finding must be decomposed into several logical propositions in order to see both why it holds in Stekram, and why it was so strikingly untrue before the Revolution. In a first stage, the citizen (out of an instinct of common humanity) declares that the minor actors and losing players deserve almost as much as the stars and the winners. (Tennis is indeed an exceptional case, since the consumer actually attributes greater value to the loser). The relevant prices on InterPrice screens go up. Second, since a decent income appears assured, members of UFUs flood into the relevant media or sporting industry. Third, Stekramers attain saturation point in consumption of media and tennis or whatever. Fourth, salaries fall to a level between those of constables and traffic wardens, with a slight premium for stars and winners (and tennis losers). It has also been suggested by SoftSensor researchers that consumers are prepared to pay high amounts for media gratification as long as a chain of intermediaries is involved (as was customary before the Revolution) but will in general fail to show any consumption appetite if payment is made (as is now usual under InterPay) directly to the bank account of a very rich person. This latter proposition is, however, more difficult to prove.

generational Security Engagements). The willingness of the Capital Banks to finance family units in this way is however conditional on ruthless credit assessment. The position of the infertile is particularly to be regretted, since clearly the younger the signatories of a paradise the greater its contribution to credit backing; promises to procreate have some value, but are made subject to fertility tests. Quite a few families cannot provide sufficient evidence to justify solvency on this basis.

The plight of the UFUs in Stekram remains a serious problem. These units may eke out an existence with employment income, but if proper finance is not available they will never attain the decent lifestyle of even the more modest FUFU[7], where investment income typically now accounts for around 90 percent of total income. Human nature has responded to this situation with prudence: even fairly well-financed units have scaled back their consumption needs wherever InterPay and the SoftSensors will allow them to do so. Living within one's means has become an obsession for all but the over-financed.

Insurance markets, however, have boomed. It amuses Stekramers to recall how little insurance they carried before the Millennium, or even before the Revolution (since most insurers were in fact bankrupted during this period, it in fact took some years for a new generation of underwriters to assemble sufficient capacity to accommodate insurance demand). The reason, of course, is that people at that time had a quaint faith in the ability of the administration to provide them with rainy day financial cover – the so-called Welfare State. But when the rains came, of course, the wind blew as well; and the umbrella was the first to be swept away. Yet the demand for a real safety net, organised competitively without ValueWaste, is huge. Varieties of personal insurance abound, but most Stekram families would be unable to sustain an economic strategy if they did not have the comfort of a few Lives[8] to fall back on.

*   *   *   *   *   *

There is no room here to describe fully the complex relationship between competition and monopoly in this economy. It should be relatively easy for my readers to understand, however, why Stekramers attribute so little importance to this subject.

The first point to grasp is that before the Millennium the administration wasted considerable resources in maintaining certain monopolies (mainly its own, but also those of some private concerns which had a good understanding with those in Government) and in preventing – or at least arguing with – others. When the administration collapsed, anti-trust bureaucracies disappeared without trace. In the upheaval which followed, many industries that could benefit from monopoly achieved it by the classic process of acquisition and merger and with a minimum of fuss, notwithstanding that Anti-Trust legislation was still on the statute books. The revolutionary Pro-Trust legislation regularised this position, and industrial structure now adopts freely the level of competition which is natural to the sector. There is a wide variety of forms of competition which citizens understand and accept without difficulty, including mutualistic structures such as InterPay, full monopolies such as Food, a variety of friendly or uncooperative duopolies and oligopolies, and even sectors of tremendous competition – the latter being primarily in areas where work

7. Fully Financed Family Unit
8. Longterm investment volatility endowed stabilisers

is still considered natural and necessary. Competition generally works through statistics rather than through alternative offers, since alternatives are so wasteful; but the intense interest shown by everyone in information analysis ensures that no company, whether Food Inc. with its twenty-five staff or a humble neighbourhood police company with ten times as many constables, can afford to be inefficient in its organisation.

Readers accustomed to archaic forms of economic organisation may have all sorts of dark imaginings about the behaviour of the big monopolies. In practice these prove largely unfounded. No-one would claim that graft and incompetence have been extinguished entirely in Stekram, but now that the flames are no longer fanned by administrative ValueWaste this is simply not a major political issue. After all, since stock market quotation is legally required for any economic activity, people can sell incompetence and buy graft as much as they wish, and in practice all FUFUs have well-diversified portfolios ensuring that they benefit from the fruits of monopoly. The stock markets ensure that, after allowance for market sentiment and occasional bouts of irrational expectation, most economic sectors offer the same Risk-Adjusted Return on Market Capitalisation. Citizens are in general well content with the performance of the main monopolies making up the Stekram economy, since these provide the backbone of their financial holdings and underlie their insurance Lives.

*　　*　　*　　*　　*　　*

Where can the UFU go from here? Unfinanceable and unfed, often with little more than a computer screen to their name, the poorest family units live in abject poverty and without benefit of policing, in squalid no-go areas and waste disposal sites which are isolated from the rest of Stekram society in order to prevent the spread of contagious human and computing viruses.

Meanwhile life for FUFUs is perhaps better designed than it ever has been. It is true that, since the promulgation of the severe laws against Financial Gambling a few years ago, when the wickedly enjoyable high-risk financial games of pre-Revolutionary times were finally eradicated, there have seemed to be few opportunities for rapid enrichment. In the absence of any really valuable employment, diligence and enterprise do not seem to provide the key. Investment returns, even allowing for a modest degree of volatility, do not provide scope for one FUFU to enrich itself more rapidly than others – since all FUFUs are essentially investing in the same range of underlying assets. Can it be that prudence and economy indicate the road to riches?

This is not to say that Stekram is an egalitarian society, nor a Calvinistic one. The fact is that, if fortunes can no longer very easily be made here – and of course the law ensures that fortunes cannot be received as a gift – they can be inherited. No wealth tax, no inheritance tax; no tax advisers; undertakers are competitively priced; solicitors, who were perceived as not adding enough value, have ceased to exist. Quite a lot of money comes through, even (or perhaps one should say above all) from a parent who had no intergenerational commitments. So what seems to have happened in the upper reaches of Stekram society is that the game of financial musical chairs has stopped, and the winners are the winners, and their children are tomorrow's winners, and so on to the last syllable of recorded time. Perhaps sharp minds will invent new ways of reaping pecuniary advantage, but mine at any rate was too dull; and people are understandably preoccupied with the old children-party questions of, Who won? And, Was the game played by the Rules?

As luck would have it the game stopped after a period of considerable social and political turbulence. Immediately prior to the Millennium, as is amply recorded by the literature of the time, it was a world of Snakebite Evangelists, Racketeers and Bigwig Financiers.[9] The period of the Happy Revolution, despite that felicitous Orwellian epithet of "Happy" coined by the Revolutionary Council to describe its activities and proclaim its intent, was little better. The people of Stekram today are understandably a little upset at the implication that fashionable society is composed essentially of former crooks, hedge fund proprietors, government ministers and parliamentarians, when all these and other undesirable activities have been largely eliminated in Stekram, or at the least brought under the strict control of the community. Not surprisingly, it is considered polite to infer that so-and-so won his or her fortune through the Lottery, since although this discredited institution is currently suspended on grounds of ValueWaste, it is widely felt that the Lottery did give valuable pleasure (albeit to fools, but does that diminish its economic value?), that a future update of InterPay might well see the Lottery back in, and that a Lottery Winner is in any case less distasteful than most other examples of success.

The situation, however, is no longer as tense as in the immediate post-Revolutionary years. The fact is that the son of a successful crook is a responsible member of society, to whom you may safely marry your daughter. It were ever so. Thus in another hundred years, as the heroes and villains of the heady Millennium-to-Revolution period gradually commit euthanasia and depart the scene, one can look forward to the end of all recrimination.

Dear Reader, by the time that you read this paragraph, the year 2000 may have followed its partner 1984 out of the future and into the past. For you, Stekram is merely a fable. Your children's computers have functioned without interruption through into the new millennium, and their prospects of employment are undiminished. Perhaps, as they say, it was all a dream.

---

9. The phrase is taken from a late pre-Millennium document of J. Mitchell, aptly titled "Dog Eat Dog".

Chapter 8

Information, culture and technology

Justin Arundale

The Author

Justin Arundale worked for several years as an information manager in the newspaper industry, where he was associated with many of the technological changes of the 1980s. He is a fellow of the Library Association and now works at the University of Brighton where he teaches information management.

Chapter 8

Information, culture and technology

Justin Arundale

'The endless cycle of idea and action,
Endless invention, endless experiment,
Brings knowledge of motion, but not of stillness;
Knowledge of speech, but not of silence;
Knowledge of words, and ignorance of the Word.
...
Where is the Life we have lost in living?
Where is the wisdom we have lost in knowledge?
Where is the knowledge we have lost in information?'[1]

Information, culture and technology

There is no shortage of people telling us about the importance of information. We are repeatedly assured that we live in an information society, that we are supported by a knowledge economy and that the future of commerce lies with the intelligent organisation. Words and phrases such as 'information superhighway', 'infobahn', 'cyberspace' and 'knowledge management' abound, all suggesting a future in which information technology will undo the damage of two centuries of industrialisation and usher in a new era of peace, plenty and concord.

It is not hard to discern the popular appeal of the notion of a society whose infrastructure is based on information technology. Computers are (mostly) small, clean to work with, cheap and easy to operate — and so have been widely adopted in our schools and our businesses, and domesticated in our homes. The popular imagery of the information society often rests upon an implicit, and usually unconscious, comparison between the friendly and liberating nature of computers and what they make possible, and the dehumanising, even enslaving, demands of the older industrial technologies.

Information technology is not, of course, concerned only with computers. Alongside developments in information processing have occurred important advances in communication. The capacity of the channels that carry information between computers — and between people — has increased enormously since the invention of telegraph, telephone and wireless. The global communications network that has been developed during the past thirty years is the technical infrastructure of the information society. There are few people in the world who are not directly or indirectly touched by it — through telephone, radio, television or the internet —

and few countries and societies that do not face the prospect of change that it implies.

Information and communication technologies, and the changes in individual and institutional behaviour that they bring with them, have thus been incorporated into the thinking of many widely divergent groups in society. Economists see them as a way of reducing transaction costs and improving the efficiency of markets. Sales managers see them as a tool for finding out more about consumers. Librarians see them as a device for bringing reality to their long dream of universal access to information. Politicians see them variously as a means of broadening democratic debate and decentralising power — or of establishing ever-tighter control over those they rule.

However, information and communication technologies are a paradoxical feature of our world, in that they simultaneously support both convergence and divergence in our social and intellectual environment. This is not a new paradox. The invention of printing and the spread of literacy had a similar effect, promoting a wider sharing of ideas, but also supporting political dissidence and revolution. In today's world, convergence can be seen in the way in which the technologies encourage the very wide spread of dominant cultures; and divergence in the increased possibilities for recording and disseminating the opinions and cultural artefacts of minorities. This makes it easier than ever before for powerful groups in society to present their messages to millions — but it makes the imposition of conventions and established views difficult, and the control or eradication of dissident opinions impossible.

These technologies, in other words, change the social environment in which the attempt is made to maintain the fine balance between conformity and dissidence, between freedom and control. Information is a central factor in modern human societies — perhaps in all human societies — and history shows the importance of the link between information and social change. The availability of information, and access to the educational structures which train people to make use of it, are highly political matters — they are key factors, for example, in the establishment and maintenance of democracy. Developments in information technology, and therefore changes in information dynamics, have great political and social significance, because information, by supporting belief systems and ideologies, is what creates and perpetuates societies and nations. It is the sharing of common cultural experiences and assumptions that makes groups of people hold together — whether the groups are small or embrace millions.

For a long time, certainly since the Enlightenment, there has been an assumption that the more knowledge people have in common, the more similar they will become, and the more mutual understanding there will be. As a result, cultural homogeneity, or at any rate cultural convergence, has been seen as an important benefit of globalisation. One of the most interesting, and potentially one of the most explosive, results of the spread of information and communication technologies is that is gives us the chance to test that idea — perhaps to destruction. The technologies have the potential to bring different people together by encouraging the adoption of common cultural assumptions; but they also have the potential to divide as never before, by enabling the formation of ever smaller social groupings, ever more introverted, exclusive and solipsistic.

Information and the internet

To this consideration has to be added another. The nature of communication within and between groups is changing; and much of the change is being driven by the particular topology of one communication phenomenon — the internet. Until the 1970s, communication technologies were either 'one to one' (such as the telephone) or 'one to many' (such as television or radio broadcasting). The internet is based on a different communication paradigm; and the metaphors implied in the words 'web' and 'net' are appropriate for describing the seamless, nodeless communication environment which the internet makes possible. On the internet, patterns such as 'many to many' and 'all people in a group to all others in the same group' are typical. These patterns are, of course, not new — they have always been part of more or less formal organisations such as neighbourhoods, villages, clubs and companies. What is new is the way in which the internet gives these patterns a global reach, removing the need for any form of physical proximity. Communication technologies have created an information environment in which space and physical location are irrelevant. This is the significance of 'cyberspace'[2] — it is an environment without locational definition, whose topography is determined only by information, where collocation is to be understood in terms of subject matter, not of physical presence. In cyberspace, communities are communities of interest, not of proximity, and one's 'cyber-neighbours' may be next door or thousands of miles away.

This dilemma is created in part by the sheer volume of information that technology has made available — a problem that has come to be referred to as 'information overload'. It is quite impossible for anyone to be aware of more than a tiny fraction of the information available in print, and in broadcast and electronic form. Individuals, organisations and societies all have to select what they wish to know. If the amount of information available to people is too great, then in that process of selection, the overlaps where information is held in common are liable to grow smaller. When faced — as one is on the internet — with a universe of knowledge that is impossibly large, it is a perfectly rational response to choose to concentrate on adding to one's existing knowledge rather than attempting to create new categories of interest. The consequence of this is a tendency for new, smaller, social groupings to develop around specific foci of interest and information, involving people who may be withdrawing themselves to some degree from the larger, more established, groupings to which we have become accustomed.

This characteristic of the information environment points society in two directions. On the one hand, in cyberspace there need be no isolation. Anyone can find and communicate with others who share the same interests, enthusiasms — and obsessions. On the other hand, there is less reason to speak with — and perhaps therefore come to tolerate — people who are different from oneself. Thus, great diversity is encouraged, but wider mutual understanding potentially threatened. At the same time, the traditional mechanisms of social control are much weakened. Communities always impose a degree of conformity, where members of the group are expected to acknowledge and, to an extent adhere to, the group's norms. In physical communities, membership of which is often determined by fairly random factors, these mechanisms tend to ease — or force — individuals towards a common denominator of behaviour which is acceptable to most members. This will tend to flatten out difference and dissidence, and inhibit behaviour which is unacceptable to a broad range of people. In cyberspace, where communities are more narrowly defined and membership is more likely to be by self-selection on the basis of common interests, group norms are more likely to emphasise difference and dissidence, and may sometimes encourage behaviour which would not be tolerated in a more broadly-based community.

In the complex and varied environment created by information and communication technologies, there are in principle no opinions, obsessions or forms of behaviour that cannot be explored and approved or validated. The mechanism whereby people with a shared interest in, say, South American butterflies can meet on the internet, in cyberspace, and exchange information and insights is exactly the same as the mechanism that allows white supremacists, aspiring terrorists and paedophiles to do the same thing. It takes only a very superficial acquaintance with the denizens of cyberspace and the information that they present as useful, interesting or true, to make one realise that this is an environment where any assertion may be accepted, perhaps uncritically, and any opinion will be validated, sooner or later, by somebody of like mind.[3]

Cyberspace is a useful word to describe an entirely new information environment, which exists as a result of the internet. The internet is thus at the heart of many of the changes, problems and paradoxes of the contemporary information and communication environment. However, what we recognise today as the internet was not planned initially as a grand project to revolutionise communication. Like many technologies that have changed society, it has evolved far beyond what was originally conceived. It had its origins in an American military project to develop a communications network which would allow data to be routed in many different ways, rather than depending on point-to-point communication. The intention was to create a system which would be very robust, capable of functioning even when substantial parts of the network were inoperative. The effect was to create a system which could in principle allow any computer, anywhere, to send data to, or receive data from, any other computer, using a simple and non-proprietary series of communication protocols.

Two key ideas were developed in the early 1960s. First, there was the idea of what was initially called the 'Galactic Network' — the concept of a globally connected set of networks, through which anyone, anywhere could access data and programs. Then there was the concept of 'packet switching' — the technology which allows data to be broken down into 'packets' and transmitted that way, to be reassembled into usable data by the receiving computer. Using packet switching means that the source and target computers do not have to be continuously connected, and only link up when a 'packet' is actually being transmitted. This in turn means that communication bandwidth — the 'pipes' through which data is passed, so to speak — can be used with very great efficiency.

These two ideas are fundamental to the operation of what has come to be known as the internet. They were initially brought together by an American military research agency, and a plan for what came to be known as ARPANET was published in 1967. The project initially advanced only very slowly, and by the end of 1969, the grand total of four host computers were linked together. An important aspect of the early years was the development of the network protocols, which are the 'conventions' or 'rules' which allow computers to access the networks and talk to one another, and without which the internet of today would not exist. The objective was that the emerging electronic environment should be 'open architecture' —

that is, not requiring participating networks and machines to conform in specific technical detail. In a sense, the objective was to develop a common language with which networks and machines of different kinds could communicate. The early protocols only regulated communication through the 'core' or 'spine' of ARPANET, and did not help with allowing communication between the networks or machines linked to it. But in 1983 a more sophisticated protocol was introduced, which came to be known as Transmission Control Protocol/Internet Protocol (TCP/IP), and which is still the basis of internet communication.

In due course, the military bits of ARPANET were split off, with MILNET taking over the military communication needs, and ARPANET remained to serve a general research community. During the late 1970s and early 1980s, a number of other networks sprang up and were connected to ARPANET. Some were designed to link together researchers in very specific areas; others were more general — USENET being the best known. In 1984 the UK joined in with the launch of JANET, which was explicitly set up to serve the needs of the entire higher education community. By the mid-1980s the internet was becoming established as the pre-eminent tool for enabling communication between academic and government researchers. But it was also beginning to develop conventions and assumptions — in particular, it began to take on the character of a somewhat anarchic forum in which freedom of expression, and the right to hold and criticise any opinion, were fiercely defended.

In the early 1990s Tim Berners-Lee (originally from CERN, later MIT) developed the ideas behind the World Wide Web. 'The web' is the development that changed the internet from a comparatively specialist technology into something of far wider and more popular interest. It uses internet protocols, but offers a graphical interface (which makes it easier to use) and makes extensive use of hypertext (which makes it simpler to find information and 'navigate' between different information sources). Although the web is the face of the internet that most people know best, it is a comparatively recent development, which would be impossible without all of the technological developments that lie behind it. But it is the web that has 'democratised' the internet and made it a mass phenomenon.

During the 1970s the number of host computers connected to the internet grew slowly, from about 20 in 1971 to about 200 in 1980. In 1986 the number was about 5000; by 1990 it had risen to about 310,000, and by 1993 to about 1,300,000. In January 1999 it was about 43,230,000.[4] This increase in the size of the internet reflects a dramatic change in its nature. From being essentially a military and research network, the internet has become an important, indeed indispensable, aspect of modern communication. It is now a world-wide phenomenon, used by many millions of people, and its influence can be seen everywhere — in education and commerce, in design and the arts, in military organisations and in television, to name only a few. It is difficult to find areas of modern life which have not been touched in some degree — and enthusiasts for the technology regularly argue that the internet is the great transforming technology which will change our culture and civilisation as nothing has done since the invention of agriculture.

In some areas of activity these extravagant claims have some weight. In business, for example, the impact of the internet has been enormous, and phrases such as 'e-commerce' and 'the i-economy' have become common jargon among managers and management consultants as well as between internet enthusiasts. One example of this perception of the importance of electronic technologies for business has been the extraordinary rise in the prices of shares of internet companies. (Between the beginning of 1998 and the middle of 1999, internet shares, as measured by the Goldman Sachs Internet Index, outperformed other media shares by a factor of three.[5]) This pattern of share performance, not generally justified by underlying profitability, is an indication of the optimistic view that investors have taken of the importance of the internet as a business tool. There are other, more sober, indications of the rising importance of electronic commerce. The number of secure internet servers — that is, those using encryption software to protect the integrity of data transfers — has been rising sharply, particularly in the USA. Since secure communication is essential for the reliable conduct of commercial activities, this suggests a similar rise in the number of organisations conducting business-to-business transactions across the internet. The measure is crude, but it bears out the suggestion that an increasing proportion of firms are expecting to do at least part of their business electronically. The belief that this is the case is certainly shared by many governments, and underlies, for example, the European Union's efforts to put electronic commerce on a stable legal and technological footing.

In social and cultural terms, the impact of the internet is harder to evaluate. The number of people connected to it is undoubtedly increasing very rapidly. What is less clear is what they are using the internet for. One of the simplest functions — e-mail — is still the most popular, which suggests that for many users, the facility of cheap, almost instantaneous communication is what is valued. The enormous quantity of information available on the web is obviously another major attraction, for schoolchildren doing homework as well as for professional researchers. The use of the internet for domestic purchasing (for example for book purchases and holiday and travel deals) is rising, and the growth in the importance of 'electronic auctions' is an indicator (particularly in the USA) of an internet-driven shift in household spending behaviour. The opportunities the web provides for recreation are also very extensive, with games, music, 'chat rooms' and so forth. However, the assumption that the internet is most used for serious educational and informative activities and innocent entertainment should be treated with caution — it is worth observing that a significant proportion of transactions appear to involve visits to pornographic sites.

Pornography and other 'undesirable' material on the internet has often been presented, particularly by sceptics in the media, as the major problem with the technology and its application. In fact pornographic sites (which for the interested person are expensive to access and for the uninterested are little more than an easily-avoided irritant) are a small part of a wider phenomenon, which includes sites devoted to racial hatred and religious extremism. The ease with which a web site can be created and maintained means that any individual or interest group can publish information — and make it available world wide. This is why almost any social, political or religious opinion, and almost any special interest, is represented and can be explored on the web.

This diversity has always been celebrated as one of the great virtues of the internet, and the dominant discourse in cyberspace has always tended to challenge any assumption that cultural homogeneity is an attractive objective. The idea of an information environment in which there are no controls is an alluring one — and many citizens of the internet community pass by few opportunities to tweak the tails of governments and other large organisations that attempt to conceal or suppress information. For as long as those connected were predominantly academics and techno-enthusiasts, this lack of controls was hardly noticed, and certainly not seen as a problem. But as the internet began to take on the characteristics of a mass phenomenon with potentially enormous social implications, concerns were raised by politicians, by the media and by users themselves.

Information and globalisation

Not all of the issues to do with diversity are directly concerned with the internet. The development and globalisation of television broadcasting have also had important implications for the diversity and integrity of cultures around the globe. On the one hand, global communication gives a voice to the voiceless, and makes possible the wide dispersal (although not necessarily a deep understanding) of information about different cultures, religions and social customs. On the other, much of the information which is easily available emanates from the nations of the West, and from the USA in particular — these being the nations that have privileged access to the global channels of communication. The language of the internet is English (about 78 percent of websites are in English[6]), and much of the output of the global broadcasting and entertainment industries is based on Western (and often specifically American) values and assumptions.

Concern about the dominance of the information environment by Western organisations and Western values, and about the imbalance between the supply and the consumption of information is far from recent — indeed, it has been the subject of a great deal of political activity for much of the past forty years. In the 1950s UNESCO pioneered research into the international flow specifically of news information, and sought to encourage developing countries to exchange information among themselves (a 'South-South' exchange) rather than remain dependent upon news exported by four or five major news agencies located in the developed industrialised nations (a 'North-South' exchange). During the 1970s, the developing countries demanded the establishment of what came to be called the 'New World Information and Communication Order' (NWICO), pointing to the Western countries' domination of the major international information channels, and arguing that the Western media give a view of developing countries which is distorted and self-interested.[7] The Western countries, the USA in particular, riposted by suggesting that demands for a 'right to communicate' and a 'balanced flow of information' were little more than a cloak for censorship and state control of news. This debate, and the heated allegations of the politicisation of UNESCO that it generated, led to the withdrawal from UNESCO of the UK and the USA.

UNESCO has moved away from grand projects such as NWICO, and in this area has concentrated its resources (much reduced for many years as a result of the non-participation of the UK and the USA) on supporting information production in developing countries, and encouraging news and cultural institutions to adopt new technology and more effective practices. However, the passions that the NWICO debate aroused is an indication of the sensitive nature of the issue of what has been called 'information imperialism'.

In this issue, as in so many others in the information society, can be seen the tension between the potential for information and communication technologies to create — or impose — a global culture, and the struggle by minority cultures to retain their difference and identity. The very considerable forces of global capitalism point us towards the erosion of difference, and the gradual convergence of markets and societies. On the other hand, there are powerful interests and ideologies that will be well served by holding Western values at a discreet distance, and maintaining a measure of cultural solipsism.

Whether, on a global scale, information and communication technologies support a process of social convergence, or whether they will enable non-Western cultures to maintain and develop their characteristic identities, will determine what kind of world we live in. We may end up with a homogenised global culture, with picturesque local variations but with essentially the same characteristics and concerns everywhere. On the other hand, we may witness a deepening of differences and an increased self-confidence and assertiveness on the part of non-Western societies, and an increased defensiveness as all cultures seek to retain their autonomy by excluding, and becoming hostile towards, what they see as threatening their own characteristic ideals and identity.

This is not only a cultural issue. There are also important implications for trade and the world economy. Because knowledge and cultural artefacts are an important part of world trade, cultural self-defence can only be practised by means of trade restriction in one form or another. An example might be seen in the EU rules about 'local content' in film and television, or in French rules designed to counter the advance of Anglicisation, both of which have at various times met opposition or ridicule from those against whose cultural encroachments they are intended to be a defence.

Information and law

Until recently, control of the internet has been exercised only with respect to technical matters and things such as the allocation of addresses and domain names. Content has not been subjected to any consistent or effective regulation (what many would call censorship), not least because of the difficulty of devising any regulatory régime that could cope with the global nature of the internet, and the fact that information can be located anywhere in the world, in any jurisdiction. However, many powerful interests in society are threatened by an uncontrolled information environment, and this is no doubt partly why there has been an increasing tendency to bring the internet within the normal regulatory controls of the law, and an increasing body of case law that constrains activity in cyberspace. The establishment, for example, of the principle that an act of publication occurs when information is accessed, and the increasingly confident use of the principle by the courts, has created a means whereby content can at least sometimes be subjected to legal scrutiny.[8]

But the internet is not the only information phenomenon presenting legal and regulatory problems. The increasing commercial significance of information has raised the profile of those aspects of the law — in particular, intellectual property law — which deal with it. Intellectual property law, as the word 'property' suggests, is based on ownership rights in things such as copyright, trade marks and patents. Trade in these things — information of one kind or another — is an increasingly important part of the modern world's economy. But trade is based on ownership — transferring ownership and licensing use — and cannot occur without clear concepts to support it. Thus, intellectual property law is the legal foundation upon which much of the structure of the 'knowledge industries' is based. The importance of this area of law in the information society can hardly be understated.

Yet information is an abstraction, which has to be distinguished from the physical object which may be required to carry it — ownership of a copy of a book does no confer ownership of the copyright. The thing owned, to which intellectual property rights refer, usually has no physical existence. Moreover, information often has no significance until it interacts with a human mind. Its meaning and value will vary from person to person and from context to context, and may well change drastically over time. A piece of information may be a single thing — a fact or a formula — yet it can be acquired by many people, and its acquisition by one person only sometimes affects its value to another, and does not prevent others from acquiring it as well.

The objects to which intellectual property law refers are therefore different from the objects which most other laws of ownership deal with. They are different in kind, by virtue of being intangible. But they are also different because of the significance of information in terms of the social, intellectual and spiritual development of individuals, and its importance in the evolution and maintenance of free societies. The concept of the ownership of information, and therefore of its regulation through law, raises important issues to do with the balance between the legitimate interests of its creators and disseminators and the rights and freedoms of its users — especially in view of the fact that copyright (in particular) is inherently a monopoly right. The balance is a delicate one — a fact that has been recognised for at least two centuries. In 1785, an English judge, Lord Mansfield, expressed the matter thus:

'. . . we must take care to guard against two extremes equally prejudicial; the one that men of ability, who have employed their time for the service of their community, may be deprived of their just merits and the reward of their ingenuity and labour; the other, that the world may not be deprived of improvements, nor the progress of the arts be retarded.'[9]

Copyright law has its origins in the sixteenth and seventeenth centuries, with attempts to regulate and protect the rapidly-expanding printing and publishing industry in Europe and America. The notion that copyright had an important role to play in protecting the interests of the creators, as opposed to publishers, of literary and artistic works came a little later, although it has since become a significant part of all arguments for supporting, extending and strengthening this area of law. During the past century or so, copyright and other aspects of intellectual property law have developed rapidly, both within national jurisdictions and by international agreements (for example, the Berne Convention of 1896 or the Paris Convention of 1883), with the objective of improving the protection offered to the owners of intellectual properties. This process has greatly encouraged the growth of the so-called 'knowledge industries' — indeed, without protection against copyright piracy and the unauthorised use of patents and trade marks, it is difficult to see how such industries could thrive.

In terms of its development, therefore, intellectual property law has been seen as part of the process of regulating the conditions of trade for a set of valuable commodities. The situation of the owners of intellectual property rights has received careful attention, and their commercial interests have been protected with increasingly extensive regulations and legislation. The argument is repeatedly and persistently advanced that, without the protection afforded by intellectual property law, human creativity will be inhibited. Artists and writers, it is suggested, will be impoverished; and scientific research will cease because the investment it represents cannot be protected (a suggestion which seems reasonable, but has never been proved).

However, intellectual property law, and copyright law in particular, faces a problem. The technology and the communications infrastructure required to reproduce and republish material have advanced steadily; and each advance has required a reinterpretation or a new, and more extensive, restatement of the law. In 1956 a new Copyright Act was passed in the UK, updating the provisions of the previous Act of 1911. In 1959 the first photocopier using plain paper went on sale, making widely available a technology which the Act had not anticipated. Less than 20 years later, in 1977, the Whitford Committee acknowledged that photocopying had made the law of copyright 'unworkable' — technology had overtaken the law.[10] The same process can be seen with the Act that updated the law to take account of photocopying (among many other things) — the Copyright, Designs and Patents Act of 1988. When the Act was passed, the total number of internet hosts was less than 10,000. Ten years later, the number had risen to about 12,000,000, and is now at least twice that.

The practical challenges to copyright law presented by the internet and the computer-based technologies for reproduction and publishing are daunting. The costs associated with the reproduction and dissemination of information have become trivial, and publishing has become an activity anyone with a reasonably powerful computer and an internet connection can engage in. Moreover, once something — text, music, picture, statistical data — is digitised (itself not necessarily a complex or expensive procedure) it can be copied any number of times and sent, by anybody, anywhere in the world.

There is clearly a public interest in ensuring that publishers, writers and designers, and all those who create knowledge and information, are protected against activities which might threaten their commercial viability. The 'knowledge industries' are economically important — as employers, innovators and disseminators of information of technical and cultural importance. The arguments of the owners of intellectual property rights for the ever-tighter protection of 'their' property are important and cannot be disregarded. However, developments in copyright law should not be seen entirely from the point of view of the rights owners. Just as there is a public interest in the proper protection of the rights of creators and publishers of information, so is there a public interest in the free flow of information. The notion that the publisher of information has extensive rights, and the user none, is coming to be seen as dangerous. The word 'publish' means 'to make public', and the notion of placing of information in the public domain so that it can become part of the common human inheritance is a powerful and important one. Control over the flow of information — whether by governments or by sectoral or commercial interests — can so easily become control over what may be expressed, or even thought.

Traditionally, copyright law has dealt with this problem by making a distinction between the idea and its expression, and asserting a public right in the idea once it has been expressed. In 1918, the American judge, Louis Brandeis, stated the proposition thus: 'The general rule of law is that the noblest of human productions — knowledge, truths ascertained, conceptions and ideas — become, after voluntary communication to others, free as the air to common use.'[11] The distinction is an important one — but it is becoming increasingly difficult to sustain and defend it. The act of sending to another a copy of a document found on the internet will probably entail a technical breach of copyright — but it is regarded by many people as part of the simple process of sharing an idea. Moreover the technologies that make it so easy to reproduce and disseminate information also make it easier to track, with the result that technical breaches of this kind are easier to identify and pursue. Many publishers are seeking to establish a strict 'pay per view' principle, using technology to monitor information use in order to generate revenue. In cyberspace, there is no ready equivalent of the principle that has always applied to books — once a book is bought, it may be used as the purchaser wishes, and the publisher can control the amount or nature of usage only when it turns to abuse, such as illegal reproduction. In its place we may see another principle — that the use of information ceases to be an entirely private affair, as rights owners track in great detail who is using their property and for what purposes.

Copyright law therefore presents us with an interesting conflict of objectives. The arguments in favour of a legal régime that offers extensive protection to copyright holders are based on economics, and are supported — unsurprisingly — by a powerful lobby of publishers, software houses and media corporations seeking to protect the fruits of their commercial research. But an increasingly vocal opposition questions this purely economic approach to copyright, arguing that the monopolistic nature of copyright ownership makes it necessary for the interests of the information consumer to be protected. At a major conference of the World Intellectual Property Organisation in Geneva in 1997, the main substantial opposition to the commercial lobby came from librarians, some of whose national and international associations have grouped loosely together to form the European Copyright Users' Platform.

This conflict — between the interests of copyright owners in strengthening their monopolistic rights and those of information users is securing free access to information — is being increasingly debated. Some legal thinkers have proposed the development of a new legal concept to lie alongside copyright — the concept of 'copy-duty'. Copy-duty would impose on copyright holders certain obligations to disseminate, license or otherwise make available their information. It would balance a public right of access to information and cultural artefacts against copyright holders' rights to exploit their property for gain. In recent years, the debate about copyright has focused mainly on the economic interests of copyright holders, and the weight of political and media indignation has been directed against copyright piracy and theft. In the near future, we can expect the terms of the debate to change, with more attention being paid to the monopolistic nature of copyright and the way in which it can be used to deny information to those whose only wish is to engage with their cultural environment. The debate is an important one, because it raises the question whether information and knowledge are part of a common human resource, or simply another traded commodity, subject to the rigours of the market. 'Copy-duty' is an idea which will receive more attention.

Among the more important owners of copyright must be included governments, who generate large quantities of information — everything from internal policy documents through census data to map and charts. This is increasingly being recognised as a category of information to which special rules should apply, and many countries are now adopting legislation to force official bodies to make much more information publicly available. This process — intended to make governments more open and accountable — is gathering pace; and there appears to be an important shift in the nature of the partnership between governments and citizens, in which there is an assumption that government-generated information should be freely accessible — or even, perhaps, regarded as public property.

The issues associated with the ownership of information assume particular importance in the context of the tendency to codify details of individuals' lives — individuals who are both required (by public authorities) and encouraged (for example, by market researchers) to turn every aspect of their lives into data which is controlled by others. This data is a saleable, valuable commodity — it is increasingly used for planning and marketing purposes; and an increasing range of important decisions (for example, to do with credit-worthiness) are made on the basis of it. This has led, during the past three decades, to the development of another new area of law — data protection.

The first legislation to attempt to regulate the relationship between the owners and users of personal data and the subjects of that data was enacted in the German state of Hessen in 1970, and was followed by similar legislation in a range of other European countries. The reasoning behind the legislation is expressed in a UK government White Paper of 1975,[12] which addressed the implications of computing technology for individual privacy. In it there was a good description of the advantages of computers, from the user's point of view (particularly the commercial user's), and the threats they might represent for individuals and their rights. According to the White Paper, computers have a number of characteristics that suggest a need for the regulation of their use in processing data about individuals. First, they facilitate the maintenance of extensive record systems and the retention of data in those systems. Then, they can make data easily and quickly accessible from many distant points, and make it possible for data to be transferred from one information system to another. They make it possible for data to be combined in ways which might not otherwise be practicable; and they hold, store, process and transmit data in a form which is not directly intelligible, meaning that few people may know what is in the records, or what is happening to them. All these characteristics confer upon those holding personal data considerable power to use that data in ways that may be contrary to the interests of the data subjects, but which those data subjects cannot discover or challenge.

The early data protection statutes relied on the use of regulation or licensing arrangements to control the activities of those holding data about individuals. However, since the 1980s, and with the dramatic increase in the number of computers in use, the focus has changed, with increasing emphasis being put on the notion that individuals' consent is required before data about them can be lawfully processed. This notion, often referred to as 'informational self-determination', is creating a new set of rights and obligations to which those dealing with information have to have regard.

These two areas of information law — copyright and data protection — give rise to similar problems and conflicts. Both involve the notion of ownership of something that is intrinsically abstract and non-tangible; and both control or prohibit the use of information where its reproduction and dissemination are technically very simple. Both areas of law can prevent people from doing what often seems, intuitively, to be their right, and certainly does not appear to be immoral or illegal. With both copyright and data protection, there is a conflict between information 'owners' (a concept created by the law) and information 'users' (a concept which reflects a very long-standing social practice). The conflict is in part between organisations and individuals, and is therefore concerned with an important balance of power within society. However, it is interesting to note that organisations and individuals are not consistently on the same side. In copyright law, organisations — usually businesses — are generally the information 'owners', and individuals the 'users'. In data protection law, they are the opposite way round, and it is individuals who 'own' rights over the information about themselves.

This conflict, between information 'owners' and 'users' is fundamental in a society where the creation, appropriation and retention of knowledge is so central, and so important for individuals' self-development as well as for the efficiency of organisations and the prosperity of businesses. The way in which the debate over these competing rights is conducted will be an important feature of the development of the information society, and its resolution will go a long way towards determining what kind of world we end up with.

Information and organisations

Some of the most startling effects of the information revolution can be seen in the structure and culture of organisations. Large organisations — whether governmental, military, ecclesiastical or commercial — have traditionally been very hierarchical. Formal structures have always been favoured, and perceived as the only way to ensure stability and avoid organisational chaos.

Hierarchies exist to transmit two things — power and information. The orderly devolution of authority through a pyramid-shaped structure — from a single person or very small group of people at the top to a large number of functionaries at the bottom — is an assumption that underlies many organisations. But just as the people at the top hold more power, so they control more information — and the control of information is part of the mechanism for maintaining that power. Thus the orderly movement of information down through the structure from the powerful to the functionaries is also assumed. Control over the flow of information is an important part of the exercise and maintenance of authority and power.

The gains available to most organisations from the introduction of information technology are enormous, in terms of efficiency, flexibility and the ability to learn and adapt. For this reason, most organisations have, over the past decade or so, embraced the new technologies. But information technology will create, or at any rate create the conditions for, new patterns of information flow, which will not necessarily support, and may even work against, the traditional power structure. Information technology is therefore strongly associated with profound organisational change — change not merely in working methods, but also in the nature and structure of authority and thus in the way the organisation is conceived and understood.

When information flows through relatively formal channels which follow or imitate the hierarchical structures of authority, then it is easy to control. But the introduction of information technology encourages the development of less formal networks, where information can move easily between any points in the organisation, and is not confined by formal structures, nor so readily controlled. Not only can information generated within the organisation flow easily, but information from outside can be acquired readily, without significant cost and without the necessity of any form of formal authorisation. The result is that all employees can be better informed and more aware of developments within and outwith the organisation. This means that the rationale for decisions and policies has to be presented and argued in a different way, with an inevitable consequence for the way in which authority is perceived and exercised. Words such as 'consensus' and 'empowerment', and concepts such as 'obedience', 'loyalty' and 'delegation' are subtly redefined.

Most large businesses have confronted this situation — and their responses tend to show a common pattern. First, the organisational structure is drastically flattened, with many fewer levels of authority and a more open culture, with a freer attitude towards communication and sharing information. Second, the organisation devolves power from the centre to the periphery, often relinquishing control over many of its functions and buying in goods and contracting in services that it once controlled directly. The management writer Charles Handy many years ago gave an account of this process, and coined the phrase the 'doughnut organisation' to describe an organisation where the centre is small and occupied by a few very well-informed people, and where the majority of activities occur on the periphery, and are undertaken by firms and individuals whose relationship with the centre is controlled by means of contracts for services, and not contracts of employment. The development of 'doughnut organisations' has been a marked feature of institutional change during the past two decades, and has brought with it other changes — social, economic and political — not all of which have been benign.

A particularly dramatic example of these changes can be seen in the film-making industry in Hollywood. Until the 1960s, the studio was supreme — a tight, often autocratically controlled, organisation that employed the full range of people involved in making films — the writers and the stars just as much as the camera crews and editors, and the caterers and the cleaners. Today, the core of people controlling the making of a film is very small. Everything is contracted out, and Hollywood is now characterised by a myriad of small, specialist firms rather than a small number of large, monolithic and very powerful studios.

It is interesting to observe that many of these organisational characteristics have been associated from the start with firms concerned with the development of information technology, and software firms in particular. Where the critical business emphasis is on innovation, then the organisation will tend to be driven by the creativity of employees, not by their ability to follow established instructions and strategies. The flattening of the organisational hierarchy is an almost inevitable consequence; and it has often been observed that computing and software development firms tend to have an organisational culture that is relaxed and informal, with a strong emphasis on achievement rather than conformity.

Most organisations — and most businesses in particular — can state and achieve their objectives within this context of change. However, there are many organisations where the transmission of authority, for whatever reason, is a significant part of their objective and reason for existence. Most religions fall into this category. The hierarchy of the Catholic Church, for example, is determined by the theologically defined need to transmit spiritual authority. The idea of flattening the organisation in anything other than administrative terms appears to be profoundly at variance with the entire purpose for which the Church exists. Yet the information revolution has affected Catholics as well, and for them the Church is no longer the only, or even the main, source of moral and spiritual speculation and assertion. There are many institutions in his sort of situation, which will struggle to come to terms with any restructuring or flattening of hierarchies — military organisations are another example. They raise the interesting question of 'what is authority?', with the long-standing answers being challenged by technology.

What are the implications of this? Institutions may be weakened and their ability to act as foci for authority and stability reduced, thus contributing to social divergence. On the other hand, institutions may be reinvented, acquiring authority from new sources and wielding power in new ways. Either way, the nature of the consent that binds individuals and institutions together will change dramatically. As communication within society adapts to the topology suggested by the internet, people are increasingly likely to define themselves in terms of what they are and what they know, and less in terms of what they do. Their relationships with institutions will therefore change, and they are more likely to be able to offer their services and loyalty because they choose to do so and not because they have to.

Information and knowledge

The sheer quantity of information available in today's world presents a host of problems — both management problems to do with controlling it and cognitive problems to do with understanding it. 'Information overload' is a major issue for most organisations and many individuals. Most of the responses involve using technology to solve the problems that technology has brought about. In organisations in many areas, more and more reliance is being placed on exploring the potential of automated systems for filtering information, and for making decisions. The automation of the collection and dissemination of the organisational memory is even associated with a special term of art: 'knowledge management'. For individuals, the options for using automated systems to cope with the information deluge are less grand. But even here, the increasing sophistication of the systems that allow users to locate information on the internet is pointing towards an automated mechanism for dealing with the information chaos in cyberspace.

However, dealing with information overload and information chaos is not necessarily the same thing as codifying information, still less creating knowledge and understanding. It is all too easy to use databases to organise information without having regard to how meaning might be extracted from it. It is all too easy to use complex search tools, employing exciting statistical algorithms, to retrieve information which will only ever be added to other information, or simply passed on, without any consideration being given to how it should be understood, or integrated into the individual's knowledge and experience. Indeed, as the tools used to retrieve and organise information become more sophisticated, and the patterns they make with the information become more beguiling, so an increasingly spurious impression can be created that understanding has been achieved and knowledge created.

The retrieval and organisation of information is not an end, but a beginning. In the complex, demanding, fascinating environment created by information and communication technologies, it is not always easy to remember that. Individuals, organisations and societies develop towards success, stability and fulfilment, not by maximising the quantity of information they deal with but by optimising the quality of what they do with it. It would be a pity if the information revolution increased almost infinitely the amount of information available to the human race, but increased knowledge and understanding barely at all.

Chapter 9

Cities as a key to sustainability

David Goode

The Author

Professor David Goode, Director of the London Ecology Unit, is a recognised authority on the ecology of towns and cities. In his 30-year career as an ecologist, he has played a key role in bringing an ecological dimension to urban planning, through his work with both central and local government in the UK, and as an advisor on the urban environment for major cities in developing countries. He is a past-President of the Institute of Ecology and Environmental Management, and a Visiting Professor in University College, London and at East China Normal University in Shanghai.

Chapter 9

Cities as a key to sustainability

David Goode

Introduction

On the wall of the council chamber in the ancient hill town of Siena is a fresco depicting the effects of good government in the town and countryside. Painted by Ambrogio Lorenzetti in 1338, it symbolises perfectly the relationship between the town and its surroundings. The values to which Siena aspired in art and culture could not be achieved without ensuring that the hinterland of olive groves, pastures, woods and streams was in good heart. For those governing Siena at the time it was a constant reminder of the symbiosis necessary for prosperity.

We are faced with a very different picture today. Our world is changing dramatically as a result of urbanisation, which has become a global phenomenon of immense proportions. World population has grown from just less than one billion in 1800 to six billion today. The growth of urban populations over this same period has been even more dramatic, from only 3 percent of the total world population in 1800 to about 50 percent today. In total numbers this means a one hundred fold increase from 30 million urban dwellers in 1800 to 3 billion today. The rate of urbanisation continues to rise and United Nations projections suggest that the world's urban population will be twice that of rural areas within the next thirty years.

We are rapidly moving to a new kind of world where the vast majority of people will live urban lifestyles. This is, of course, already true for most developed countries which went through the demographic transition relatively early. These countries have become accustomed to the fact that the majority of people, commonly 70-80 percent of the total population, live in towns and cities, and the proportion is still gradually rising. But it is in the developing world that we are seeing the greatest changes. In both Asia and Africa the level of urbanisation rose from about 15 percent in 1950 to 35 percent in 1995. Asia now has almost half the world's urban population, and both South America and Africa have larger urban populations than North America. The total number of people living in towns and cities of the developing world now vastly exceeds those of developed countries. The difference is that *per capita* levels of consumption are still extremely low when compared with European or North American lifestyles.

It is perhaps the scale of urbanisation itself which poses questions of sustainability. London seems to have been the first city to reach a population of one million in about 1800. World-wide, there are now nearly 300 cities with over one million inhabitants. A further 60 exceed five million and a small number have reached megacity proportions with populations of 10 to 20 million. United Nations' estimates suggest that by 2015 the average population of the ten biggest cities will be 22 million. So it is not surprising that attention has increasingly focused on the sustainability of cities. Finding ways of making modern cities vibrant and liveable, and at the same time reducing their global environmental impact to levels which can be maintained in the long-term, is perhaps the greatest challenge facing us today.

The picture of Siena is a perfect metaphor, reflecting the dual responsibility for human well-being and for the wider world on which this depends. The implication that this can be achieved through good governance has more than a little relevance to our search for sustainable cities.

Sustainable development

The human race has a remarkable ability to compartmentalise its activities. Even in environmental matters where one might expect integrated thinking as the norm we have all too often failed dismally to make the necessary connections. I say this because, looking back over the past thirty or more years, it is remarkable how little cross-fertilisation has occurred between people engaged in solving the problems of cities, and those developing principles of sustainable development. With some notable exceptions these two fields of human endeavour have proceeded in parallel and it is only in the past ten years that the two strands have really come together. In his recent book on sustainable cities, David Satterthwaite comments that the tendency for debates about sustainable development and sustainable cities to develop independently is clearly demonstrated by the fact that so few individuals have worked in both these fields.[1]

In some ways it is simply a matter of different disciplines ploughing their own furrow. It was ecologists and natural resource biologists, such as foresters, who first developed concepts of sustainable use, whilst planners, designers and public health officials have been responsible for solving, or sometimes creating, the ills of the city. But I suggest the separation goes deeper than that, in the sense that many environmentalists have viewed cities simply as a problem, preferably for someone else to solve. The possibility that cities might offer ecological solutions is a relatively new idea which is only just beginning to permeate the professions.

One of the notable exceptions to this compartmentalisation, was the Stockholm Conference of 1972. This first United Nations Conference on the Human Environment was notable for the remarkable synthesis of environmental issues by Barbara Ward and René Dubos who clearly understood the ecological consequences of urbanisation. Since then we have seen a series of global initiatives drawing from each other but often following separate courses.[2]

Themes dominating the first UN conference on Human Settlements at Vancouver in 1976 were almost entirely driven by human needs. Housing, poverty and health were paramount. With hindsight it is clear that Vancouver failed to envisage the sea-change that was soon to engulf cities of the developing world as population growth accelerated and environmental problems exploded on a scale never experienced before. However, the critical issue of human health, combined with rising numbers of urban poor, led to new initiatives for cities during the 1980s, especially the healthy cities campaign spearheaded by the World Health Organisation.

Two things were notably absent from Vancouver, the role of the community and the whole question of sustainability. Solutions were still seen firmly from a top-down perspective. Governments were expected to take the lead in planning urban

126

development which itself was largely seen in terms of roads and buildings. The idea that community-based initiatives might have a role to play was hardly appreciated and the concept of sustainability was even less relevant to those grappling with the problems of cities at that time.

The World Conservation Strategy produced by IUCN in 1980 was in fact the first international initiative to expound the principles of sustainable development.[3] Though narrower in focus, it set the scene for the subsequent Brundtland Commission statement that 'sustainable development is about meeting the needs of the present without compromising the ability of future generations to meet their own needs'.[4] Inherent in this statement are two key principles which together form the basis for sustainable development. On the one hand is the 'meeting of human needs' and on the other is the 'need to do this in ways which will not reduce the earth's natural capital, either now or in the future'.

This duality formed the basis of the Earth Summit of 1992 which, in addition to agreeing crucial conventions on climate change and biodiversity, precipitated an extraordinary surge of activity by cities across the world under the banner of Local Agenda 21. Many cities in both North and South have introduced innovative long-term programmes to improve environmental conditions, reduce their use of resources and minimise waste. The whole approach has gained recognition from governments and city authorities to a remarkable degree.

By the time the UN held its second conference on 'Human Settlements' at Istanbul in 1996, the question of sustainable cities was firmly on the agenda, in contrast to Vancouver twenty years earlier, when the concept was virtually unknown and certainly not seen as relevant. By the time the next 'City Summit' is held in 2015 the world's urban population will have passed the four billion mark and there is no doubt that issues of sustainable development will dominate the agenda.

Ecological footprints

Since the present phenomenon of urbanisation is so closely bound up with global trade, any consideration of sustainability must take account of the demands made by cities on ecosystems far removed from their natural hinterland. Throughout history, the most wealthy and powerful European cities, such as London, Paris, Amsterdam and even ancient Rome, had the capacity to draw on lands way beyond their immediate surroundings. Indeed, it can be argued that their growth and prosperity depended to a large extent on their ability to utilise the natural wealth of distant lands. The process which was set in train through imperial and colonial rule continues today, on a vastly greater scale. World trade has increased dramatically over the past fifty years as a result of cheaper transport and exponential growth in the global economy. The predominantly urban populations of developed countries have increasingly used the natural resources of other countries to maintain their high living standards. In the new wave of urbanisation in the developing world the more prosperous cities are following suit, drawing on a multitude of countries to supply raw materials, food and energy supplies. In this new scenario it is almost impossible for individual cities to consider the ecological consequences of their actions because their hinterland is spread far and wide. Whilst the city fathers of Siena could easily check the state of the environment on which they depended, it is a very different matter for city governments today. Yet, if we are to take sustainable development seriously it is crucial that we are able to assess how present and predicted demands relate to the natural carrying capacity of these 'distant' hinterlands.

The problem is not new. In its Man and Biosphere Programme, UNESCO promoted two decades of studies looking at cities as ecological systems. John Celecia points out that the 'long roots of the city' were recognised as a crucial element in the detailed ecosystem model of Frankfurt as long ago as 1981.[5] Subsequent studies in Tokyo, Hong Kong and Rome identified the global demands made by these cities on distant ecosystems, but their overall ecological consequences were as yet unquantified.

The most significant step came ten years later, when William Rees devised a method to calculate what he called the 'ecological footprint' of any given community. This he defined as the total land area required to provide the necessary food and raw materials and also to absorb the output of waste products such as CO_2. Working in Vancouver, Rees calculated that the lower Frazer Valley of British Columbia, including the city of Vancouver, has an ecological footprint about 20 times its own land area.[6] He and his colleagues went on to examine ecological footprints on a *per capita* basis and found that the average citizen of the United States or Canada required approximately 4-5 ha of productive land to support his or her present lifestyle. He states that 'extrapolation of the present North American lifestyle to the entire world population of 5.7 billion would require about 24 billion ha of ecologically productive land, using existing technologies. Since there are only 8.8 billion ha of such land on the planet we would need at least two additional Earths to bring just the present human population up to north American ecological standards'.[7] Lest Europeans feel complacent, I hasten to add that the equivalent figure of 3 ha for European countries means that one extra Earth would be required to raise lifestyles world-wide to European levels.

Global figures such as these bring us up sharply against the realities of ecological limits, and also demonstrate very clearly the enormous disparity in use of natural resources between countries of the developed and developing world. But the concept of ecological footprints can be applied at any level, right down to local communities. The technique has already been used to compare the ecological efficiency of different kinds of housing schemes, commuter patterns and various types of products. At a domestic level it has been shown that, in North America, high income households in rural or suburban areas have a larger ecological footprint than equivalent households in cities, mainly because city dwellers use less fossil fuel energy.

This approach could have considerable potential for comparing the performance of individual cities. Indeed, some attempts have already been made. Undaunted by its size and complexity, Herbert Girardet examined London's ecological footprint and came up with a figure of 125 times the land area of the capital. He points out that this is almost equivalent to the entire productive land of the UK, though the footprint is, of course, dispersed throughout the world.[8]

Comparisons can be sometimes be misleading. The calculation for London was based on an area of 1580 km², most of it heavily urbanised, with a population of 7 million. Rees chose a region of 4,000 km², which included an extensive rural hinterland in addition to the city of Vancouver, supporting 1.8 million people. No doubt the quality of the statistics varied too. But, despite all that, progress is being made. For the first time it is possible to quantify the ecological performance of a city in its global context. It will not be long before we have an ecological performance rating alongside traditional economic indicators for every major city. Assessment of the ecological footprint is a vital step in this direction.

Urban metabolism

But understanding the ecological footprint is not sufficient in itself. We need to look in greater detail at the ways in which a city functions in order to find out where consumption can be reduced, or re-use of resources maximised. Using the metaphor of urban metabolism could well be a vital key to find ways in which urban systems can be made more sustainable.

I have already alluded to the series of studies promoted by UNESCO which considered cities as ecosystems.[9] These multi-disciplinary studies aimed for the first time at examining the metabolism of a whole city as if it were a natural system. Inputs and outputs were quantified and energy levels measured in great detail. Some researchers went to extraordinary lengths in quantifying all the components and tracking their paths through the urban ecosystem. What they found demonstrated a critical difference between urban systems and natural processes Whilst natural ecosystems have a series of inbuilt circular processes, the metabolism of a modern city is almost entirely linear. This is particularly the case for affluent cities in developed countries, where vast quantities of material are imported daily for human use, and waste products are discharged from the city as unwanted residues. This contrasts with natural biological systems, where food chains, decomposer cycles and mineral recycling ensure that no components are wasted.

Although there are differences in the problems faced by the cities of the North and South, the basic components of their metabolism are the same. On the input side is the daily requirement for food, water and energy to sustain the human population in all its activities. The more affluent the society, the greater the associated inflow of paper, glass, plastics and other materials to meet daily expectations, as can be seen in loaded shopping trolleys anywhere in the developed world. Less obvious perhaps, but no less significant, are the vast quantities of material imported by business, industry and public institutions to meet their daily needs. Entirely unseen and hardly ever appreciated is the daily combustion of oxygen through use of fossil fuels to power all the processes of the city.

The outputs are the waste products of society, which are disposed of every day in huge quantities. They include sewage, contaminated water and a vast array of solid wastes. The majority of these solid wastes, which are a special feature of developed countries, where affluence has resulted in a throw-away society, are dumped in landfill sites outside the city. Again, mostly unseen, are the gaseous emissions resulting from use of fossil fuels for heating, lighting and cooling of buildings, and also, of course, for transport. Carbon dioxide and carbon monoxide, nitrous oxides and sulphur dioxide are all produced in vast quantities as a result of present energy and transport policies.

But the city's metabolism is not restricted to these daily functions. Growing cities also require enormous quantities of aggregates, glass, timber and metal products for construction of new buildings, roads, subway-systems and all the paraphernalia of our modern built environment. Rapidly growing cities make demands on the wider environment for these materials, which are then locked into the urban system. When eventually replaced by the need for more up to date structures, the original components all too often simply join the stream of unwanted residues – the city's waste.

Numerous attempts have been made to quantify individual elements of city metabolism with particular emphasis on the day to day functions. As a result, it is now possible to compare particular aspects of the ecological efficiency of individual cities. Take energy for example. The need to reduce emissions of greenhouse gases has led to a plethora of studies throughout the world examining the energy efficiency of cities. Girardet used the London Energy Study to demonstrate that London uses the equivalent of two supertankers of oil each week to supply its energy needs and, in so doing, discharges some 60 million tonnes of carbon dioxide into the atmosphere every year. He points out that the *per capita* energy consumption of Londoners is almost the highest in Europe, exceeded only by Brussels. Comparisons with Helsinki and Amsterdam demonstrate that the technology already exists to reduce London's energy use by 30-35 percent without affecting living standards.[10] Yet energy consumption in European cities is relatively modest compared with most North American cities. The average Canadian city dweller uses twice the amount of energy as a citizen of Amsterdam or Helsinki. The profligate use of energy by North American cities reaches its peak in the sprawling low density desert cities of Arizona and California, where the potential for solar energy is all too obvious but hardly used, and levels of fuel consumption per person for travel by car are amongst the highest in the world.

The reasons for these major disparities in energy requirements are now well understood. London is less efficient than Helsinki partly because it relies on inefficient methods of electricity generation and long distance transmission, whilst Helsinki gains from local power generation and greater efficiency obtained through combined heat and power which supplies much of the city's needs through district heating systems. Efficiency could be further enhanced by using renewable energy to a much greater extent. The UK government has set a target to generate 10 percent of electricity from renewable sources by 2010, and much of this will be achieved in major cities. For, although renewable energy is often assumed to be a rural technology, it is fast becoming a major factor in the ecological efficiency of cities. Use of solid waste as a fuel could meet 10 percent of London's energy needs over the next 10 years. Although there is still widespread public concern about the air pollution resulting from incineration, new technologies are now widely used in Europe to convert solid waste into energy. Nearly half the municipal waste in France and Germany is utilised in this way, and the resulting ash is used as an aggregate for construction purposes.

Architects are also placing increasing emphasis on the design of buildings to maximise the use of solar energy and reduce the need for either heating or cooling. Passive solar design could well become a major factor in reducing the ecological footprint of cities in the future.

In the UK the Building Research Establishment has developed a scheme to assess how far an individual building scores on sustainability. The scheme (known as BREEAM) was first introduced in 1992 and is now achieving international recognition. It covers many different environmental issues in the design and operation of a building, ranging from energy use to recycling of construction materials, transport costs and potential ecological impacts. The city of New York has just published guidelines on 'high performance buildings' which do precisely the same job.

Catering as it does for US business interests, such procedures are justified in terms of increased financial profit and human well-being; and it is no surprise to find the phrase sustainable development conspicuously absent.[11]

Understanding the city's metabolism is equally important when dealing with waste. Waste reduction plans are becoming an essential part of city management, whether in the developed or developing world. In London, a detailed study of the sources of 14 million tonnes of solid waste per year found that only 25 percent comes from households.[12] Three quarters of London's waste comes from the commercial, retail and business services sectors. London's prominent service sectors in finance, entertainment, communications, hotels and catering produce a quarter of all the waste; and businesses such as advertising, PR and marketing produce about 1.5 million tonnes or 10 percent of the total. The business sector is now being challenged to develop plans for reducing the amount of waste it produces; and this will inevitably be a major issue for the new Mayor of London, who is required to produce a waste management strategy for the capital. In parallel, is the need to find effective uses for all kinds of waste. Innovative schemes are happening all over the world, one of the most extraordinary being the production of new forms of asphalt for road surfaces, using recycled glass mixed with discarded car tyres.

One of the best examples of innovative solutions to waste management comes from Curitiba in Brazil. Before the visionary Mayor Jaime Lerner came on the scene, most of the city's rubbish was dumped along riverbanks on the edges of shanty towns, where it polluted the water courses and led to chronic health hazards. Lerner mobilised the shanty town dwellers to take rubbish to organised landfill sites by paying them with transport tokens. Children were given books and food. An effective local economy was established, and the urban poor were able to embark on new commercial enterprises by selling crafts in local markets.[13]

The great value of the metabolic model is that it allows us to identify where savings in resource use can be made; and then, of course, we have to find local solutions to make these changes. Instead of viewing the city as the source of all the problems, we can look at it as the most efficient way of contributing to sustainability. It is in the cities that the greatest opportunities exist to make the necessary changes towards sustainability. Rees points out that there are enormous potential savings to be made through the economies of scale inherent in high urban densities In addition, it is at the local level that this can best be achieved, through partnerships between local government and local communities. What this means is that effective policies for sustainable development gain their greatest leverage in cities.[14]

Urban design

What we need now is a manual for high performance cities. Just as the green building assessment, referred to earlier, looks beyond the internal functions of the building to its wider environmental context, so we need to look beyond the metabolism of the city to the wider issues of its geography. Density is one of the clues to sustainability. All the evidence points to higher-density cities being more efficient, not only in energy, but in many other aspects of their functional metabolism. The way that people move about the city is the key issue, in particular the extent to which people are dependent on cars as the predominant mode of travel.

Newman's classic study of 32 major cities in North America, Europe, Australia and Asia showed that cities could be divided into distinct categories of car dependence. At one end of the spectrum are extremely low density cities, such as Phoenix, Arizona, where fuel consumption per person is five times that of London. Most European cities have moderate densities but low levels of fuel consumption, whilst the Asian cities examined, such as Tokyo and Singapore, have a significantly greater density and lower levels of fuel consumption. Newman emphasises that wealthy, prosperous and desirable cities can have low levels of fuel consumption, reflecting less car dependence. Cities such as Vienna, Copenhagen and Stockholm are among the most desirable cities in which to live, yet they have one fifth to one tenth the level of fuel consumption of American cities.[15]

It is perhaps instructive to look at some of the extremes in a little more detail. I am told that the original attraction of Phoenix, set in the desert landscape of Arizona, was its clean air and perpetual sunshine. For those who had suffered a lifetime of bronchial problems in heavily polluted industrial cities, it was an attractive retirement prospect. Now, with a population of seven million people, almost entirely dependent on cars, Phoenix suffers the same acute problems of air pollution that have long afflicted places like Los Angeles, which is famous for its photochemical smog. In terms of sustainability Phoenix has other problems too. One can hardly imagine a less sustainable environment for a major city. Water supply will inevitably become a critical issue within the next ten years, and could well be the one constraining influence that limits its future growth. Yet the fastest growing places in the USA today are these sprawling patchwork suburbs of the so-called sunbelt. Few people outside the USA will know their names; places like Chandler and Scottsdale are expanding satellites of Phoenix; Henderson, an extensive desert suburb of Las Vegas, holds the record for growth in recent years. These are places without a centre, where scattered shopping malls and office parks, accessible only by car, substitute for the downtown of the older industrial cities; cornershops are in the filling stations, and no-one walks except on the golf course or along the hiking trails in the surrounding desert mountains.

At the other end of the scale are the rapidly developing cities of China, where transport patterns are changing overnight. Densely populated cities, such as Shanghai and Nanjing, still have the highest levels of bicycle use in the world; but the policy to increase the use of cars as a means of promoting economic growth is changing these cities beyond recognition. In Shanghai, a network of raised motorways, or 'viaduct roads' as they are called, has been superimposed on the existing lay-out of city streets, which were previously used predominantly by bicycles. The effect is an extraordinary clash of transport cultures which don't easily mix — with the inevitable daily toll of accidents to both cyclists and pedestrians. As the numbers of cars increase, Chinese cities are now experiencing serious traffic congestion. The response in some places, including Beijing, has been to close certain roads to cyclists! Given the high density of most Chinese cities, the most effective way of improving travel will be through more efficient public transit systems, alongside cycling and walking. By positively encouraging car use we see the pattern moving progressively away from what was a reasonably sustainable solution and, in the process, creating a whole series of new problems.

Recent events in Nanjing illustrate the point rather well. The characteristic tree-lined streets are a special feature of this ancient city, once the capital of China. Plane trees, pollarded so that their branches form broad canopies, provide continuous shade along cycle routes on one side and over busy pavements on the other.

These shaded pavements are the heart of social and commercial activity. In a city, which is so hot in summer that it is known as the furnace of the Yangtze, the street trees are vital to the well-being of the city. They are so much a part of life that they tend to be taken for granted until, that is, the city authorities decided to chop them down so that the roads could be widened to accommodate more traffic. Removal of the trees provoked public reaction on a scale rarely seen in China, with the result that the widening was restricted to specific streets, and the authorities agreed to plant replacement trees. For me, the trees of Nanjing now have a special significance as a symbol of sustainability.

The benefits of high density are plain to see in most traditional Chinese cities, but they are in danger of being lost in the headlong rush for economic growth. As Richard Rogers (now Lord Rogers) pointed out in his 1995 Reith Lectures, Shanghai had the opportunity to create a modern example of a sustainable city in the development of its new commercial and financial district in the heart of Pudong.[16] The Shanghai authorities decided that the new district would be a magnet for major international companies to locate their offices and would be primarily accessed by car. A complex road network was designed to cope with maximum rush hour demand, with the result that road coverage of the site was three times greater than in New York, but with less than half New York's building density. Most of the land would be allocated to car parking.

In contrast, Rogers proposed a diverse commercial and residential quarter enhanced by a network of parks and open spaces and accessed primarily by public transport. 'Above all we aimed to establish sustainable local communities, convivial neighbourhoods that would also consume only half the energy of their conventionally planned counterparts and would limit their impact on the environment.'[17] Six large, compact neighbourhoods were envisaged, each supporting 80,000 people. Each neighbourhood would have its own distinct character and all would lie within ten minutes' walk of the central park. Rogers aimed to reduce energy use by half compared with conventional schemes; but found that it would, in fact, have been possible to make a 70 percent saving, owing to the compact nature of his design. Sadly, his vision never came to fruition; but it remains a challenge for others engaged in major new developments of this kind.

But there are success stories, which demonstrate more sustainable solutions. Again, we go to the Brazilian city of Curitiba, where a remarkably successful public transport system has led to enormous social and environmental benefits. Growth and development of the city was planned around a system of main transport routes served by cheap and efficient express buses. This system, which has been developed over the past 25 years, is now remarkably sophisticated and is used by well over a million passengers every day. Nearly one third of these previously travelled by private car, resulting in savings of up to 25 percent in the city's overall fuel consumption. Air pollution has also been significantly reduced, and Curitiba now has one of the lowest air pollution levels in Brazil. As a result of these changes, it was possible to return the centre of the city to predominantly pedestrian use, with tree lined plazas in place of streets full of traffic. The city has received international acclaim for its achievements and has become a model for cities in all parts of the world. As so often is the case, its success depended on strong leadership, but the government also promoted considerable public participation, which was also a key element in its success.

Curitiba is not alone, of course. There are numerous examples of cities, particularly in developed countries, which have substantially reduced car dependence through improved public transport and direct control of car usage. These include some well publicised examples such as Rome, Zurich, Singapore, Freiburg and Toronto. The fact that this has been achieved in some of the wealthiest cities in the world shows that people are willing to change. Dependence on cars can be reduced in major cities even in societies where a high proportion of people are used to owning cars.

However, as Newman points out, on a world scale there is still a rapid growth in car dependent cities and overall use of cars continues to increase.[18] This has clear implications in terms of future levels of greenhouse gas emissions. So if we are to move towards more sustainable cities we need the models such as Curitiba and Zurich. Curitiba is particularly important because it demonstrates how rapidly developing cities in the south (where the most rapid urbanisation is now occurring) can find financially viable long-term solutions to reduce their dependence on cars, while also enhancing their prosperity and quality of life. It is the people of Curitiba who are gaining as a result of their visionary city, and they still have the highest level of car ownership in Brazil. They just use them more wisely.

Zurich reduced its car dependence by providing a wonderfully efficient tram system, which allowed the city centre to be transformed into a new public realm with pedestrianised shopping malls and outdoor restaurants. The story is repeated in many European cities, and even in Portland, Oregon — the only place I know where a freeway has been converted into a park. With the introduction of a light rail system, Portland has revitalised its city centre, and is one of the few American cities which, by imposing a green belt boundary, has pushed new residential developments back into the centre of the city.

As a general rule, the more compact the city, the more it can contribute to sustainability. It isn't just a matter of being more efficient in the use of resources in all the daily metabolic processes — crucial as these are; the benefits go much wider. I sometimes ask people to name cities which are nearest to a model of sustainability. They talk about places which have retained a human scale, where everything is within easy reach and where there is a vibrant sense of community. These are places where people can feel comfortable, where life can be enjoyed. Some of the most vibrant and attractive neighbourhoods are found in the old city centres which grew up long before motor transport came on the scene. Mixed neighbourhoods such as these, where people can live within walking distance or a short bus ride of their work, or can walk their children to the local park, provide the human element of sustainability which is sadly lacking in many modern cities.

The idea is gaining momentum that we plan cities in order to re-create such neighbourhoods. There are already examples in many towns and cities throughout Europe, where people have learnt from past mistakes, and are reinventing some of their past successes. The concept of the compact city is now being actively promoted by both architects and environmentalists. Richard Rogers is one of its strongest advocates. It was a central theme in his 1995 Reith Lectures and, not surprisingly, underpins the whole philosophy of his recent Urban Task Force report to the UK government, entitled 'Towards an Urban Renaissance'.[19] Attitudes are changing; and this became evident at the launch of his report, when a spokesman for Friends of the Earth admitted that it was far more radical than their own proposals for British towns and cities published almost ten years before.[20]

On a recent visit to Paris, I was encouraged to see that small shopkeepers provide rather attractive carrier bags for their customers with a message reminding them to use their local neighbourhood shops — 'they are the life of your community'. Connections at this level are crucial; they determine whether a city is sustainable; whether it is liveable. It is here that we have to start in our search for sustainable solutions when we address the other dimension of sustainability — namely 'meeting human needs'. So far, I have concentrated almost entirely on just one of the two key principles of sustainable development — 'the need to order our affairs in ways which will not reduce the Earth's natural capital, either now or in the future'. My argument is that cities should take responsibility for their actions in the context of finite global resources. But cities also need to take responsibility for 'meeting human needs' — indeed it has been the role of cities to do just that ever since the first towns and cities were created.

Nature in the city

One of the human needs which often gets forgotten is the need for contact with nature. I have argued elsewhere that one of the consequences of urban growth on the scale occurring today is that city dwellers may become completely divorced from the natural world.[21] There is abundant historical evidence from the industrial cities of the 19th century to show that nature was almost eradicated from many towns and cities at that time. The popularity of linnets and canaries as cage birds in Victorian back-to-back slums was perhaps symptomatic of a deeply felt loss. Even today we are building cities in ways which exclude nature from our daily lives. What contact is there with the seasons in a modern air-conditioned shopping mall with its built-in trees? It seems that, for most city dwellers, nature is something to be encountered in exotic locations whilst on holiday, or perhaps only through wildlife programmes on television.

But times are changing; and, over the past twenty years, hundreds of new initiatives have been developed to encourage wildlife in towns and cities. Until now, this has largely been a phenomenon of developed countries. Concern to re-establish links with nature seems to have been a feature of post-industrial society throughout Europe and North America. Maybe this concern is a natural response to increased urbanisation; it may be a natural part of the demographic transition. What we are seeing is a host of local projects providing opportunities for people to enjoy nature in the urban setting. It is a process which seemed to develop spontaneously in many different places and is well documented.[22]

Planning and designing for nature in the urban environment is now widely accepted in many cities in Europe and North America. Berlin, London, Toronto and Washington all have their urban wildlife programmes. Most major cities in the UK now have a strategy for nature conservation whereby significant remnants of natural habitats are protected from development and new habitats are being actively created in urban areas deficient in nature. In London, over one thousand wildlife sites have been identified for protection in local plans, and numerous new wildlife areas have been created. They vary in size from tiny school-yard nature gardens to the extensive new 40 ha wetland being created alongside the Thames, only four miles from Westminster.[23]

One of the most effective programmes in the USA is the urban wildlife refuge system of Portland, Oregon. To see a colony of white egrets only a stone's throw away from the city centre is one of the great pleasures of such a scheme. These are the living resources of the city which will help future generations to maintain their crucial links with nature.

With notable exceptions, such as Cape Town and Durban, both of which have well established wildlife programmes, towns and cities of developing countries have not generally seen nature conservation as a priority issue. But ideas are now spreading more widely. The first urban nature trail in China was opened earlier this year at Nanjing Botanical Garden, and the Metropolitan Government of Shanghai is proposing to create an ecological park on an area of landfill — again the first of its kind in China. Santiago, in Chile, is leading South American cities in having a biodiversity plan for the city,[24] and I am working with the government to create two environmental awareness centres in the capital. Curitiba already has its famous centre for environmental education, and I am sure many cities in other developing countries will follow suit.

Achieving solutions

Cities in the developing world, such, as São Paulo, Calcutta, Mexico City and, indeed, Santiago, face totally different problems from the affluent cities of the north. When we talk of sustainability in Mexico City, people do not think immediately of global warming or biodiversity. Rather, we are dealing with crisis management in a city which has acute problems of water supply, sanitation, and adequate provision of housing and food. People there are concerned with the basics of human existence — problems exacerbated by the enormous size of the city and the long-term effects of environmental contamination. The priority is to ensure that the city can function effectively and that people can live their lives without daily exposure to acute environmental stress.

Cities in the north and south have one thing in common — that sustainable solutions will be most effective at the local level. Radical solutions need vision and political courage to carry them through. They need strong leadership; but they also need the support of the local community. Some of the most imaginative and successful schemes have depended on visionaries such as Pasqual Maragall, the former Mayor of Barcelona, or Jamie Lerner, the Mayor of Curitiba.

Links are also being forged between towns and cities to achieve the objectives of sustainability. In Europe, the Aalborg Conference in 1994 resulted in the 'European Cities and Towns Campaign' based on the Aalborg Charter. This was an attempt to identify sustainability issues common to all towns and cities in Europe as a means of fostering new solutions.[25]

But there is also a crucial role for citizens themselves. Agenda 21 has mobilised citizen action throughout the world and we are seeing new forms of participative democracy emerging as a direct result.[26] Local sustainability forums and 'round-tables' are springing up in many towns and cities — sometimes posing challenges to political leaders, but, more often, providing the support for what might otherwise be regarded as unpopular policies. Sustainable solutions depend on having strong political leadership combined with effective action at the community level. What we are seeing is a new form of governance involving all sectors of the community. How this will develop in the future is difficult to predict, but it could be one of the most effective ways of ensuring that cities meet their responsibilities both globally and locally.

Future scenarios

So what are the future possibilities for cities? One thing of which we can be fairly certain is that, during the first half of next century, the total world population will continue to rise and that the proportion of people living in towns and cities will also rise, but at a faster rate. Whether we take the highest or lowest of the population projections of the United Nations, we can expect to see a significant increase in the world's urban population by the year *2050*. The increase could be anywhere between 500 million and 4 billion, and virtually all of this will be in towns and cities in the developing world. The question is whether continued growth of towns and cities can be achieved in ways that are sustainable in the long-term.

Taking the worst-case scenario: if cities of the developed world continue to use resources at current levels, and the growing cities of the developing world aspire to this as the norm — as many do at present — then we are in for a tough time. Competing demands on the world's finite resources are likely to become a continuous source of conflict. Increasing numbers of urban poor in the developing world will place additional strains on already severely under-resourced cities. It is likely, too, that ecological insolvency will lead to further deterioration in environmental conditions in and around all but the most affluent cities of the south. The effects, in global terms, will be continued depletion of the earth's natural capital, with significant loss of biodiversity, vastly increased levels of greenhouse gases contributing to global warming, and continual incremental losses of productive land on an unprecedented scale. It must be said that present economic trends continue to lead in this direction, particularly as a result of world trade agreements and the unwillingness of rich nations to reduce their share of the cake. Whilst the economic value of the earth's natural capital is becoming more widely appreciated, it seems unlikely that attitudes will shift sufficiently quickly to avoid the stark realities of current economic demands and expectations.

At the other end of the spectrum is the admittedly idealistic picture of cities that have maximised their metabolic efficiency, where all waste products are turned to good use, and where use of fossil fuel is reduced to the absolute minimum. In this ideal world, every city would take responsibility to ensure that its ecological footprint was reduced to a level consistent with the earth's natural carrying capacity — thereby ensuring global equity in the use of resources. Such a scenario would have immense implications for northern cities; it could be achieved only through radical approaches to fiscal and taxation policy which would reflect full social and environmental costs. It would require major changes in lifestyle for people in developed countries and would necessitate different expectations on the part of developing countries. Given current economic pressures, it is highly unlikely that a social and economic change of this magnitude could be achieved within the next fifty years, but it is worth considering how far we might move in this direction.

What is far more likely is that action for sustainable development will occur in a piecemeal fashion, as it has in the past, resulting in a scenario somewhere between these two extremes. Many northern cities will gradually become more resource efficient, as the social, economic and environmental advantages become apparent. We can expect to see major cities promoting innovative new ideas, and legislating to preclude non-sustainable activities. Policies for renewable energy and waste reduction will become an essential part of life for city planners. City-wide plans for biodiversity will have equal weight with strategies for transport and economic development. Mayors will be found cycling along their new urban greenways, and children will learn again the pleasures of walking to school.

For cities in the South, priorities will be more immediate for many years to come. The need to overcome poverty and environmental stress will be the driving force for change. It will need imagination, creativity and strong political leadership to mobilise local communities in ways that can solve the day to day problems of the world's poorest cities.

But, as Anne Spirn demonstrates in her wonderful scenarios for future cities,[27] most of the ideas are already there somewhere, in some shape or form. All it needs now is the political courage to ensure that the growing cities of the South develop in ways which are less demanding on global resources and yet can still provide a high quality of life. Who knows? Perhaps we shall see them giving a lead to the North in their search for solutions which are more truly sustainable. The metaphor of Siena remains the key to their success.

Chapter 10

Sustainable livelihoods
The one-point agenda for planetary survival

Ashok Khosla

The Author

Ashok Khosla is the President of Development Alternatives in New Delhi. His primary professional concerns include appropriate technology, environmental management and institutions for effective governance. He has held posts in the Government of India and in UNEP. He has been on the boards of IUCN, WWF, the Earth Council, SEI, IISD, WETV and EXPO 2000, and is a vice president of the Club of Rome.

Chapter 10

Sustainable livelihoods
The one-point agenda for
planetary survival

Ashok Khosla

For more than one half of the people on this planet, the turn of the millennium —
or for that matter the turn of the century — will be no different from any other day
in their lives. Full-time preoccupation with matters of survival and subsistence
tends to distract one from appreciating the significance of such milestones in history, even when one is actually present at the event. For such people, it will be just
another day or, at best, another year signalling yet another cycle of poverty, deprivation, marginalisation and exclusion.

These three billion people, each a person in his or her own right, are both an
integral part of the global economy and yet completely outside it. In the terminology of ecological science, they lead the front and bring up the rear of the economic
'food chain': the low-wage producers and the unpaid scavengers. In these roles,
they are very important to the world's economic flows, as shown by the large numbers of foreign, 'guest' or landed immigrant workers in any industrial country and
the huge volume of low cost commodities and products its businesses import from
the South. The good things in between — the products and services generated by
the economy and the purchasing power to acquire these — go, of course, to the
well-fed consumers in the middle. When it comes to the modern amenities of life,
most of the three billion are definitely not in the game.

Unquestionably, the world today is a better place to live in than it was, say, a
hundred years ago. Progress in medicine has eliminated many killer diseases,
devised treatments for a wide range of disabilities and made it possible for people
to live longer than ever before. The revolutions in materials and energy technologies
have opened dramatic arrays of options for satisfying basic human needs and for
extending our capacities to shape our destinies in many new directions.
Transportation and communication have opened opportunities for work and leisure
that could not have been dreamt of earlier. We now have more, control more and
know more than ever before.

But few societies today have escaped the widespread scourges of growing pollution, waste accumulation, social alienation, drugs, climate change, and a wide
range of generally unsustainable production and consumption patterns. Rampant
unemployment and accelerating inflation; growing supplies and depleting resources;
stagnant and unmet needs — these are the paradoxes and hallmarks of many
economies today, no less in the North than in the South.

In a natural ecosystem, all niches are crucial to its survival and functioning — and each receives its due share of importance in the grand design that gives it viability and resilience. In the world of human beings, on the other hand, this is not necessarily so: politics, economics and, more recently, technology have all been appropriated by the privileged few and worked to create vast disparities in social standing and access to societal resources. Decision systems that determine the direction of the whole economy are largely controlled by the rich for the benefit of the rich and the day-to-day reality of the poor rarely enters into their calculations.

The more fortunate upper half of the three billion poor subsist on an income of under US$2 a day, working endless hours in factories and fields to produce cheap products for an insatiable market. They themselves can never hope to get the benefits — the goods or services — widely flaunted by this market. The remaining 1.5 billion survive, under conditions as inhuman as any known to history, by collecting and processing the detritus of the modern economy. The star earners among them get at most one dollar a day and have virtually no meaningful contact with the monetised economy.

At the top of the food chain, some one billion people live in the unprecedented splendour and luxury made possible by the dynamite combination of modern technology and concentration of wealth. Dynamite: first, because of the sudden and explosive growth of this transnational class over the past one hundred years; and second, because of its in-built, inevitable destiny in due time to self-destruct as a result of the narrow, short-sighted pursuit of its self-interest. Spread throughout the world, they form a rich and powerful network that now controls the economy of virtually every country.

Equity and ecology

What, one may well ask, have the extremes of wealth and poverty that manifestly exist in the world today to do with planetary survival and sustainable development? The answer: a very great deal. Both extreme affluence and extreme poverty —wherever they exist, whether in the North or the South — are highly effective destroyers not only of societies but also of nature.

The affluent reside in largest numbers in the industrialised countries but certainly not only there. They are also to be found in significant numbers in every nation. In addition to having a great deal of money to spend, their major distinguishing characteristic is the virtually total control they establish over the structures of governance and business in their countries. It is through these structures, by influencing legislation and organisational strategies, that they collectively capture the major part of the economy's rewards. Both to symbolise and to demonstrate their status in society, they acquire an insatiable desire for material goods and physical services and the means to generate an unlimited demand for these. These demands inexorably concatenate through the economy into the natural resource base, producing tremendous pressure on the earth's biosphere.

One major category of impacts is, of course, the depletion of 'non-renewables' including fossil fuels and minerals.[1] A second group of impacts is the destruction, often irreversible, of the regenerative capacity of 'renewables' such as water, timber and gene pools. The third set of impacts is the disruption of biogeochemical cycles and global life support systems such as the stratospheric ozone shield and the greenhouse effect. These impacts are not small: the anthropogenic movement of material world-wide is, today, approaching in magnitude the natural, geological flows of material – a trend that is clearly untenable for long without major break-downs of critical life support systems. They also foreclose many options for a decent and healthy life not only for future generations but even for a large part of humanity today.

The consumption patterns and production systems adopted by today's rich also produce large quantities of waste and pollution. Much of this is in the form of chemicals and toxic substances that are expensive to dispose of safely — and in many cases accumulate in dumps that will continue to endanger health for many generations to come. Even with the exemplary efforts of some industries over the past decade to adopt cleaner production methods, the emission of pollutants world-wide continues to grow rapidly.

The affluent in any country create other, more subtle and indirect, impacts too. By a combination of money muscle, political power and social supremacy, they reg-ularly manage to appropriate the best and most productive lands — squeezing the original inhabitants further and further into marginal, and ecologically fragile, areas such as forests, deserts and uplands. The 'reasons' for displacing them come in many forms: factories, dams, power stations, roads, bridges. Or just mechanised agriculture and plantations — always in the name of economic efficiency and social progress. Except that the poor, who keep having to pay the bulk of the costs in terms of personal loss, never seem to get any share of the wealth and progress cre-ated. They only get blamed for the ecological disasters into which they constantly find themselves squeezed.

In fairness, it would be difficult to deny that the poor also create significant impacts on the resource base. The exigencies of survival and the lack of options available sometimes force them to undermine their own futures by using their resources in a non-sustainable way. In villages, fuelwood collection certainly con-tributes to the loss of trees and the creation of wastelands; excessive pumping of water has in many areas drawn down water tables to levels that are now uneco-nomic to use. In urban areas, slum dwellers contribute their share of pollution haz-ards and social disruption.

Nevertheless, the poor can never hope to compete with the affluent in the mag-nitude of environmental destruction they cause. Moreover, the distribution of impacts in the two cases is quite different: the rich can escape from the problems they cause, the poor cannot. In the one case the environmental costs are largely externalised, in the other they are borne by the people who cause them, however involuntarily.

Consumption patterns and population growth

The current debate on environment and development, particularly at the interna-tional negotiating table, has thus become quite polarised. Some — primarily the affluent and most of them from the North — feel that population growth poses the greatest threat to planetary survival. Others — mainly the poor — claim that it is the consumption patterns of the rich that are the primary cause of environmental destruction. As in many such debates, there is some truth in both positions. Rapid population growth certainly places intolerable pressures on the resource base as does mindless material consumption, though there are, of course, differences in the precise nature of the impacts each produces. But what distinguishes this from many other debates is that the underlying causal factors, and many of the solutions for both sets of problems, lie largely in the hands of one side: the rich. They also have the means to solve these problems — though not necessarily the inclination to do so.

In a world, like ours, that has the resources to meet every individual's basic needs, it is intolerable that such needs are not met. But we do not need moral, ethical or ideological arguments to prove that extreme poverty or disparity is unacceptable. Eradication of poverty, at least of involuntary poverty, is as much an ecological imperative — a matter of self-interest, and possibly of survival, for everyone, rich or poor. Unless the pressure on our natural resources is urgently reduced the very basis of our economies will be irreversibly lost, a disaster from which no one will be able to escape.

The urgency to eradicate poverty is further heightened by the longer-term implications of every new baby that is born. Unless we quickly bring down human birth rates, which are highly correlated with the degree of poverty, population growth will certainly continue well into the second half of the twenty-first century before it levels off. Each additional person added will, of course, be entitled to the same standard of living as everyone else and this can only add further to the already high pressures on the life support systems of the planet. The best investment that the affluent can make today, purely as a matter of self-interest, is in measures that accelerate genuine, sustainable development in the third world and thus help speed up the demographic transition from high birth rates to replacement fertility.

The present increase in the quantity of material now being physically moved by industrial economies clearly cannot continue for long without massive disruption of the natural processes on which life depends. Yet, the largest part of the current material and energy production is appropriated by far less than half the world's population. It would appear we have two basic choices. *Either* we accept a future that is even more unjust and inequitable than the world of today *or* we drastically cut the *per capita* use of materials and energy by adopting alternative lifestyles and reduce the material intensity of our production systems.

The Factor 10 Club has shown that material intensity in the industrialised North must be reduced by a factor of at least ten if we are to avoid massive ecological catastrophes. It has further concluded that it is technically possible to reduce material and energy use in the industrialised countries by an order of magnitude without loss in quality of life or access to 'the services' people have today. At the same time, to prevent the 'rebound' effect of growing populations, the countries of the South will need to obtain access to sufficient material and energy resources to help accelerate their passage through the demographic transition. And they must do so NOW. This apparent paradox is resolved through the principle of 'convergence': everyone, and particularly over-consumers, must now reduce the material intensity of their lifestyles and under-consumers may continue to raise their access to resources until the two converge at an acceptable level that lies within the limits set by nature. This is probably the only short cut we have to a sustainable future for the globe as a whole.

Production systems and human development

Our concepts today of human development and what constitutes well-being are not the same as they were at the turn of the last millennium – or even at the beginning of the century. In a world dramatically shrunk by technological innovations in transportation, communication and information processing, the opportunities available to people to find fulfilment have exploded in terms of range and variety. And so have their expectations, embedded in such concepts as 'progress' and 'economic development'. Short of some unforeseen catastrophe, few societies will ever accept on their own to go back to the way people lived in earlier times.

In any case, one thing seems clear to most societies: economic progress is determined by the ability of a society to make investments — in infrastructure, in scientific research and technology development, in utilisation of resources and in the skills of the people. The ability of society to make investments depends on the surpluses and savings it generates. In a modern democracy, where feudal types of human exploitation are not permissible, such surpluses can only be generated by raising the productivity of workers. In a viable ecosystem, the productivity of resources must also be raised. Raising the productivity of labour and resources needs technology and enterprise.

Many contemporary economies have demonstrated how much 'progress' is possible by adopting an aggressive use of technology and enterprise. But their experience has also pointed to the need for careful selection of the types of technology and forms of enterprise, to avoid wholesale destruction of human, social and environmental values that are also important. It has become clear that *how* something is produced, *where* it is produced and *for whom* it is produced are issues as important as *what* is produced.

For the poorer half of the world's population, living mostly in the 'developing countries', exposure to technology and enterprise is more recent than for those who travelled this road earlier. While they start with the obvious handicap of limited means, however, they also have the relative advantage of being able to learn from the earlier experiences of others and to leapfrog over expensive intermediate stages.

Human development appears to be closely related to the degree of control people have over their lives and over the decisions that affect them. In today's monetised economy, such control is, in considerable measure, linked to the degree of financial autonomy a person has. And in a world that is rapidly adopting the work ethic throughout, financial autonomy for the average person comes largely from the income one gets from one's job or livelihood. Work also offers other rewards such as status in society, self-esteem and a focus for one's life.

The modern economy would appear to be creating a world where cheap machines produce ever-cheaper products for other cheap machines to use. As a consequence, human beings have less and less to do. It is common to see more and more automation in the face of more and more unemployed people. With each day that passes, we have more and more products chasing less and less purchasing power. Today's labour-displacing technologies and mechanistic economic structures can only lead to growing supply and stagnant demand — until, perhaps, we reach the catastrophic environmental transition when supplies collapse altogether and both human populations and their demands collapse with them.

Ever increasing consumption and ever more 'efficient' production systems also spell ever greater demands on the resources of nature. Under particular threat from this run-away consumption are those that are non-renewable or capable of being lost irreversibly — like minerals, life-support systems or biodiversity. Rapid growth of population, largely associated with poverty, has its own impacts on natural resources, particularly those that are classified as renewable — such as forests, rivers and soils. Analysis of production systems shows, again, that extremes of affluence and poverty inevitably lead not only to economic and social breakdown but also to ecological and biospheric catastrophe.

The false promises of globalisation

In a developing country, as in any other, a job is the most basic need of all, a means to generate income with which to meet the other basic needs. The third world needs to create some 50 million jobs each year if it is to accommodate the needs of all the new entrants into the job market, plus the backlog of unemployed people, within a reasonable time frame of, say, fifteen years. India alone needs to create 12 to 15 million jobs per year. Partly because of improvements in farm labour productivity, and partly because of the natural limits to agricultural expansion, no more than 25 percent of these can be in the agricultural sector. Off-farm industries, tertiary sector services and other activities must produce the remaining ten million jobs.

The capital investment needed to create one job in a modern industry is significant. In regions like North America, Europe and Japan the average cost of a making a new workplace is about one million dollars.[2] To compete in the global economy, even a country like India needs more than US$100,000 to create an industrial workplace. To start bringing the unemployment rate down, India would therefore need to invest, each year, some US$1,500 billion — 5 times its GNP — just in creating new workplaces. This simply cannot be done; it is not a coincidence that during the past decade during which the Indian economy has been progressively liberalised, the rate of industrial job creation has also come down substantially. The country is currently creating not even 3 million industrial jobs a year (out of the 10 million needed), steadily losing the race against unemployment. The answer to job creation for sustainable development clearly lies elsewhere.

'Global competitiveness', 'comparative advantage', 'economies of scale', 'environmental externalities' and other such shibboleths — the ultimate being the 'free' market — based on simplistic (and entirely unrealistic) assumptions are concepts of neo-classical economics that do not easily translate into the language of sustainability. In fact, they do not translate at all, since economists have been unable to recognise the issue of sustainability in the first place — presumably because it would complicate the mathematics of their elegant models.

The theories of global trade and comparative advantage have no meaning unless the full environmental and resource costs of transportation are included in factor and product prices. Till today, such costs have been ignored, as have the social and human benefits of widespread employment. To complicate these calculations, barriers to trade in various guises today (under such pretexts as human rights, child labour, low wages, lack of environmental standards) distort international transactions to an extent that was not envisaged even in Mr. Ricardo's rich, original framework.

The economies of scale depend directly on the technological, organisational and infrastructural choices available to a production unit. It is easy to show that with a small change in any of these choices, the economies of scale can be quickly turned into diseconomies of scale. Recessions in some of the industrialised countries have forced many large corporations to learn this lesson the hard way, a process Japanese and American companies these days often refer to euphemistically as 'downsizing'.

Unfortunately, the assumptions underlying current economic theory — and the machinery of the modern marketplace that they naturally lead to — are not sufficiently solid to support the common platforms of human values on which societies must stand to benefit collectively and equitably. Growth, they have claimed, must come first, even at the expense of distributive injustice and human misery. Efficiency above equity. Machines over people. The rich before the poor.

The limits to self-employment

Over the years, we have gradually learned that for any development process to be sustainable, it must be equitable, efficient, ecological — and empowering.

The term empowerment, despite its impending nose-dive into the sea of meaningless jargon as it rapidly becomes the buzzword of the development and gender sets, is a valuable, highly integrative concept that brings together many of the desirable goals of social and economic development. It signifies activities that help the people who are most marginalised in society — particularly women, but also the poor and the handicapped – to gain a reasonable degree of control over the decisions that affect their lives. A person becomes more empowered when he or she is able to participate effectively in family and community processes. To be empowered, one must have access to basic knowledge, health and social status. Successful routes to empowerment therefore include access to education, adequate nutrition and health care, and information about one's rights.

Nevertheless, it is difficult to imagine that, in the increasingly materialised, commoditised, and monetised world of today, anyone can feel truly empowered without access to the income or status that, for the poor, comes only with a job. But, in the real world, jobs are usually created by capitalists more concerned with their own financial ambitions than with the welfare of their workers — creating conditions of exploitation that quickly undermine any 'empowerment' that the job might have led to in the first place.

It is basically this reasoning that has led to the widely held belief among development workers that 'self-employment' is the most effective route to genuine empowerment. Several initiatives in recent years have dramatically shown how much can be done by simply providing access to small amounts of (commercial but fairly priced) capital in getting women and other disempowered people back on their financial feet, thus enabling them to stand on their own legs in the community. Indeed, the success of such programmes as the Working Women's Forum (WWF) and the Self Employed Women's Association (SEWA) in India, and the Grameen Bank and the Bangladesh Rural Action Committee (BRAC) in Bangladesh, have led some actually to equate 'empowerment' with 'self-employment'.

Although some of these ventures have achieved outstanding results, their impact in the long run can only be limited unless the financial inputs are supplemented by a variety of other support systems. Genuine, durable development means that the structures of production and distribution must be transformed – though admittedly not in the mould of the North. This means that households and local economies must create surpluses to be able to accumulate savings and thus make investments that will in turn enable them to continue to improve their material well-being. It is not possible, simply with one's hands and without any amplifying tools, to do more than survive and subsist in the modern world.

Thus, in addition to the boat of financial capital, people need the engines of technology, management expertise and marketing skills to be able to sail towards a better future. And this means enterprise.

But entrepreneurship is not for everyone. An entrepreneur must have very special traits: to be able to mobilise resources; to manage people, money and machines; and, above all, to take risks. Not everyone in society has these attributes. In fact, very few do. Most prefer the security — and sanity — of a steady job. Is it fair, then, to expect village (or even city) women — illiterate, vulnerable and without any safety nets to fall back on — to take risks that others are not prepared to take?

To avoid this trap, we need to evolve new kinds of enterprises that will mobilise resources and create steady jobs for local people.

Sustainable livelihoods

The central issue facing society, North and South, East or West, is the need to create sustainable livelihoods. Large numbers of sustainable livelihoods.

Any simple solution proposed for a complex set of social and economic problems must inherently be suspect. Yet, if there is a One Point agenda for sustainable development, it surely has to be the large-scale introduction of sustainable livelihoods.

Sustainable livelihoods create goods and services that are widely needed in any community. They give dignity and self-esteem to the worker. They create purchasing power, and with it greater economic and social equity, especially for women and the underprivileged. And they do not destroy the environment.

In short, a sustainable livelihood is a remunerative, satisfying and meaningful job that enables each member of the community to help nurture and regenerate the resource base.

Sustainable livelihoods, and the human security they engender, underlie the one set of issues that is common to all nations and societies, at all stages of development. They provide a powerful synthesising, unifying concept that can bring the most disparate interests together to design more viable economic systems for the future in any country, rich or poor.

Economists often oppose any alternative approach with the objection that the primary need of poor countries is 'economic expansion', meaning that, unless the growth rate goes up, there will be no wealth to spread around. One cannot take exception to this view — only that the quickest pathway to such expansion lies in placing a heavy emphasis on sustainable livelihoods. While development economists debate whether GNP should grow at 7 percent or 8 percent per year, eradication of poverty within a reasonable time will in fact need growth rates in the double digit region. China has demonstrated that such a growth rate is not only possible, but that it can be sustained over long periods. And the only way to do this when a country has limited capital resources is to put everyone to work.

Clearly, a better mix of large, small, mini and micro industries is now needed. Given the continued failure of policies to address the needs of the small, mini and micro sectors, a proper balance will require greater encouragement and incentives to such industries. There are, of course, sectors for which the economies of scale favour large, mechanised production units. These probably include steel making, oil refining, petrochemicals and automobile manufacture. But there are many sectors where economies of scale are not relevant. Most industries producing basic goods for rural populations are commercially viable even at quite small scales. And because of the low capital requirements and short gestation periods, they can have high returns on investment – in some cases even double those of their larger counterparts.

By definition, sustainable livelihoods bind people to their communities and to their land. Not only do they thus have a positive impact on health, fertility reduction, migration and other demographic behaviour, but they also permit a far more effective use of resources for the benefit of all.

But without improved productivity and better management and marketing systems, they can never lead to the quantum shifts in lifestyle that people everywhere now desire. For this, the large-scale success of sustainable livelihoods will depend on our ability to design:

- sustainable technologies;
- sustainable enterprises;
- sustainable economies; and
- sustainable institutions of governance.

Sustainable technology

With the evolution of societal perceptions, aspirations and conditions, and with recent developments in science, design, new materials and production processes, technological innovation is becoming increasingly important for solving the problems of poverty. New products and technologies, many with significant, positive social and environmental spin-offs, are now possible for mass distribution as a result of the application of sophisticated scientific and technological knowledge.

Technology that serves the long term goals of development is defined as 'sustainable technology'.

Sustainable technology springs from endogenous creativity, in response to local needs and possibilities:

- It is relevant and ready for use by the common people, and aims directly to improve the quality of their lives.

- It derives maximum leverage from the local cultural environment by drawing upon the existing managerial and technical skills and providing the basis for extending them.

- It uses the physical potential of an area, and maintains harmony between people and nature.

Sustainable technology is the offspring of the marriage between modern science and traditional knowledge: a method, a process, a design, a device or a product which will open up new possibilities and potentials for improving the quality of life. It requires frameworks for innovation and delivery very different from those that exist today, either in the global economy or in the village. Throughout the Third World, there is an evident and pervasive need among both rural and urban poor for a whole variety of technologies ranging from cooking stoves and lamps to gasifiers and windmills.

Moreover, tens if not hundreds of designs are available for each such technology, scattered in laboratories, workshops and archives throughout the world.

Why have these needs not led to a more widespread demand? and why has the existing technical capacity not led to supply?

The answers to these two questions are complex, and interlinked. A combination of economic, social, political and cultural — not to mention scientific, technical and institutional — factors have greatly inhibited the supply and demand for sustainable technology. They apply, in varying mixes, to all rural technologies. The more important among these factors are:

- capital/operational costs;
- efficiency of the technology;
- evidence of improvement over traditional methods;
- ease of operation and ergonomic design;
- availability of spare parts and ancillaries;
- ease of repair and maintenance;
- problems of production;
- adaptation to local conditions;
- existence of marketing organisations;
- availability of information;
- promotion, training and extension services;
- management skills and social organisation; and
- social, class, political and cultural attitudes.

Above all, the 'appropriateness' of a technology must be measured by how well it satisfies the needs of the end client and with what success it takes advantage of the opportunities and constraints of the production and marketing processes. Contrary to past development understanding, sustainable technologies need to compete in the marketplace. To design technologies that can reconcile the conflicting requirements of the market, nature and people, requires systems for innovation and delivery comparable in sophistication with those of the most successful multinationals.

Many technologies for such enterprises already exist. So does the demand for their products. What prevents the poor from setting up such enterprises is their lack of access to these technologies and their inability to put together the financial capital required. What prevents them, once set up, from becoming profitable is the absence of entrepreneurial and management skills, infrastructure and marketing channels. Much more public investment is needed to provide these, but probably not nearly as much as is being made today for the benefit of large, urban industries.

Governments, the private sector and universities have shown little interest in the innovation of appropriate technologies. Most of the work in this area has been done by independent-sector organisations. Development Alternatives is one that has had some success with a wide range of technologies. Examples of Development Alternatives' products include building materials, water pumping and water purification systems, recycling of waste materials, handmade paper, energy from renewable fuels and local infrastructure such as sanitation, communication and transportation.

Sustainable enterprises

Sustainable livelihoods using sustainable technologies will require sustainable enterprises. Sustainable enterprises produce goods and services that are needed to better the lives of the great majority of people, including those who have been left outside the mainstream economy. At the same time, being environment-friendly, they minimise waste, use renewables and residues and generally conserve resources.

To break out of the present poverty-pollution-population trap, we need to create new kinds of corporate institutions that integrate considerations not only of economic efficiency, but also environmental soundness and social equity into business decisions. Neither the current policies for national development, nor the activities of the corporate sector are geared to achieve this kind of goal.

The way forward that is needed lies far outside the imagination of the planners and decision-makers of most countries. Much of it lies in small scale, decentralised industries of a new kind. Such (mini-or micro-) industries use good technology to raise productivity and local resources to make products and services that satisfy the needs of local people without destroying the environment. To be viable, they will need to evolve substantially modified market mechanisms that can take account of full-cost pricing and social impacts.

Traditionally, corporate response to social issues has largely been driven by fear — the fear of jail, of markets lost or of financial liability. This must change, and the broader social good has to be internalised in decision making at a level no less than the bottom line of cash profit. To bring about this change, the sustainable enterprise will have to strike a radically new synthesis across sectors and institutions, either by redesigning itself or through partnerships with other entities that have complementary strengths.

Taking the complete cycle from biomass generation to end-product use, entire jobs can be created at costs of a few hundred dollars, the environment can be enriched at no cost at all, and the basic needs of whole communities can be met through the additional purchasing power created. Hand-made, recycled paper manufacture demonstrates the possibilities in this direction. In comparison with a large-scale paper mill, a small paper enterprise has many environmental, social and even economic advantages:

Cost of creating a workplace	=	1/10
Capital investment per kg of paper	=	1/3
Energy consumption per kg of paper	=	1/4
Water consumption per kg of paper	=	1/2

If the full economic and environmental costs of the processes and resources used in manufacturing and delivering products is taken into account, and no 'perverse' subsidies are allowed for energy, transportation, financial and other services, small scale production can become quite competitive.

All that the small enterprise needs to beat the large corporation at its own game is better access than it has today to technology, finance (not necessarily cheaper finance), and marketing channels. The primary role of the public sector in facilitating these is to provide basic infrastructure for financing, communication and transportation.

The myth of the 'economies of scale' that justifies the bulk of national invest-ment going into urban infrastructure and large industries is as hollow as it is deeply embedded in a bankrupt theory of development economics.

As evidence of this, 'small and medium enterprises' already form the backbone of the national economy. For example, in India, they account for more than 60 per-cent of the industrial production in India, and for more than 65 percent of industrial exports. More important, they employ more than 70 percent of the industrial work-force. When adjusted for the vast subsidies and infrastructure that large scale indus-try can take advantage of, their real contribution to the economy is even higher.

Sustainable enterprises are usually quite small. They have between one and 100 employees, with an average of around 20. They are generally informal and flexible, and quite labour intensive. They may properly be called 'mini-enterprises', since they bridge the gap between what are usually referred to as micro-enterprises on the one hand and small scale industries on the other. However, being small, dis-persed and largely unregulated, mini-enterprises can often have environmental and social impacts that are fairly negative. To overcome this, they need access to better technologies as well as other supports.

The design and operation of rural enterprises is a complex, still unfamiliar busi-ness. They have to master the technology-environment-finance-marketing linkages, while keeping their overhead costs low. They must do this without access to highly qualified engineers, management specialists, marketing experts — or to friendly bankers or market infrastructure, either for buying raw materials or for selling prod-ucts.

An interesting solution to these seemingly insuperable obstacles lies in building franchised networks of small, private enterprises capable of growing and processing biomass to manufacture products for both the urban and local markets. To be suc-cessful, the franchise arrangement will have to provide high technological and mar-keting inputs and access to capital. Technology and Action for Rural Advancement (TARA), the commercial wing of Development Alternatives actively franchises mini-enterprises based on appropriate, sustainable technologies.

Sustainable economies

The possibility of improving equity, efficiency, ecological harmony and self-reliance — and thus of achieving sustainable development — rests on how quickly and effectively innovations can be introduced into the economy. Given the size, spread and poverty of the rural population, which must now comprise the primary target of any effort aimed at sustainable development, it becomes immediately clear that any viable approach must be:

- highly replicable;
- locally accessible;
- self-financing.

These criteria imply that the strategies of development must now turn many of the earlier paradigms upside down: technologies must be economically viable, insti-tutions must be decentralised, and the environment's capacity to supply resources must be conserved. To achieve these attributes, we will need whole sets of new concepts: participation, networks, appropriate technologies, the diseconomies of scale, environmental and social appraisal of projects, rapid resource surveys, corporate research and development, and non-governmental action.

Yet development of sorts has taken place in some parts of the world, and management approaches that have succeeded can yield valuable ideas which, appropriately generalised and adapted, can also be made to work in the rural business environment. Among these, the most important lessons for effective organisation of rural technology efforts come from:

- organisation of innovation in high technology industries, such as the manufacturers of electronics components;
- effective decentralisation of production and marketing through franchising, such as the fast food chains; and
- the management of complex systems and projects, such as space programmes.

The urban markets of any developing country have so far provided highly attractive business opportunities to entrepreneurs and have prevented them from fully appreciating the possibilities offered by the rural areas. However, the huge and untapped potential for profits in lower income 'peripheral' or village markets will soon necessitate and produce a more systematic corporate approach to generating both supply and demand through innovative technologies and marketing systems.

Several mechanisms are now evolving to help enterprises overcome the barriers to obtaining technology, to using effective transport and communication facilities and to introducing modern management methods. However, as the table shows,

Size of Enterprise (Investment in plant and machinery)	Potential for Livelihoods (Investment per Job)	Transformation	Reach	Impact	Annual Loans In India (estimated)	ROI & Recovery
Micro enterprises ($20 to $250)	Highest ($20 to $200)	Survival to subsistence	Household, local neighb'd, village	Family self-sufficiency	>$1 billion (from formal sector)	High & very, good
Mini enterprises ($250 to $25K)	Very high ($200 to $2K)	Subsistence to security (for workers) and surplus (for owners)	Local neighb'd, community, village, town	Local self-reliance	<$ 10 million	Very high & excellent
Small enterprises ($25K to $250K)	High ($2K to $20K)	Surplus to savings and productive assets	Town, region, OEM*	Resilient industrial base	$2.5 billion	High & good
Medium enterprises ($250K to $2.5M)	Medium ($10K to $50K)	Assets to major capital investment	Region, OEM, exports	Quality, standardised products	$7 billion	Medium & good
Large enterprises (Above $2.5M)	Low ($25K to $250K)	Capital investments to private wealth	Buyer of OEM, national market, exports	Global competitive-ness	>$70 billion	Medium to low & good

Note: K = Thousand (1,000) ; M = Million (1,000,000)
*OEM = Original Equipment Manufacturer

credit continues to be the key missing link. Currently, finance is fairly easily available in some developing countries to 'small and medium enterprises' that have capital requirements of US$25,000 or more. Also increasingly available is finance to micro industries that need capital of less than US$250.

Unfortunately, this is not so for mini-enterprises that fall in the range between these two categories, with capital investment of US$250 to US$25,000. It is precisely this sector that optimises the twin objectives of sustainable livelihoods and returns on investment. They are small enough to be responsive to the local economy yet large enough to get high productivity through technologies and skilled workers. At the same time, they are big enough to take advantage of public infrastructure, credit facilities, technology support and marketing channels, provided these are available. There are numerous technology based mini industries in this range that could be set up today and run profitably.

Such enterprises can create, directly, several workplaces, each at a capital investment of US$200 to US$1,000. In addition, they indirectly lead to the creation of several more jobs, upstream or downstream, usually at an even lower capital cost. Such workplaces, in the village or small town, yield incomes for workers whose purchasing power is comparable to, if not better than, those created at a hundred times the cost in large urban industries. At the same time, they permit very high returns on investment, sometimes with payback periods of less than a year.

The potential clientele for mini credit, accompanied by proper technology, marketing and policy supports is very large, certainly in the millions. Empirical studies by the Government of India, the World Bank and others show that among these potential clients, a significant percentage has high levels of credit worthiness. Carefully designed lending programmes can therefore be both financially profitable and socially worthwhile.

The paradox of our economy is that there is virtually no source of funding today that can actually deliver adequate financial credit in this intermediate range (which might properly be termed 'mini credit') where it has greatest potential impact, both on the generation of employment and on the national economy.

These initiatives will have to come from the non-government sector and, more widely, the 'Independent Sector', hopefully with direct encouragement and support from Government.

Sustainable governance

For development to be sustainable, people must acquire a sense of ownership and responsibility for their resources — economic, social and natural. And they must be able to oversee and correct the actions of their elected representatives on a continuing basis. Such a sense of ownership can in the long run come only from actual ownership — enshrined in institutions of local governance involving the entire adult population. Such bodies should collect revenue from local resources, decide on local priorities and authorise higher level institutions to co-ordinate activities that involve other jurisdictions or skills and knowledge not available at the local level. And for any such citizen oversight to be effective, it needs certain basic prerequisites —transparency, accurate information and the right to be consulted in all matters that affect the citizen.

In the minds of many people, democracy is synonymous with elections, held periodically at the national or state level. By themselves, however, elections are not sufficient – or even the most important features of democratic systems. The success of Jeffersonian democracy lies largely in the existence of strong institutions of local governance, hawk-eyed media and a flourishing civil society, all of which America has, painstakingly, built up and nurtured over the past two hundred years.

The institutions of the marketplace, which are both the products and the prime supports of this democracy, also underpin much of its success. But in their present, capitalist form, they are also the cause of its greatest failure — extraordinarily profligate and wasteful consumption patterns on the one hand, and the consequent, unprecedented inequity and poverty on the other, which now threaten the very survival of the planet.

As the work of People First, the advocacy wing of Development Alternatives, continues to show, a true democracy, of the type that can support sustainable development and generate sustainable livelihoods, is actually an inversion of today's highly centralized, top-down systems of government. People First aims to bring about, in India, a bottom up form of democracy that assigns the primary decision-making responsibility to the local community — the entire adult population of a village or city neighbourhood. The local community will retain the portion of the tax money collected in its jurisdictions to implement these decisions. These bodies and the ones above them devolve successively upwards only those decisions and activities that cannot be handled at a given level of governance, together with the residual funds to implement these.

Changes in systems of governance are not easy to achieve in any society, much less in one that has a traditionalist's pride in its existing Constitution. The political and administrative leaders who have come to be the major (if not, together with the rich, the only) beneficiaries of the national economy will do all they can to prevent change. Nevertheless, if the massive spread of poverty and the equally massive degradation of the environment are to be slowed down, people's movements can help to bring about the changes needed.

Chapter 11

The limits to sustainable development

David Fleming

The Author

David Fleming is an economist. During his early career through engineering, textiles, detergents, advertising, the financial services and business school, he worked in the environment movement as economics spokesman of the Green Party and chairman of the Soil Association. His recent research on economic conditions in the 21st century, under the project title of The Lean Economy Initiative, has been supported by Elm Farm Research Centre.

Chapter 11

The limits to sustainable development

David Fleming

'Sustainable development' has been the guiding principle of environmental policy for the last quarter of the twentieth century.[1] During this time its effects have been positive. The neat, widely-accepted phrase has allowed environmental issues to be painlessly inserted into business and government policy. It has spawned a business culture of environmental awareness, and the practical consequences of this have been good. It is now less likely that the claims of the environment will be ignored in business and government planning decisions. You no longer have to be a fanatic or visionary to give consideration to environmental matters; the big-decision makers do it, or at least claim to do it, all the time.

However, I shall be arguing in this chapter that the time has come to move beyond the concept of sustainable development, that the concept is no longer appropriate to the circumstances that actually face the market economy, and that it is now doing more harm than good. I shall argue that, in the longer view, the main significance of the concept may turn out to be its function as an instrument of denial. The challenge facing environmental protection now is to lay aside the easy reassurance provided by sustainable development, and to face up to the urgent need for a re-think about the relationship between the market economy and its environment.

* * * * *

As Karl Popper reminds us, there are two ways of setting about the business of definition.[2] The first is from-left-to-right: 'This word/phrase has the following meaning . . .' The problem here is that there is no frame of reference within which to reach an agreement about whether that definition is the correct one or not. Different people will have different interpretations, and there is nothing to stop them falling out over it, perhaps even to the point of war, or fudging the issue until there is no meaning left there at all.

The other way is from-right-to-left: 'To this meaning, which I have explained at leisure and in specific detail, I shall attach the following label . . .' This gives the label a secondary position in the argument; if anyone disagrees with you about the meaning, and prefers a different meaning, they have to find another label to attach to it. The result is that different meanings are clearly identified with different labels; it is clear what you are talking about, and you avoid the heroics of having to defend an icon, with all its symbolism, emotional investment and subtexts.

Sustainable development belongs to the first of those two methods, and the consequences are immediately evident in the form of a litter of definitions. No less than twenty-six definitions are paraded before the bemused reader in one of the first books to make the claim to integrate economic and environmental policy;[3] since then, newly-formulated definitions of sustainable development have become almost a mandatory element in the opening-paragraphs of articles on the subject. This is inevitable in a discussion whose starting point is a word/phrase rather than a proposition, so the mandatory definition is needed here, too: 'Sustainable development is development that meets the needs of the present without compromising the ability of future generations to meet their own needs'.

Now, that definition comes from the World Commission on Environment and Development (the 'Brundtland Commission').[4] Actually, in the Bruntland Report, it is combined with an assertion that 'Humanity has the ability to make development sustainable', and the whole of the report is committed to arguing the truth of that proposition. Whether it succeeds in that argument is another matter, but it is at least an argument with which you can agree or disagree: there is a case to answer. What the Brundtland Report's implanting of sustainable development into public discussion has done, however, is to introduce a phrase which has escaped from its immediate context, which has evolved a life of its own, and which presents, in terms of logic alone, three fatal problems.

First of all, the relationship between the two words is not specified. 'Development' can be taken to refer to business-as-usual, with the market economy advancing as a provider of jobs and higher living standards; the development process builds institutions in economies where they are still poorly-developed, elaborates them in economies in which development is already well established, and draws its energy from the sustained advance of economic growth. 'Sustainable', can be taken to refer to a pattern of behaviour and economics such that the environmental endowment is maintained in a stable, fertile condition, abundantly yielding its resources for the indefinite future. While these interpretations of the two words are reasonable enough, it is far from clear how to proceed when they are placed together because, unless you can be sure of achieving both objectives completely, it is necessary to make some kind of a trade-off. More 'development' will very often mean less 'sustainability' – so the question is, where do you strike the balance? Since this is not specified, the natural solution is to strike it wherever you believe is most convenient – which may be at either end of the spectrum of possible combinations, but which is quite likely to be at the extreme which favours maximum economic development with nothing more than some politically-correct lip-service to the notion of sustainability. Any doubts you may have about whether you can defend your choice of development to the virtual exclusion of sustainability may be calmed by appealing to the well-attested fact that the wealth earned from development can later be reinvested in the technology, regulations and conservation programmes required by sustainability. You can also invade and colonise the sustainability agenda by claiming that job-creation is as legitimate a part of sustainability as is (say, with easy parody), protecting newts. In other words, it is trivially easy to back the 'development' horse entirely, while claiming that you are building up the necessary resources for sustainability. (As an added refinement, you can use government-approved sustainable packaging in the form of phrases about 'extending existing best practice', 'integrated thinking', and delivering goods and services 'in more innovative and resource-efficient ways').[5]

The second logical problem lies in the ambiguity about what kind of a statement 'sustainable development' actually is. Technically, it is not a statement at all, since it has no verb. But if we do now endow it with some syntax and upgrade it from a mere definition into a statement, what should it be? An assertion that sustainability and development are mutually consistent? This is the form in which it appears in Brundtland – and the problem with this interpretation, of course, is that it can be simply denied, or qualified out of existence with statements such as: 'I disagree that they are mutually consistent, unless we define "development" as . . . and "sustainability" as . . .' – which opens up the whole debate afresh, and re-defines sustainable development itself virtually out of existence. Is it an imperative (i.e. thou shalt develop sustainably . . .)? If it is, then it begs the question around which the argument is being conducted: *of course* a combination of sustainability and development would be a very nice idea if it were possible, but the whole point at issue is that it is not clear how they can be made mutually consistent, if indeed it is possible at all. Is it a form of reassurance to be uttered by people struggling with a tricky problem?[6] If so, it joins 'unsinkable ship', 'perpetual motion' and 'virgin birth' in that unreliable little list of incantations whose purpose is to fill in awkward gaps in the logic.

The third problem is that sustainable development is a pre-packaged answer to a complex set of issues, and it has within it, in a well-intentioned but still disturbing way, some aspects of authoritarianism. Authoritarian regimes, which want to secure support for themselves while obscuring the facts, devise a characteristic form of teaching. They train children to learn the answers, and not to collect the building-blocks of information, intelligence and independence which would enable them to work out answers for themselves. They are happy with the idea of reducing ethics to a code and of saying what has to be in that code. Sustainable development has become one of these secular doctrines, and in this form it asserts that economic growth and environmental protection for future generations are mutually consistent, and often indeed mutually dependent. In the hands of a religion, such certainties can be valuable: the 'noble lie' can be a certainty expressed in mythical terms with no practical content; it can help to build personal confidence and a sense of identity, forming a power-base from which the individual can explore and think independently. But when such pre-packaged certainties are captured by a political regime, they block independent thought, and in the process they become weak and rootless with nothing but authority to back them up. Ultimately they invite the straight, refreshing, rebellious response, 'No!' Faced with that response, political correctness – even in its benign form of genuine concern about the damage inflicted by development on the environment, and by the environment on development – has no reply: the phrase is revealed as empty, leaving no real guidance for action to take its place.

* * * * *

Those three logical difficulties with the concept of sustainable development should be more than enough to alert us to the possibility that there are also practical problems associated with it, which could inflict damage in the most literal way – and, indeed, that is how it turns out. The decisive practical weakness of the concept is the proposition, explicitly or implicitly contained in it, that environment protection and economic development are mutually consistent. Except under special definitions for each term, that proposition is not only wrong but irrelevant. It is totally out of touch with the issues that actually confront the market economy at the turn of the century. The developed market economy, world-wide, is on the point of having to fight for its life in the face of environmental degradation and resource depletion, and many of the less developed regions of the market are already having to do so. That is to say, the idea that we should try to develop in a way which does not damage the environment is the inverse of the truth: the agenda that faces us now is to find those forms of economic management which remain open to us when the effects of the environmental damage that has already been inflicted begin to bite.

Starting in the early years of the century, we shall have to try to cope with the damage which environmental degradation is inflicting on the market economy. 'Development' in the broadly-based sense in which it is used now, will be obsolete; 'sustainability' will have been transformed into a tough fight to derive what value we can from an environment which has finally – and not without warning – turned mean. It is the role of sustainable development in helping us to avoid facing up to this, and in thereby ensuring that we shall waste the vital period of preparation, that is the critical and decisively damaging effect of the concept. If we were to approach the coming transformation with a sense of uncertainty, with no ready-made, reassuring talisman to hold, with a proper sense that there is no coherent policy in place at all, we would be in a less exposed and vulnerable position. We would be allowed to think. The starting-point for the transformation that lies ahead is not a comforting sense that everything is in order, but an acknowledgement of a discontinuity in the evolution of the market economy, for which it is – materially, emotionally, politically, totally – unprepared.

When a complex living system grows so large that it gets into trouble, the problem is usually expressed in many different ways at the same time. The reason is that, over the years, and with every new challenge and problem, it has become clever at adapting its behaviour and opening up alternatives. In the past, it has gone systematically through all the options, using up its environmental resources with ever-increasing efficiency, and becoming confident in its ability to meet every challenge. Only when every avenue has been explored does the problem hit. By that time, all possible substitutions have already been substituted; the detail has already been developed; ingenuity has had its day, and the crisis takes the form, in a sense, of confirmation that the ground has been comprehensively used – that, in the poet Gerard Manley Hopkins's words, 'the soil is bare now' – not just here-and-there, but everywhere.

This widening of the damage, as the market economy approaches its crisis, has naturally produced a broad array of responses: reform of taxation is now a central theme of environmental policy; company accounts have to make their environmental costs explicit; the double agenda of sustainable development has now grown to a 'triple bottom line' with the addition of social justice.[7] The environmental policy profession is creative, elaborating its solutions in step with each new problem. Its attention is distracted by complex strategy ideas such as the 'seven-dimensions' in which it sees the world that it has invented.[8] It is therefore less than fully-prepared for the single, devastating dimension of the environment issue that is about to be made plain: the coming oil shock.

* * * * *

Extraction of the total recoverable world resource of crude oil, from first barrel to last, is now close to its half-way mark. Most of what remains belongs to the five

Middle East OPEC countries, Iran, Iraq, Kuwait, Saudi Arabia and the United Arab Emirates. Production from all other producers is now on a plateau; its half-way mark takes place around the turn of the century and, from 2001, the flow of crude oil from all producers other than the Middle East OPEC starts its long decline towards depletion.

This has the following implications. The Middle East OPEC Five group is now beginning to find itself with a large and growing share of the market. The decisive share – that, is the share which allows the group to influence the scale of global production and oil prices substantially – is about 30 percent. It was a share of 30 percent-plus that enabled it to determine supplies and prices so effectively during the 1970s, when the price of oil rose from US$2 a barrel to a brief peak of around US$50 a barrel. Its share fell below 30 percent at the end of that period because major new provinces were opening up, notably Alaska and the North Sea. Since then, oil prices have been low. However, its share is now back at the 30 percent level and rising; there are now no new oil provinces to be found, so its dominance of the oil market will not be challenged again. This opens the way for the Middle East OPEC producers to control world oil production at the margin – which is where prices are set. As effective control over the price of world crude oil is surrendered to this group – and this would be true for *any* group endowed with monopoly control over a wholly indispensable but declining resource – the price can be expected to rise far and fast.

At first, the price movement will be within limits that the market economy can take in its stride. The other producers, who still account for nearly 70 percent of the market, can still (for some five years or less) manipulate their supply to offset action by the Middle East OPEC. But many of them, too, would dearly like to see a rise in the price of oil. In any case, they are beginning to become aware of the depletion of their own reserves, and of the rise in value of the oil that is still in the ground, so that they have an incentive to hold back their own supplies. At the end of that five-year transition period, however, the decisive power in the oil market will be concentrated unambiguously with the Middle East group, who will, themselves, be moving to within a decade or two of their own half-way marks. They will have no reason to increase their supply to compensate for the fall-off in supply from the other producers. They will not need to do anything differently from what they are doing now: if they maintain a constant flow, or if they invest to increase their production to a modest extent, or indeed whatever they do – short of a fantasy world in which they make a massive investment in order to keep the price of oil as low as possible for fifteen more years – the result is the same. The concentration of the oil resource in one area, along with declining production from the rest of the world, will bring on the decisive mid-life crisis in the oil market; this is the moment when the market shifts from the cheap and easy options available in its early middle years, to the high costs, volatility and tensions which will develop as it passes its peak.

Now, of the questions raised by that analysis, the three key ones are these. First, on what authority is it based, and what justification is there for setting it out with such apparent confidence? Second, what are the implications? Third, what should be done?

The source of the information is the International Energy Agency (IEA), a body that was set up by the OECD in 1974 to monitor world oil supplies and to provide advance warning of any supply problems. That job is, in fact, an extraordinarily difficult one. Much (but not all) of the data on discoveries, reserves and flows in the oil industry comes from within the industry itself, and it is coloured by the political or strategic agendas of the companies and nations that produce it. There are incentives to understate and to overstate reserves, and to refer to political considerations when making estimates of the oil that is yet to be found. At the same time, techniques of estimation, exploration and extraction are improving all the time, so that reported reserves can increase, while the actual significance of this lies in the relentless fall in the quantity of oil held in basins which are as yet unconfirmed.

It is therefore only to be expected that estimates of the future reserves of oil should include some surprises – that occasionally a large quantity of information should fall into place at the same time. This is what has recently happened in the case of information on the prospects for oil. Unpredicted surges of new data place any official body in an awkward situation. New information, however sound the research on which it is based, has some uncertainty about it; this particular new information also has explosive implications which, if widely understood, could affect stock markets and world trade very profoundly and rapidly. Under these circumstances, the only option open to an agency with the weight and authority of the IEA is to trail the data in a somewhat guarded form. This allows it to be picked up, discussed, re-examined, tested and in due course filtered into the political process, by which time responsibility for the unwelcome news will have shifted to some extent from the original source and towards the much wider constituency of people who have taken up the story and followed it through.

It seems clear that these are the considerations that have influenced the IEA's solution to the problem presented to it by the sudden emergence of this new, hard (though not unanticipated[9]) data on the market for crude oil. The data about the shift of power into the hands of the Middle East OPEC producers is contained in a chapter, more or less without comment, in the October 1998 edition of the IEA's report, *World Energy Outlook*.[10] The chapter tells the story in the form of a thought-experiment, explaining what would have to happen *if* the world supply of oil were to continue to grow at its present rate for another fifteen years. It shows that this would require Middle East OPEC producers to more than double their production in record quick time in order to avoid troubling their customers with any inconvenient increase in price. It then leaves readers to draw their own conclusions. This is how one of the most critical pieces of information in the history of the oil market has come into the public domain.

Certainly, the authority of this source will not deter scepticism, and certainly there could be some slippage in the moment at which world energy prices can be expected to take off, and some softening of the impact as a result of reductions in demand imposed by the world's response to climate change. But the scale of the coming shortfall in oil supplies (relative to demand) is immense. By comparison with this, doubts and reservations with respect to estimates of reserves and timings are insignificant. Indeed, even if the most optimistic estimates of future reserves were correct, then a massive, one-way oil price shock would be postponed only for some ten years. That postponement would be helpful, but the nature of the problem would remain unchanged: the market economy still faces the shock of having to adapt, at very short notice indeed, to life after cheap oil.

In this way, the environment bites back. This is no longer a world in which the environment is regarded as something we ought to feel responsible for, in order to protect the interests of future generations: it is the environment, not the market economy, that now has the upper hand. The first member of the environmental family to swap places with the market economy, calling the shots instead of passively receiving them, is the global oil resource. Other members of the family (relations of the kind that wear wrap-around dark glasses, heavy stubble and suits with bulging breast-pockets) will be along very shortly. In the meantime, it would be a good idea to take stock of the implications.

* * * * *

The critical issue is the mismatch in timing between the constraint in the supply of oil, and the rate at which the market economy can adapt by radically reducing its energy-dependency and switching to renewables. It is probable that the transformation in the oil market, followed by a progressive worsening of the volume and security of supply, will take place within the first decade of the century. By contrast, the sheer scale of the programme of shifting the market from its present oil dependency onto a new energy-efficient basis might normally be seen as requiring some fifty years or more.[11] It can be expected that, for the first fifty years of very high oil prices, there will be inadequate alternative energy supplies. The consequences of this will be severe, and at worst they could lead to the deconstruction of the market economy, especially when other environmental issues such as climate change and food supplies are factored-in. The strong assumption that must underpin any useful discussion of the matter, however, is that society can find solutions; in the longer term, there is a prospect of stability. The focus of attention must therefore be directed to the period of critical and extreme stress in the early decades of the century.

The scale of price rises may be guessed at, if only in order to make the problem specific enough to give some meaning to a discussion of implications: there might be (say) a 5-fold rise in prices by 2005, a 10-fold rise by 2010, and a 20-fold rise by 2020. The sequence of cause-and-effect that follows is clear: from much higher energy prices, to a fall in real incomes, to a sharp upward trend in unemployment. That will be moderated by new jobs in the conservation and renewable energy industries – but not enough to compensate for the loss of jobs across the economy as a whole. The recessionary effects, world-wide, are therefore likely to be profound.

Food prices will rise, roughly in pace with the rising cost of oil, unless governments intervene. In effect, we eat oil: the industrialisation of agriculture between 1950 and 2000 has made food production oil-dependent, from fertiliser-production to tillage and transport; agriculture is now less a process of cultivation than an industrial process for converting oil into food. High food prices will, in turn, add to industrial costs and reduce consumers' real incomes, deepening the recession.

Transport prices will rise, with implications not only for the front-line companies such as airlines, but for all the tourism industries. Airlines are already in a financially-sensitive situation, since their aircraft-leasing arrangements tend to be based on ambitious growth targets; if air-traffic growth were to falter, business failure would ripple through all the industries with interests in air and road transport, hotels and tourism. A recent study which estimates that transport accounts for some 50 percent of GDP in a typical European economy,[12] and the observation that tourism is the world's biggest employer,[13] are not irrelevant in this context.

Taken together, business failure and job losses on this scale could have the effect of tipping the world economy into depression. As oil price rises continue on their upward path, there will be no easy way back: the trend of rising costs for oil and food, and the knock-on effects, in a positive feedback, from unemployment to reduced demand to unemployment, would mean that time would not be on the side of economic recovery. There are potential consequences here in terms of civil unrest and insurgency. The economic and social systems of the global market economy are so interconnected that no economy would be insulated from the secondary effects: recession breeds recession; it cannot be forecast how deeply this sequence will damage the essential structures of market economies round the world.

* * * * *

So, what should be done about it? The broad outlines of a policy response can be described quite easily. First of all, it is necessary to move on from the market economy's accustomed complacency and, particularly, from that leading manifestation of it – sustainable development. A speedy verification of the IEA's figures and of the conclusions that have to be drawn from them is needed; what is not needed is the more habitual response – the two-step routine of denial followed by panic – the panic being an inevitable consequence of the fact that the whole of the time which could have been spent in preventing the problem reaching its crisis has been used up in denial. A clearly-defined political vision is the indispensable condition for an effective response. The substance of the response itself falls into place, once that leadership is present. There is expertise readily available in all the technologies and logistics that will be needed; the political leadership itself does not have to be expert in anything except in getting people to co-ordinate effort in a focused programme. Once that has been established, the resources and ingenuity of the market economy, so often used as reasons for not recognising the physical limits within which it must work, will take over with massive effect, and with a degree of success which could expose the debate between pessimists and optimists as dated and irrelevant.

A brief glance at those practical policy options themselves is justified, if only to point the contrast with the arthritic processes that pass for environmental policy. Society approaches the coming transformation with some immense resources in its favour. They include the conservation technologies, whose leisurely but useful development over three decades opens up the possibility of delivering heating, transport, lighting, industrial production and other energy services at a level of energy-dependency as little as a tenth, or less, of the energy required at present.[14] This is the massive energy-service resource which brings the post-oil economy within the range of feasible options on a time-scale of two generations. Its technical development and application should now be accelerated to a pace which bears a serious relationship to the need, comparable to the level of urgency and political energy applied to the development of technologies such as radar and the Manhattan Project during the 1930s and 1940s.

Equally (and, of course, the two technologies overlap in many ways), the renewable energy technologies are a rich and essentially undeveloped resource. The technology, the development, revisions to the electricity grid and to voltages, new standards and equipment, and the training that will be needed to install and maintain all this, represent a clear agenda. A significant threat to both the conservation

166

and the renewables systems is present in the prospect of heavy investment being diverted into nuclear power, despite the evidence that, dollar for dollar, the yield in terms of available (i.e. saved or generated) energy that can be obtained from the 'soft' conservation and renewables technologies is greater, on a scale of 7-fold or more; the soft technologies also yield immediate energy savings and do not have nuclear power's intense demand for energy for construction, mining, management, waste-disposal, and decommissioning.[15]

For all the promise of the soft technologies, the job of shifting the market economy off its oil-dependency cannot be bounded within the limits of the technical fix. The deep agenda is to get to grips with the logistics – the patterns of distribution, land-use planning and transport, which have developed on the presumption of unlimited cheap oil. The mature market economy has lost, or abandoned, a sense of space, distance, proximity and locality. That will have to be rebuilt, in detail and quickly. Much of this will consist of undoing half-a-century of serial error in land-use planning, whose way has been smoothed so effectively by appeals to sustainable development. Grossly energy-inefficient industrial agriculture will have to be replaced by organic production; food-miles (the distances over which food is transported) will need to be drastically shortened; the logistics of firms and households will need to be transformed; a process of 'accelerated incrementalism' – similar to the Japanese concept of *kaizen* in the process of step-by-step production improvements[16] – will have to be applied. Every stage in the transformation will need to be a feasible, liveable option, along with intensive pressure for change.

Beyond these baseline policies, a complex of solutions will need to be developed, including the potential of electronic technology as the infrastructure for an energy-rationing system.[17] It will be necessary to reassess some of the core principles of the market economy, since in a post-oil world, dependence on sustained economic growth – the precondition for stability – will no longer be an option.[18] One of the principles that will need to be part of this solution is a greater responsibility for the conduct of economic and social affairs at the level of the household. As the material patronage which the market and the government can provide weakens, the household, whose present passive role is witnessed by the very word 'consumer', will need to call on reserves of resourcefulness and skill which have not been needed to date, but which are appropriate for a developed market economy as it learns to cope with the crisis of its maturity.

* * * * *

This chapter has indicated some difficulties with the concept of sustainable development, and it has drawn attention to the imminent failure of one of the essential conditions for the stability of the market economy. The argument is one with which many readers will disagree, not least because there is an implication here that much of the agenda of environmental protection for the last quarter of a century has been built on the wrong set of assumptions. *The Limits to Growth*, published in 1972,[19] and perhaps one of the most-rubbished and vilified books to have been published since the great heresy debates of the Middle Ages, pointed to the prospect of resource limits; that message could have been coolly appraised and set, for instance, in the context of oil: the declining rate of discoveries at that time was a very clear warning of the problem that was to come. If that had happened, we would now have an energy and agricultural system far better adapted to the conditions of the new century, and the security of the market economy would be very much more firmly based. In the event, however, the environmental movement switched its attention, doing a lot of useful work by improving the standards of industry's pollution control, but ignoring the matter of resource limits, which now turns out to be the first of a family of deep environmental problems with the potential to destabilise the market down to its core.

It would be nice if the coming debate about the prospects and extent of an oil supply problem could be short-circuited. Ten years of pleasurable controversy on the matter would use up the whole period which distinguishes a typical optimistic assessment of crude oil supply prospects from a more pessimistic one. Even if we had ten extra years to prepare for the oil shock, that would be forty years less than is actually needed to achieve the transformation with any real degree of confidence and comfort. The difference between five-years warning and fifteen years warning is, however, not trivial. Of course, debate cannot easily be short-circuited; if you try to do so you only end up with a debate about whether to short-circuit it or not. On the other hand, a lengthy debate is itself a form of short-circuit: those who argue that nothing should be done win the argument by the mere fact of extending the debate.

If there is a solution to this, it can only take the form of a dawning realisation that something very practical, economically-sound, employment-creating, energy-saving, inspiring, can be done now. The energy/logistics programme that is needed is totally consistent with the goals of a competitive market economy: it improves its efficiency; it creates jobs; it builds an export market in the goods and services it has pioneered; it provides a logical integrity that underpins the whole framework of social and economic policy. It provides evidence that there is political leadership with a clear vision, which knows what it is trying to do, and is not so fazed, confused and frightened by the forces of scepticism that there seems to be nothing to be done other than to plead ignorance and walk away.

And yet, ignorance is also a central character in this story. As soon as the transition that lies ahead is acknowledged, we are immediately plunged into a world of ignorance. It is an obvious paradox. Sustainable development, steadfastly ignorant of any particular problems that may lie ahead, disingenuously appeals to the argument that the future is unknowable, or vaguely refers to the amazing last-minute rescues that could be imagined. These solutions include unimaginably huge reserves of gas, oil sands and shales,[20] a robustly-ignorant confidence in the potential of fusion, energy-collection from space,[21] and unspecified applications of human ingenuity. A veil of ignorance thus becomes magically converted, under the spell of sustainable development, into a knowing confidence in business-as-usual. That sense of confidence inspired by sustainable development is, of course, why it is so magnetic. That is why the concept, those two words, now polished and smooth by repeated contact with the lips of the faithful, is so robust. That is why a demolition of sustainable development, with an urgent call to wake up and look at what is coming, derives its strength from ignorance. This new situation, the phase-transition which lies ahead, takes us straight into the unknown. Ignorance is our entry ticket to the subject. We shall, all too soon, need to learn hard and fast. But we have to start uncluttered. Step one is to say, with W.B.Yeats, 'I would be – for no knowledge is worth a straw – ignorant and wanton as the dawn.'

Acknowledgements

Thanks are expressed to the International Energy Agency, Miriam Polunin and Nick Robins for comments; and to David Astor and to Lawrence Woodward, director of Elm Farm Research Centre, for the financial support that made it possible for me to write this chapter.

Chapter 12

Contemporary order, peace and conflict: the balancing of opportunities and risks

James O'Connell

The Author

James O'Connell is Professor Emeritus of Peace Studies, University of Bradford and continues to work as a research professor in the Department of Peace Studies. He has taught in Nigeria, the United States and Britain. He has written extensively on global and ethnic politics. He has also written on philosophy and theology.

Chapter 12

Contemporary order, peace and conflict: the balancing of opportunities and risks

James O'Connell

Introduction

This chapter falls into three broad sections. It begins by considering technology and its social implications. In crucial ways technology – in the broad sense of scientific attitudes, the multiplication of tools and techniques, and bureaucratic organisation – is the most dynamic socio-economic factor in the contemporary world. It has intensified production, increased trade and influenced social organisation. It pervades inter-state and intra-state activities; and it constantly shapes and facilitates, and on occasion impedes, communication and co-operation between groups. Technology, however, functions within received and acquired cultural values; and it offers opportunities and advantages to its possessors. It also rapidly uses up natural resources; and it both enhances and threatens our human condition and the environment.

After looking at the uses and abuses of technology, I examine actual and potential conflict between countries and groups. For almost forty years after World War II, the world in military terms appeared bi-polar through the global confrontation of two superpowers, the Soviet Union and the United States. With the break-up of the Soviet/Russian empire and a gradual move away from policies of nuclear deterrence, a multi-polar military world has emerged and has begun to reflect a complex social and technological world environment. The weapons generated by the Cold War (and the period immediately before it) have, however, brought an historic change in military affairs: countries in an age of nuclear, biological and chemical weapons can now no longer rely dominantly or exclusively on military defence, and on the attitudes that traditionally accompanied it, to provide adequate security. Yet governments have not so far found other understandable, practical, and readily acceptable ways of thinking and acting in relation to new problems of security and conflicts of interest.

In what is now an era of one superpower, the military capacities of a technological age have begun to be further developed and, in measure, to spread to middle-ranking and even smaller powers, including those in politically unstable regions. In consequence, a global nervousness over security and survival remains. In the case of strong powers, this also gives rise to frustrations when they face unprecedented and apparently intractable situations that inhibit the safe use of full military strength. This mixture of power and frustration is surfacing in various areas of the world from Bosnia and Kosovo to Angola and Sierra Leone.

I want, finally, to argue that, by drawing on the benign features of our new technological capacities, and while taking account of their dangers, we may seek to create peace in tangled parts of the world and underpin new forms of global and regional unity.

The parameters of global technology

In suggesting interpretative paradigms or categories that organise the understanding of the role of technology within world society, one needs to take into account how quickly technological innovation has occurred in recent times; and how we are still using thought forms, value attitudes, and patterns of social organisation that are linked to earlier stages of the industrial revolution. Furthermore, technology cannot be separated from values either in its genesis or in its use. While it affects inherited values and sometimes harms them – some traditional values and belief systems are incompatible with advanced technology – it also develops within the values of, and in great measure for the benefit of, those who control it.

Potentialities and organisation of technology

It seems clear that the potentialities of science and technology imply continued, though not necessarily uninterrupted, economic growth as well as more powerful control over human and other living organisms, and over the environment. Moreover, though the logic of technology in its productivity, communications and trade is global, it is shaped also by regional and national attitudes. Furthermore, advances in technology are likely to generate opportunities for new forms of work and new attitudes to work and leisure, especially but not only in the developed world. All these potentialities occur within a pace of change that not only raises social and personal aspirations but also creates insecurities and some dislocation in nearly all societies. Not for nothing has Fukuyama called the 'sixties' and the decades that have followed the 'Great Disruption'.[1]

In reconciling the potentialities and organisation of technology, the last century mostly eliminated as inhumane and unworkable a purely *laissez-faire* or entirely unregulated economic system. The last sixty years have seen, instead, a clash between central planning and social market systems. With the failure of the centralised planning systems of the once Communist countries, there is however a convinced recognition of the role of the market (competition regulated by authorities) as the correlative factor to technology in economic growth; but there is less agreement on the mix of competition and regulation that economies need for both investment efficiency and social distribution. Moreover, in the developed countries, there is going to be continuing tension – though it need not necessarily be a malign tension – in the allocation of the fruits of economic growth between more propertied or more skilled and less propertied or less skilled groups. The Swedish, German and Austrian models of social democracy, which in good part laid a foundation for economic success and which were in turn buttressed by economic success, have formulated and applied the concept of the social market; they have generated close liaison between government, industrial and commercial corporations and labour unions in economic expansion; and, through a universal and generous system of welfare benefits, they have reduced the travails of technological change, softened the lot of those marginalised in skilled societies, and protected a sense of community.

Returning to the central theme of the considerable scientific and industrial achievement of our times, two observations are in order. On the positive side, this achievement has led to health and comfort, communication between groups and control over the environment to an extent undreamt of in pre-industrial times; and it makes peoples, including those peoples who do not have ready access to them, globally aware of new economic and social possibilities. On the negative side, it has led to enormous pressure on available resources; it has wounded the planet in various places and in various ways; and it has put the greater part of the world's resources in the hands of a minority of the world's population. There is little enough hope that the contemporary human majority will reach, within their lifetime or that of their children, the living standards of the minority.

Unevenness and ambivalence in economic growth and interchange

Trade and communications relate directly to the capacities and interests of the industrially advanced countries; and these, moreover, trade mainly with one another. As a result, the use of resources is not only ill distributed but the terms of trade are weighted against the poor. In other words, though general economic growth may take place, such growth is, and will continue to be, uneven in its geographical, ethnic and social class distribution. Also, in a world grown small, patterns of resource use that draw on a narrow range of materials may well create increasing tensions among competing groups. Yet, if there are dangers in uneven and ill-regulated growth, there are also opportunities for all countries in the development and self-correcting capacities of modern technology and its organisation. Opportunities will, however, best become available only if the well-to-do share through sensible aid policies with the less well-to-do. I return to this theme in a later section.

In the interchange between developed and developing countries, particular stress needs to be put on the asymmetry of communication; the image of the world that emanates from the media of the First World is so dominant and pervasive that the perceptions of both rich and poor countries are distorted. In the event, the rich are confirmed in their right to existing levels and patterns of consumption, and see them as reflecting social and economic ability as well as the rights of historic possession. The poor, in turn, are tantalised by the glittering prosperity that they glimpse in the new global proximity and are also moved by a revolution of rising expectations that can easily go sour in continuing frustrations. Developing groups are led to judge their own progress and the achievements of their leaders by standards that are borrowed from abroad; they are tempted to employ technologies that are ill-adapted to their situations; and they are weakened in their self-confidence by the obvious gaps of achievement between themselves and the economically advantaged. In this context, groups within the developing countries have scrambled desperately and competitively for greater power and prosperity and, in the process, they have most often drawn on ethnicity to organise the seizure of power and the build-up of wealth. In consequence, poor countries have often looked almost condemned to instability; and such instability has been increased by disruptive outside interventions from interested governments and commercial groups. It is, therefore, not surprising that the great majority of wars since 1945 have taken place in the Third World; and military spending there takes up undue proportions of many paltry national budgets.

175

Convergence of development and ecology

Technology has had, and continues to have, certain malign effects. Not only has damage been done to the environment – chemical pollution, erosion of the ozone layer, atmospheric warming and dangerous radiation, among other effects – but resources that took cosmic evolution millions of years to put in place are being prodigally used up in the space of decades. These resources would be used up even sooner, were more peoples in a position to exploit them with current technologies. Although, in rich countries, there are many people who are convinced that they have to use resources in a non-polluting way, it is still difficult to convince most that resources need to be used more economically; and even more difficult to convince them that resources need to be used in a way that will make them available to those who do not presently possess them. In most poor countries, it has, in turn, proved immensely difficult to create the kinds of skills and organisation that are needed to underpin technology; but, at the same time, there is no reversing the hopes awakened by technology. Moreover, poor countries cannot be convinced to put their meagre resources into caring for parts of the global environment that fall within their jurisdiction unless rich countries compensate them for such difficult investment.

The industrial countries seldom grasp that they can go on using existing materials in their present volume and style of use only because some three quarters of the world's population do not have equivalent access to those materials and cannot use them in the same way. If it were not so, there would neither be enough resources nor materials, within the limits of contemporary technology, to go round among all countries; nor enough clean capacity to absorb their wastes. For such reasons, the challenge for those who control technology is to re-think how we use the resources of the planet in ways that are economical, interdependent and wholesome – the last term taken in the broadest sense to include health, a user-friendly and aesthetic environment, and respect for other species as well as humans. They need, also, to carry out research that will enable their contemporaries to use existing resources more frugally; to work on approaches to add value; and to create new resources in skills, equipment and materials, especially of renewable energy, that can be expanded and made available for global consumption but that, at the same time, respect and enhance the care of human beings and the good use of the earth. If they do not do so, enormous tensions will build up, as new revisionist powers come on the world scene, seek to share world resources, and threaten those who may not wish to share.

Economic viability and aid

Aid motives at their best generously relate to a common humanity and justice. They are also functional in heading off future conflict for resources. In a utilitarian sense, aid for development also recognises the potentiality of great Third World markets that replicate the growing markets which the Western industrial proletariat furnished internally to its countries at the end of the 19th century. Understandably, poor nations mostly prefer to put economic growth and social improvements ahead of planetary ecology. For such reasons, in their bargaining on ecological issues, governments of developing countries want economic benefits. In exchange for preserving biodiversity and for restricting the cutting down of their forests, they seek financial aid, debt remission, more advantageous trade conditions and access to clean technology without restrictions on intellectual property.

If the environment is not for many countries the most immediate problem, it has brought together the issues of development and ecology. Groups in the rich countries, who would not have readily taken on the issue of development, have had to confront it, once it had become clear that under-development was harming their own environment. Well-to-do countries which worry about the effects of global warming, the holing of the ozone layer, and tropical deforestation cannot cope with the consequences without calling on the majority of the globe's inhabitants, at least as auxiliaries, in the struggle. In other words, while the poverty of the developing world touches the rich psychologically, the ecology of the poor touches them structurally.

World technology and the urgency of peace

Developed countries have now largely realised – the European Union which set aside age-old rivalries is the prime example of a new and co-operative political structure – that, while they may compete with one another, their interests still converge. Such convergence is much less clear in developing countries, whose economies seldom complement one another, and whose newly formed states, new ethnic proximities within them, and changed relations between groups create no small temptation to conflict.

In examining the issues of peace and war in our time, we do best if we consider not simply manifestations of the breakdown of peace but the sources of present and future conflict. Broadly speaking, there are three such sources. First, many in developed countries are driven by a passion for forms of growth that comes out of attitudes engendered by the historic human struggle against hunger, disease and natural disaster; human desires have not yet come to terms with the ease and abundance created by contemporary achievements. While advances in health as a subject, and computing as a technique, may reasonably continue, it is not, however, sensible to go on ravaging the environment for relatively peripheral advances in transport, cosmetics and comfort. In Western countries, what this anachronistic passion for goods and luxury may do, especially, is to lead peoples to get the priorities, timing, and style of desirable future progress wrong. Above all, it prompts Westerners to put their own immediate interests before those of the developing countries.

Second, contemporaries neglect at great peril to their peace the consequences of the spread of technology and the communications explosion of the contemporary world. The expanding élite groups of developing countries grow impatiently aware of what the developed countries have and what they themselves have not; and they are under pressure from the lumpen élites[2] of their countries for faster rising living standards. In urging reconsideration of such issues in industrially developed countries, the central problem is that, not only is it difficult for governments and peoples to readjust economic perspectives, but they seldom face up to longer term and international issues until they become acute, including the allocation of resources between peoples. Even in solidly democratic and well-educated countries the future has few votes. Since, however, the poor are not going to press their noses indefinitely against the plate glass windows of the rich without threatening and trying to break through, and since the spread of military technology is now such that even the poor can threaten the rich, the rich, if they want to safeguard their peace, need to gear technological progress more towards sharing prosperity.

Third, we have in nuclear – and biological and chemical – weapons the capacity to destroy, or at least greatly to harm, our civilisation militarily. The only way in which we can avoid doing this is to set aside age-old over-reliance on military security and accept that, in our time, conflict between powers best admits of non-violent resolution. We must create world political structures that match, contain and control the global reach and transforming impact of technology, and develop an understanding of our common human belonging that flowers in global fellowship. For the developed countries, the alternative to such policies is to set up a siege society, walled off against the poor majority, that would make the erstwhile white South African laager look simultaneously trivial and rational.

The tensions and complications of a multi-polar world

If the technological and social factors sketched in the previous section set broad paradigms for global peace, prosperity and ecology, what are the immediate political situations that we need to take into account? There are, it seems to me, four such broad situations:

a) Expanded opportunities for regional integration are arising in Europe in the wake of the passing of the confrontation of the superpowers and the ending of the Soviet grip on countries in Eastern and Central Europe; but moves towards integration are flanked by threats from new forms of ethnic and regional competitiveness, not least in post-Yugoslavia.[3]

b) Continuing instabilities and hostilities bedevil the Middle East.

c) The progress, vicissitudes and aspirations of the developing world, including China and India, are beginning to impinge on the whole world. In looking at Europe, the Middle East, and the Third World, we need also to take into account a theme in political development that has become salient in recent years – namely the questioning of the legitimacy of particular states, and of actions of governments, as well as diminished sovereignties and the growth of international bodies that cut across the historical ways in which states have associated with one another. Such questioning arose initially in Africa but has now also become acute in Asia and Europe. A challenge ahead for international organisation is to work out ways of intervening across national boundaries in instances when governments have oppressed or have failed to protect peoples. It becomes a question of working out responsible decision making about where interventions should take place, taking on board the gravity of disregarding sovereignty and putting in place safeguards that prevent the users of force from primarily following their own interests.

d) Looming over world politics is the lonely position of the United States as the only current military superpower. Both its military outreach and its economic linkages lead the US into playing a global role, while playing a specifically regional role in the Americas. (Japan plays a global economic role as well as an Asian regional one, but it has avoided wider political and military commitments.)

The European scene

Europe has aspirations to maintain a multi-polar political world and to protect its global trading interests. But it has, in measure, turned in on itself politically, while it works out the conditions of its unity and deals with the aftermath of the break-up of the Soviet empire for which it was little prepared psychologically.

The Union and others: The grouping that most co-ordinates social standards and underpins peace in Europe is the European Union. In its overall implications, the European Union offers the best contemporary witness that nation states are obsolescent in social and economic terms and interests and, for that reason, also in military terms and interests. The measure of the success of the Union is the way in which its citizens and others now take for granted that war is as impossible between Berlin and Paris as between Ottawa and Washington. The Union has not only constrained traditional enmities, but has channelled political co-operation positively towards the future. It has generated both economic capacity and made use of cultural strength with which to offer understanding and aid to Eastern Europe and elsewhere.

In the evolution of European politics the smaller continental countries have known that they live – in military terms at least – on sufferance from the three great powers of the continent, France, Germany and Russia, or from a balance of power between those states. They gained this knowledge from old memories of the movement of peoples; from the power play of politics since the 18th century; and from the new liquification of social structures resulting from the industrial-linked wars of the 20th century and the cross-frontier nature of new technologies. Unfortunately for the European peoples, a political stability, which had, in the 19th century, buttressed growing prosperity, has, in the 20th century, broken down in two great wars. These were, initially, European civil wars, which arose as the leading states developed nationalist industrial and welfare policies – socialist nationalism – that united élites and workers. These policies set state against state, as each, unaware of the now immense and growing possibilities for production, competed for markets that seemed static to politicians and others, while they were, in fact, increasing and could even be made to grow faster. The genius of Monnet and Schumann was to understand that, in recoiling from war, they could, in the years after 1945, use the new industrial technology to create and to share wealth among hitherto divided countries and to draw them together within a quasi-federal organisation.

Nothing better exemplifies the new Europe than the transformation of Germany. Germans, who before 1871 lived in small states that were endlessly threatened by neighbours, feel insecurity in their bones, as many of the smaller European states still do.[4] Germans had acquired a certain *hubris*, as the Prussians showed their military prowess in humbling Austria and France, and as the industrial clout of a united Germany began to gain them influence. That this *hubris* was fragile was brought home to them by defeats in two life-consuming wars, as indeed their defeats by Napoleon had earlier foreshadowed. It is a mistake that many commentators make, to believe that the Germans want, through the Union, to be made safe from their own power. In fact, they want to be made safe from a simultaneous condition of power and weakness. If the smaller states have wanted a community within which their identities and roles could be safeguarded and their futures made not only in peace but in friendship, Germans, in their turn, have wanted a community within which they could exercise the strong role to which they believed their capacities entitled them.[5] But they wanted to play this role in a community within which they would be spared a search for hegemony, and in which the conditions of association would give them security and influence, but not require overall control.

They, too, wanted a place in which, through their positive and generous contributions, their strength would remain welcome; and in which not only they themselves, but also their neighbours, would avoid the tantalising and destructive temptations of the past.[6]

Moreover, with the pulling down of the iron curtain that artificially divided Europe and Germany itself, the German commitment to the Union is enabling Western Europe to cope with the re-emerging concept of Central Europe, and with remaking the old heartlands that have, over the centuries, held the continent together. In the process, not only do Western and Central Europe draw more closely together, but Germany, from within the Union, is already taking the lead in Central European industrial investment and in helping, with others, to reactivate the old Hapsburg lines of transport and communications. In the process it is also assisting the growth of social democracy in those countries.

Obviously, the Union itself faces proximate organisational challenges: creating federal or quasi-federal structures; putting right the 'democratic deficit' in its decision-making; ensuring that monetary union proves successful; and planning, and acting on, the further enlargement of a Union which is working, through various stages, to reach much further east as well as north. This expansion is obliging governments, as a start, to rework the Union's Common Agricultural Policy (CAP), since it would be impossible to retain the existing funding for farmers in a Union that embraced Poland and Hungary. But, more fundamentally, the Union is faced with taking decisions on regional support policies, that would benefit its existing poorer members, but that the richer members hesitate to extend to new members. Yet it is difficult to conceive a healthy Union that tolerates undue disparity in living standards between its members or within its individual states. Finally, Europe is faced with devising not only economic, but also security, arrangements with the United States. Such arrangements should not only link both regions in a common security policy, but, in a revision of the North Atlantic Treaty Organisation, should create a European decision-making centre that possesses its own relative autonomy.[7]

After the Soviet Union: The post-Soviet regime in Russia is trying to survive the vicissitudes of liberalisation and democratisation, deal with the intractable remnants of an industrial command economy and agricultural failure, and overcome general social resistance to bureaucratic and social restructuring. Tensions abound in a country where civil society is weak, where there is not enough trust or respect between rulers and ruled, where bureaucrats remain insensitive to popular needs, and where mafia groups amass wealth, terrorise business people and distort law and order.

In the evolution of Russia several phenomena continue to puzzle observers. First, a relatively educated population remains passive and, so far, shows little evidence of reacting violently against its unhappy lot. Secondly, the military, in spite of worsening living standards for the officer class and uncertain and low salaries for all ranks, have stayed relatively apolitical. Thirdly, though they are desperately short of resources, Russian technicians, both military and civil, have continued to maintain nuclear and related facilities and to provide basic safety measures. Dangers however remain in relation to Soviet nuclear weapons technology. For one thing, one cannot rule out the possibility that unscrupulous and/or irresponsible military personnel may sell nuclear weapons to certain states that are seeking to create or enhance a nuclear capacity; and, for another, in the present depleted economic situation of the weapons industry, that nuclear experts or teams of experts may seek to market their expertise in order to gain or improve their economic circumstances. Fourthly, though nationalist parties and groups have sought to build on suspicions that Russian political and economic decisions are dictated from outside as well as on resentment of Russian loss of influence in world politics, they have not so far been successful. Yet resemblances with the German Weimar republic are too strong to be ignored in a country that is floundering economically but that still possesses weapons of immense destruction.

Western countries will run great risks if they do not monitor carefully the evolution of Russian politics and society, and do not go on providing Russia with enough economic means to stave off mass unemployment or bankruptcy. In the longer run, they will not only profit from a market created by growing Russian prosperity, but will avoid the military consequences of ultra-nationalism and maverick leadership. Yet, while existing frustrations in Russia arise in good part out of continuing economic failure, all the comparative evidence suggests that even if, with time, a degree of relative economic success is achieved, aspirations will rise faster than possibilities of fulfilment. Western countries have no option during the coming decades but to manage actively their relationship with Russia.

Added to economic woes, many nationalities within the former Russian empire are demanding autonomy, or at least greater autonomy, as well as beginning to quarrel among themselves. Some, such as the Ukraine, have unresolved territorial problems with Russia, as Moldova has with Belarus. Many such tensions mirror the struggles that took place in Africa from the 1960s onwards, as the autocratic grip of colonialism eased. This relaxation left ethnic and other groups in new competitive power relations with one another; and also stranded groups, that had, during the colonial peace with its commercial opportunities, moved into parts of countries away from their places of origin.[8] In these situations the Russian government has been made extremely uneasy by NATO's intervention in the Serbian/Kosovan war; on the one hand, it does not have the political or economic resources to bolster Slav solidarity; and, on the other hand and more important, remembering its Chechen conflict, it fears precedents that legitimise interventions across national boundaries – and in this reaction it is joined by the Indian and Chinese governments.

Finally, one may sympathise with the desires of nationalities and other groups to win freedom for themselves. But, if a considerable disintegration of the former Soviet empire should take place, new political insecurities may well appear, as curbs are lifted from old hostilities. Moreover, immense waste may be involved in casting aside, or not making proper use of, a once shared infra-structure in administration and communication, a huge market, and considerable collective bargaining power.

Post-communist Central and Eastern Europe: Alongside the successor states of the former Soviet Union are those other Central and Eastern European countries – Poland, Hungary, the Czech Republic, Slovakia and Slovenia – that have shrugged off the shackles of communist political control and rigid economic structures; and that have had some success in establishing responsible and secure governments, building up competent bureaucracies, coping with economic liberalisation and resolving environmental difficulties. The situation in other Eastern European and Balkan countries is, however, much less healthy. The Yugoslav situation indicates internal problems that these other countries may encounter. For they are not only struggling with difficulty to build up civil societies and bureaucracies, but they are also tottering into disputes over issues of ethnic minorities, the environment and resources. In the course of these developments, Western governments may help in working out, through the European Union and in conjunction with the United Nations, mediation and sanctions policies to cope with a spate of ethnic and border disputes; and they may be grateful for the role of Russia in helping to broker and preserve the peace among those countries which border it and whose stability is important for its security and economy.

The European Union, if it acts with foresight and shrewdness, may be able to offer aid in ways that bolster the more valuable elements of Tsarist and Soviet structures. Arguments are often put forward that the Europeans should initiate a plan similar to the Marshall Plan that helped Western Europe after 1945. The problem is that American aid was then channelled into countries which had civic systems and bureaucracies that could integrate aid, while Russia and other post-Soviet countries, with exceptions such as Latvia, Lithuania and Estonia,[9] have not yet been able to go far enough in building civil societies which embody the trust and organisational capacity that would enable them to benefit properly from aid. Essentially the same problem arises with most developing countries.

Japan – political and economic asymmetry: Ironically, it seems best to consider Japan – about whom many countries nourish military as well as economic uneasiness – in the Western context; for Japan is an integral part of the economy of Western countries. It avoided the exorbitant economic costs of the Cold War by relying on American protection, but repaid its debt in good part and protected its American market by investing and lending in the United States. It is through such linkages that Japanese power may most peacefully be contained. Yet there will be great loss if Japan does not share the results of its economic achievements; in the process, it should work with countries such as Singapore, South Korea, Malaysia, Taiwan and other Asian countries, and it should develop a symbiotic economic relationship with China. Japan may also be crucial to the technological growth of Russia. And, finally, it offers the example of a country that has successfully modernised – a modernisation which is too often misunderstood as westernisation – while retaining its traditional cultural heritage and forms of social organisation.

The cockpit of the Middle East

Middle East problems prompt a division into two great categories of countries, the second of which is far from being internally coherent in either its linkages, alliances or hostilities. These are Israel and the Arabs; and Arabs and others.

Israel and the Arabs: Israel was established after World War II, with indigenous Jewish and Western Jewish inhabitants; it drew for support on Jews throughout the world; it benefited from a Western bad conscience over the relatively unopposed Nazi destruction of Jews; it gained sympathy as a Western-type state that appeared to introduce efficiency where there had previously been Arab inertia; it expanded territorially through successful wars and the displacement of the Arab population, and it has grown demographically through the intake of Oriental and Russian Jews (the former mostly driven out of Arab lands) and some continuing migration from the United States and other Western countries. In a fundamental sense Israel represents the last – and potentially the most enduring as well as the most provocative

– Western colony.

Israel has, in fact, reached in some forty years a political identity and structure that plantation or colonial settlements took over two hundred years to achieve in the North and South Americas and in Northern Ireland. Hence, there is a reasonable case for accepting that, in spite of the shorter historical period involved, Israel has achieved a territorial legitimacy in its region and is entitled to endure and that its existence be recognised by Arabs as well as by others. That does not, however, mean that the international community has to accept new displacements of Palestinians through settlements or expulsions, or that the flouting of United Nations resolutions and the methods being employed to maintain Israeli rule in the Occupied Territories and in Southern Lebanon have to be condoned.

Yet it is important to recognise that Israel is torn between three different concepts of itself as a state: (i) as a homeland for Jews, having historic roots in Palestine but remaining especially a community and place of refuge for a people who have been savagely attacked at different times over the centuries; (ii) as a restoration in their 'promised land' of the Davidic kingdom which was the high point of historic Jewish state-making; and (iii) as a power state that, like any other such state, seeks advantage at the expense of its neighbours and that justifies territorial expansion in terms of security. Clearly there are many Israelis who combine elements of all three concepts in their thinking.

The concept of homeland, which was the dominant motive of many of the state's founders, obtained enormous support for Israel both in Europe and the United States in the aftermath of the Nazi holocaust. Those who move beyond the concept of a homeland to support the concept of Davidic kingdom argue from history – and appeal to many Jews and fundamentalist Christians – to justify displacing Palestinians from the West Bank. History is, however, a two-edged sword in arguing for possession of this land, since centuries-long history up to 1948 leans to the Palestinian case rather than to the Jewish. Those who, finally, support the concept of a power state, stress security and want extra land for new citizens; but, in doing so, they choose to live dangerously by the sword in a small state which is threatened by the hostility of vastly more numerous neighbours.

External attitudes towards Israel have undergone changes over the last two decades. If Israel has moved towards conciliating Egypt and Jordan, much Western, including some American, opinion has been influenced adversely by the Israeli invasion of Lebanon, by the handling of the Palestinian *intifada*, and by a growing awareness of the general plight of the Palestinians. Opinion has, in measure, come to believe that Israel, with its present policies, can neither enjoy internal order nor hope for external security.

If the Israeli-Palestinian issue is pivotal to world peace, it is for a set of complex and often interlocking reasons. The conflict is located in an oil-rich Middle East that supplies energy to the main economies of the world. Israel enjoys immense economic assistance and committed military protection from the United States; and it possesses nuclear and other powerful weapons which it would not hesitate to use were its existence threatened. Arab populations, generally, see Israel as an anomaly in the Middle East and the Israelis as interlopers; and many look on the Jews with a pathological hatred that has found expression in, and been reinforced by, three bitter and losing wars. In this outlook, they are joined by Muslims in most parts of the world and by anti-colonial élites throughout the developing world. If Israel itself is at present militarily more powerful than its neighbours, its superiority is gradually being eroded by modernising advances in Arab and other Middle Eastern countries. The latter are acquiring the civil and military technology and organisation that once made Israel so distinctive in the region; long range missiles make redundant the fortified defences in the Golan Heights and reduce the worth of Israeli ground forces; and the gradual growth in the technical skills of anti-Israeli terrorist groups is not to be discounted.

The future of Israel in its region obviously depends on US foreign policy in the years immediately ahead. Yet, to protect its own democracy, to continue economic growth without excessive military costs, and to preserve its long term security, Israel has little choice but to accept a phased establishment of a Palestinian state that will almost certainly with time merge with Jordan. Not least, if Israel is to cope with its water resource problems, it has to seek an accommodation with its neighbours; and its neighbours could greatly benefit from Israeli expertise in this area. More than anything else, however, were Israel to arrive at an agreed peace settlement with Palestinians and were it to work with them towards mutually beneficial economic arrangements, it would have found the one way possible to come to terms with its location and to disarm regional and other hostility. Furthermore, were Palestinians broadly united in accepting a settlement with Israel, other Arabs could, with more equanimity, concentrate on their own internal problems. If, however, intransigent and overreaching Israeli governments continue in power and the siege mentality of many of Israel's population prevails, and if the United States is immobilised by its domestic politics and does not use its overwhelming influence and strength to move Israel towards a settlement with its neighbours, it will leave in place a potential for conflict that the United States and the world may well rue.

A powder keg of countries: In the Middle East, political stability within states and peace between states face an uncertain future, not least in the uneasy truce between Iran and Iraq and in the barely contained tensions between Syria and several of its neighbours. Iraq will take decades to recover from the destruction of the Gulf war and its bombing aftermaths.[10] Future tensions appear inevitable as Iran tries to stake out a major role within the region; it was, after all, the possibility of this major role that led the Western powers as well as the Soviet Union to support Iraq against Iran in the seven year Gulf war. The truce between Iran and Iraq is likely to continue – a truce that has not yet been turned into a peace – but both will keep up a mutually harmful and isolating competitiveness. Both countries will also have to live with unresolved internal issues of political legitimacy and social discontent. Iraq faces a difficult future in integrating ethnic and religious groups and moving from autocracy towards a form of democracy, while Iran has to deal with the legacy of its revolution, by moving political power away from clerics and by seeking to modernise Islamic law while maintaining religious devotion and social cohesion.

The overall Middle East political situation is complicated by Syrian ambitions to create a Greater Syria and by the troubled Lebanese power vacuum which also prompts Israeli involvement. There are further dangers and risks inherent in the lack of political legitimacy and demographic balance in the oil emirates as well as social fragility deriving from the incompetence and corruption of their governments. The Kurdish issue in Iran, Iraq and Turkey – and it is not the only ethnic issue in the region – will not easily be resolved or contained. Moreover, disturbances in newly independent states, and in the Muslim post-Soviet independent states and Russian republics, may well have repercussions in Middle Eastern countries – not least in Iran and Iraq – and even further east in Afghanistan and Pakistan. While the United States guarantees the external security of the Gulf oil states, the situation could become much less clear in the case of dynastic and social conflicts within those states or if Iran or Arab countries were to intervene in such conflicts. Finally, while the problem of human rights remains unresolved in the Arab and Middle Eastern states, the quality of social and personal living in those countries will continue to be harmed; and the openness will be missing that is necessary for more efficient and responsive government and for self-generating economic activity.

India and China

These two great countries are currently struggling with the internal problems of economic development. The latter are compounded in India by the troubled relationships of its many nations and religions as well as by its tense confrontation with Pakistan. Both countries have, in a worrying way, complicated relations with one another and with the rest of the world by their acquisition of nuclear devices and ballistic missiles. For such reasons, the major military and economic powers need, in particular, to support moves within India and Pakistan to resolve the festering problem of Kashmir which, more than any other issue, prompts an arms race between two relatively poor countries and threatens the peace of the entire region.

In China the problems of development are compounded by regional disparities as well as by the difficulties of political and economic liberalisation that seem to portend almost inevitably a degree of political instability in the near future. Moreover, Chinese foreign policy, especially in dealing with Cambodia, Vietnam and Tibet, is hampered by old-fashioned concepts of geo-political strategy that derive from Chinese history rather than from contemporary security and economic conditions. More immediately significant than any other external policy matter, is the situation of Taiwan from where, only a few decades ago, a defeated and expelled regime laid claim to the whole of China, and which now has a government that effectively repudiates integration with China. To keep this conflict unresolved, mainly through American power stacked on the side of the Taiwanese government, is to leave in place a conflagration point that future developments could cause to explode, draw in major powers and threaten world peace.

Yet, to make crucial cultural and political points: India and China have the advantage of having cultures with values open to the integration of technological modernisation; they possess considerable human resources and energies; and they contain large internal markets. As they make economic progress, they will also inevitably question contemporary patterns of global resource allocation and make strongly revisionist bids, possibly with military overtones, to change those patterns.

The struggling developing world

In global terms, for the moment, most of the Third World falls into this category; and it is there that two thirds of the world's population live. There has been a dreadful failure of development in many, if not most, poor countries since around 1960. Although developing countries have been damaged, because terms of trade have moved against them and they have encountered protectionism exercised against new countries arriving on the industrial scene, their greatest problems have been internal. These problems are manifold. They include difficulties in creating and imparting skills; in establishing stable political cultures in which rising and impatient aspirations do not place intolerable burdens on governments; in developing an efficient and upright bureaucracy; in avoiding expensive and irrelevant prestige projects; and in building a sense of the commonwealth of the state or political association over against ethnic and local community allegiances. Countries such as Nigeria and Algeria, for example, which suddenly became oil-rich, showed that they did not have the political will, the administrative capacity, the political cohesion or the social integrity to make use of the accrued capital. Overall, while aid from rich countries remains pressingly urgent, it has never been close to adequate; and aid agencies have faltered and hesitated in taking proper account of the cultural integrity of poorer peoples, in promoting human rights, in differentiating between the interests of donors and receivers of aid, and in finding the most appropriate organisational and technical means for working on the problems of development. Yet for all its inadequacies there is no alternative to large-scale aid for many poor countries if they are to make faster economic progress.

Third World countries are racked not only by extreme poverty – in Bangladesh and Burkina Faso, for example – but in many cases are torn by war. In Latin America, social class liberation struggles remain unresolved. In Africa, inter-state and ethnic conflicts cost lives and waste resources in Angola, Mozambique, Ethiopia, Liberia, Sierra Leone, Eritrea, Somalia, Burundi, Rwanda, Sudan and the Western Sahara. In South Africa, the aftermath and patterns of apartheid are still being uncertainly unravelled. In Asia, ideological and ethnic struggles intermingle in Afghanistan, Burma, Cambodia, Indonesia and Sri Lanka. The United States, through its single minded anti-communism throughout the cold war period, has, time and again, intervened in Latin American conflicts to protect an enlarged version of its own security. In Africa and Asia – in countries ranging from Vietnam to Angola – American and Soviet rivalries have contributed arms and expertise to worsening local disputes. Now, when such foreign political rivalries have disappeared, they have left behind the gaping divisions that they helped to widen; and in the cases of countries such as Angola, Sierra Leone and the Congo (Zaire), private companies have gone on supplying arms in return for mineral concessions.

Some transnational issues

While, in this chapter, I have mainly gathered observations into regional categories, there are other relevant categories – terrorism, not least – that cut across them. To expand on these would take up too much space here. What, however, needs to be stressed is that current political hostilities and tensions are likely to be powerfully conditioned by three continuing military developments: the acquisition of long-range missiles by various countries; the spread of chemical and bacteriological weapons; and the proliferation of nuclear weapons. The possession of all these kinds of weapons by certain countries may, in military terms, offset their lack of general military influence and of economic resources; and they may create horrific regional complications in the not too distant future.

There are two further complications that arise out of the achievements of the modern world that need special mention. First, we may well be witnessing a new and explosive migration of peoples that mirrors the great migrations of history. The push from poverty and political insecurity and the pull from riches and stability will go on creating a determination among many groups to move from Eastern Europe, North Africa, Latin America and parts of Asia to Western and Southern Europe, North America and Australia. Once arrived, they are likely to meet uncertain and hostile receptions and to prompt adjustment problems as societies become multi-cultural. Second, the explosion of Aids – some forty million persons may be infected with the HIV virus by the year 2000 – the spread of which has been facilitated by contemporary communications, may create human and development havoc throughout Africa and in Asian countries such as India and Thailand.

The US: the sole military superpower

What needs to be said about the United States is that it now has had the time, since the start of the 1990s, to build up a far-sighted foreign policy. In particular, it has, through the destruction of Iraq's influence, settled for the immediate future the issue of Western and Japanese access to oil supplies. Yet the US has sufficient experience to know that, while it can reach to all parts of the globe militarily, it has no hope of dictating unilaterally to a multitude of countries, or of intervening constantly in the internal affairs of any country. It intervened benignly in Bosnia, after the European powers had been immobilised by their own divisions and hesitations; and it has latterly used the NATO alliance to move against paranoid Serbian intransigence in Kosovo. In the process, it has had to ignore the UN Security Council, where the opposition of Russia and China impeded decisions. Yet, it is only because of the super-power capacity of the US and its position on the Security Council that UN moves have been prevented against a similarly paranoid Israeli intransigence in dealing with Palestinians. Again, the Western intervention in Kosovo underlines the contrasting unwillingness to intervene against Turks attacking Kurds (though not Iraqis attacking Kurds), Russians devastating civilian Chechen groups and Hutus murdering Tutsis (and vice versa). Furthermore, the US has singularly failed to impede India and Pakistan, or Israel and North Korea, from developing nuclear weapons. These different situations and reactions underline the need for power in the United Nations Organisation to become more representative and more effective. Anything else will leave any superpower open to accusations of selective interventions in its own interest and of carrying on a modern version of gunboat diplomacy.

The central weakness of US foreign policy has lain with its domestic preoccupations. It is a huge country that, far more than most, is sufficient to itself; and the greater number of its public opinion activists evince little interest in foreign affairs or they see them through a highly restricted American prism. In consequence, US politics is preoccupied with US security, an attitude that stems from nervous factors in American history: the threats on the frontier from the Amerindians, the fear of blacks in old Dixie, the integrating Anglo-Americans' resentment and dislike of the waves of polyglot 19th century immigrants, and the corroding competition of immigrant ethnic groups in the cities. In the process, the US has had a tendency to create and to demonise enemies in 'evil empire' or equivalent concepts; and, on occasion, to move too soon towards military action. Pushed by its insecurities, it has supported corrupt regimes and thwarted decent aspirations in Latin American countries. It is tempted, too, to seek technical solutions for its own security, not least in nuclear weapons and anti-ballistic missile programmes, that set it outside any collective security system. At the same time, it has sought to create a global economy within which its trading and financial groupings could remove barriers, no matter how crucial the latter were to the economies or stability of individual countries or regions. Within such policies, particular American economic interests – communications groups, fruit companies or meat producers, mining companies, and others – have been able to exercise influence over federal policy and to warp foreign relations, even with friendly countries.

The size, energy and international tentacles of the American economy mean, however, that other countries have to go on taking American internal and external political and economic policies seriously into account. Moreover, in spite of weaknesses and rigidities in its foreign policy, most economic relations with the United States create prosperity for the countries involved; huge admiration exists throughout the world for its dynamic and free, if flawed, politics; and its cultural impact is immense in areas ranging from music to the internet. If, for the immediate future, a crucial challenge for the American intelligentsia is to leaven national understanding and improve dealings with other parts of the world, other countries have equally to learn how to reach into American opinion formation so as to deal cogently with a world power that cannot be avoided and that has to be met in its complexity.

Another future? The primacy of vision and the search for global unity and peace

In the previous sections of this chapter I have, on occasion, suggested proposals for action on unity and peace. In this final section let me expand a little on ways of moving forward.

Vision and means of unity

Useful as individual proposals may be, it is much more important at this stage to sketch tentatively a broad way forward towards a new ordering of global politics and society, while only suggesting, reluctantly, a framework for action or any possible technical detail. Caution is of the essence of the venture. Had Monnet and Schumann, for example, begun with a precise blueprint for European unity in the post-war period, they would never have achieved the European Economic Community, but would have been rejected at the outset as utopian planners. What they started with was a vision and a set of convictions; and they moved forward in constructing political and technical agreements that came out of their convictions but which, nonetheless, took account of political and economic realities. They and those allied with them – politicians, bureaucrats and commentators – convinced others of the worth of unity; they drew up and tested the acceptance and workability of a sequence of agreements; and they used political good will to generate the patience and trust that were required to accomodate adjustments that were initially painful.

In effect, it can be argued that what is needed for a secure global order are vision, outline arrangements and technical agreements. The vision is the heart of the venture. It draws on the fellowship of peoples who own a sense of belonging to one another and who share values and attitudes; it also draws on the realisation that peoples need a sense of belonging if they are to pool resources in supporting research and education, to fund and co-ordinate technical development, and to provide markets within which to exchange goods on a basis of some equality and fairness. In short, the vision is given heart by a determination to search for unity, peace and economic development. The eventual outline arrangements for global unity may look, for example, to the structures of the European Union; for these have been relatively successful in gathering together diverse countries and in drawing up arrangements that are already tried and tested. Finally, in order to tie the vision and the outline arrangements together in practical operation, technical proposals may include agreements on commodities, prices and international aid.

While there are no ready-made formulae for world unity, it still seems reasonable to predict that the paradigm for global unification might lie with the 19th century unification of Germany and Italy and the 20th century unification of Western and Central Europe; for these territories contained member states which eventually responded not only to political history but to the distance-overcoming technologies of the times, even though they had previously inflicted immense damage on one another in recurring wars. Obviously, global unification confronts problems vastly greater than those of individual European territories or even of the whole of Europe, so great are the problems created by immediately unresolvable tensions arising out of historic bitternesses, competitive production and trading, unequal standards of living, uneven distribution of goods and skills, cultural differences, and complications of communication and understanding. The magnitude of these obstacles emphasises that unification will not easily happen. It is a critical task for us is to make it happen.

The most that can be striven for, and hoped for, in the immediate future are efforts towards unity on regional and equivalent levels – which may gradually merge with global initiatives – and co-operative relations between any groupings that may arise in the process. Meanwhile, one must hope that political leaders, civil administrators and social theorists will recognise those political processes that are benign and advance them. Politics is process – ways of finding immediate answers to problems that are ultimately insoluble as they stand, but that change as more resolvable problems are worked on. Moreover, attitudes towards security, justice and freedom are likely to continue to adapt to changing technological conditions and, in adapting, to begin to shape our contemporary political structures and potentialities. An ethos must surely emerge that seeks to eliminate, in a pre-emptive way, the competition for resources that so often underlies war and which, by acknowledging that war with contemporary technology is no longer a practical option, finds other means of resolving conflict and dealing with differences.

Making peace

While it remains appropriate to have a general concern for the unity of a world order, peace requires special emphasis. There seem to be three sensible means in which future peace may be assured: by policing, by mediation and by sharing.

First, if peace is to be established and sustained in the near future, there needs to be an agreed form of world policing in which states pool their sovereignties and in which the great military powers have a dominant but not exclusive role. In this entire process, the single most strategically placed institution is the United Nations Organisation. With the ending of the Cold War, it has already come more into its own; but, to move towards greater achievement, it will need careful restructuring, not least in the composition, role and powers of the Security Council.[11]

Secondly, there is a need for careful institutional arrangements, efforts at mediation by various powers and groups, and judicious and relatively disinterested applications of influence by the larger powers, if the sources of conflict are to be eradicated, wars are to be prevented or stopped, and the arbitrary and intermittent acts of violence and terrorism that disturb national well-being and international order are to be eliminated. In other words, threats of the use of power in international policing need to be preceded by carefully built mediation efforts and flanked by widely agreed uses of economic sanctions.

Thirdly, beyond policing and the underpinning, sharing, and acceptance of mediation, there needs to be a gradual growth of accepted world structures that organise the new close neighbourhood of peoples with one another. The richer countries need to think ahead in terms of establishing global political structures, creating new material resources, sharing skills and resources with poorer countries, and engaging in political conciliation. Unfortunately governments and peoples are difficult to persuade that problems that are not now acute may become so. Consequently, a prophetic task awaits those who are aware that the time is limited for using technology more judiciously and more frugally, for rectifying imbalances of wealth, and for avoiding the dangers of conflict. The task is urgent.

Conclusion

In the last resort, what is most needed in the modern world is a conviction of community within states and across states through which a sense of belonging, collaborative methods and shared goals are used to deepen security, freedom and justice; and in which global technology assumes the tasks of production for human welfare and care of the environment and turns away from the intent of destruction. In fundamental ways, those concerned for peace, development and the environment are working with the tide of the times. If the recognition of a common humanity linked to balanced communication, mutual knowledge and social trust is accepted, if the search for a vision of the interdependence of peoples is pursued, if policies of fair trading and sensible aid are implemented, and if a commitment to the care of the earth through a disciplined technology is maintained, then the future may well achieve more stable peace, create deeper respect for human rights, and produce greater shared prosperity than history has hitherto known.

Chapter 13

Improving the world's governance

Sir Shridath S. Ramphal

The Author

A lifelong internationalist, Sir Shridath (Sonny) Ramphal has been Guyana's Foreign Minister, Commonwealth Secretary General, a member of several international commissions, Chairman of the West Indies Commission and a Co-Chairman of the Commission on Global Governance. Currently, he is Chief Negotiator for the Caribbean on International Economic Issues. He wrote *Our Country, The Planet* for the Earth Summit.

Chapter 13

Improving the world's governance

Sir Shridath S. Ramphal

The Cold War, in which the relationship between two heavily armed, suspicious super powers was the dominant feature of the world system, ended, after several uneasy decades, in the late 1980s. We have since been living in what is nominally a multipolar world, but in reality with one super power exercising a predominant influence, the other having been much enfeebled. Besides being downsized by the Soviet Union's break-up, Russia has also been affected by the political and economic turbulence it has encountered. Other power centres – the European Union, Germany, Japan, China – have not been able or not wished to offer themselves as new poles challenging the dominance of the United States.

The sole superpower's influence extends well beyond politics. The crumbling of the communist system along with the Soviet Union has been followed by the globalising advance of the capitalist market system, of which the United States is the leading exemplar and champion. The liberalisation of markets and the communications revolution that have been part and parcel of globalisation have helped not only the world-wide advance of American capitalism but also the spread of American media and culture – movies, Madonna, McDonalds and all.

The world is poised to enter the third millennium with the pre-eminence of the United States as a defining characteristic of the time. In the words of one American observer not wholly enthralled by the spectacle of his country as behemoth: 'Probably not since classic Rome or ancient China has a single power so towered over its known rivals in the international system'.[1] American leadership is not a wholly unfamiliar phenomenon. There have been times in the past when it has earned the world's regard by enlightened, internationalist policies and initiatives. Its post-war pressure for imperial withdrawal, the launch of the Marshall Plan, and its role in establishing the United Nations and Bretton Woods institutions were high-points in post-war US leadership.

But the US has since often failed to live up to its own high standards. During the Cold War, its cynical support for a string of abominable despots abroad was in sharp contrast to its espousal of democracy at home. It has often behaved like a bully in several unilateral interventions in its hemispheric backyard. It has stood against efforts, supported by a majority of countries, to extend the global rule of law, as in negotiations on the Law of the Sea Treaty and, more recently, to set up an International Criminal Court. The world's richest nation, it has not only failed to pay its dues to the United Nations, but has sought perversely to exploit its position as a major debtor to impose changes of its choice on the organisation. It has tried to use its muscle with supreme arrogance to beat down trading partners. According to one American commentator, the U.S. has, since 1993, 'imposed new unilateral economic sanctions, or threatened legislation that would allow it to do so, 60 times on 35 countries that represent over 40 percent of the world's population'.[2]

The international system needs the support of all major powers if it is to function well in the service of the world community as a whole. An isolationist United States that shuns global involvement would harm the system, but a U.S. with hegemonic ambitions would harm it even more. This is the situation that the world community has to contend with, as it seeks to shape the architecture of global governance to meet the problems and pressures of a new century.

The institutional arrangements we have for dealing with world problems and issues are largely those that were put in place more than half a century ago, when the end of World War II prompted a spurt of institution building. The principal international institutions were then designed to meet what its victors perceived to be the main issues facing the world community as the war ended. There have been some modifications over the years, but they are far from sufficient to take account of the many changes that have taken place in the world setting, or to meet the many new problems the passage of time has thrown up.

Patterns of conflict

At the time these institutions were established, the predominant concern was collective security. The focus was on the dangers of war between states; the first priority was to put in place arrangements that could prevent a third world war. This concern, logical and understandable given two hugely destructive world wars within 25 years, was reflected in the design of the United Nations and particularly of the Security Council, its most powerful organ. But it would be foolhardy today to proclaim an end to war between countries, although there has been no major conflagration in the past five decades. The wars between India and Pakistan, Britain and Argentina over the Falkland Islands, and between Iraq and Iran, as well as the Gulf War, precipitated by Iraq's attempt to take Kuwait, and the flare-up between Ethiopia and Eritrea are all evidence that war is not an extinct phenomenon. Nations have not fully outgrown the habit of going to war against other nations and several unresolved international disputes remain that could trigger armed conflict around the world.

The world's recent experience has, however, been more of a proliferation of wars *within* countries, including, occasionally, violence of a genocidal nature as in Cambodia, Rwanda and the former Yugoslavia. Over twenty-five internecine conflicts have been going on at any one time in different parts of the world, with some of them – Afghanistan, Angola, Colombia, Kashmir, Sri Lanka – dragging on for many years. While, in earlier periods, war was generally an instrument of annexation, unification or national enlargement, it is now more often instigated by those seeking secession, fragmentation or dismemberment.

Disintegrative tendencies are commonly associated with the assertion of ethnic or religious identities, and Kosovo is a very recent theatre of violence resulting from the failure to reconcile the interests of two communities with different ethno-religious affiliations – Serbs who are Christians and ethnic Albanians who are Moslems. Some have looked to economic progress to provide a solvent for ethnic grievances and to blunt the appeal of separatist nationalism, but there is evidence that prosperity does not necessarily suppress old animosities or ensure ethnic harmony. For example, in Spain the Basque region is among the more prosperous; while, in Italy, the Northern League seems to resent being linked to the poorer south. In India, the state of Punjab, where some Sikhs took to violence in a bid to create an independent Khalistan, benefited prominently from the Green Revolution in farming. It is as if those who have progressed feel they can do even better if they are not bound to others less prosperous or weaker. Whatever the factors behind the rise in internal conflicts in the second half of the twentieth century, those who founded the United Nations and wrote its rules could not have envisaged this shift in global patterns of conflict.

The Nuclear menace

Nor could they have anticipated that so many countries would become capable of making nuclear as well as other equally horrific weapons. The five powers – US, Russia, France, UK and China – who assumed the role of custodians of the post-war global security system as Permanent Members of the UN Security Council and who became the first five nuclear powers, may have hoped to remain an élite with a nuclear arms monopoly. That hope has been shattered, as a few other countries have acquired, or are strongly suspected to have acquired, the capacity to make nuclear weapons. Both India and Pakistan set off a series of explosions in 1998, defiantly announcing their graduation to the status of nuclear powers. The nuclear escalation in the sub-continent provoked specially grave concern, because the two states are suspicious neighbours sharing a long border, have gone to war with each other three times, and have a long-standing quarrel over Kashmir; proximity increases the dangers of conflagration. On the plus side, several countries – Argentina, Brazil, South Africa, Sweden – having started on the road to nuclear capability, have voluntarily given up their nuclear ambitions, and the former Soviet republics of Belarus, Kazakhstan and Ukraine agreed to give up their nuclear weapons when they became independent states.

India refused to sign either the Non-Proliferation Treaty (NPT) or the Test Ban Treaty, arguing that they were discriminatory and that the five nuclear powers were determined to perpetuate their privileged position. It maintained, therefore, in conducting its 1998 tests, that it was not breaking any international treaty undertakings. Indeed, its action served to challenge the sincerity of the nuclear powers' professed commitment to nuclear disarmament which was a foundation principle of the NPT. These powers have flagrantly ignored their responsibilities, and their hypocritical behaviour has been an incitement to proliferation. Adding to the world's capacity for self-destruction can never be a good thing but, in bringing the issue of nuclear proliferation back into public attention and by underlining the fragility of the nuclear equilibrium, the South Asian explosions may actually prod the world community into giving a new impetus to efforts to end the era of nuclear weapons. It may also remove the threat of nuclear catastrophe that continues to hang over the world, though it seemed for a while somewhat less menacing after the end of the Cold War.

A perception that the world's safety would best be served by barring national control of nuclear development of any kind, and by vesting complete control in a new global authority, emerged in the anguished reflection prompted by the devastation of Hiroshima and Nagasaki in 1945. This led to the 1946 initiative known as the Baruch Plan, with which, besides Bernard Baruch, such prominent US figures as

Dean Acheson, David Lilienthal and Robert Oppenheimer were associated. It sought to create an international authority under the UN to own and operate all nuclear facilities world-wide. But movement towards international control was blocked by Moscow's suspicions of Washington, then the sole possessor of nuclear bomb technology and, therefore, likely to have a big role in any international management.

The anxiety then was that the lethal technology would spread among nation states. Fears that nuclear weapons might also pass into the hands of terrorist groups or criminal gangs has emerged more recently, with the spread of nuclear technology and the growth of terrorism and organised crime. The end of the Cold War has not eliminated the danger of a nuclear catastrophe, whether by design, miscalculation or accident. Some of the political changes following the end of superpower confrontation may have arguably even increased the risks of miscalculation or accident, and also of nuclear weapons being acquired by terrorist groups. That some states have nuclear weapons is also a constant invitation to some others, at least, to acquire them. Besides the expectation that nuclear arms will increase national security, countries can also be influenced by the machismo effect. The belief that these weapons enhance a nation's standing in the world, making it a bigger power than it would otherwise be, is perhaps sustained by the coincidence between permanent seats in the Security Council and the status of a nuclear power. Any increase in the number of states possessing nuclear weapons adds to the chances of further proliferation, if not necessarily with the same inevitability as that Pakistan's explosions would follow those of India. The only guarantee against a nuclear catastrophe is the complete elimination of nuclear weapons.

The end of bipolar tensions, if it has not significantly diminished the risks of nuclear holocaust, has created a better climate for progress towards nuclear disarmament, and the world community must not let the opportunity slip. The major responsibility is squarely on the big five nuclear powers, and particularly on the United States and Russia which have by far the largest stocks of weapons. But all states, large and small, must see the elimination of the nuclear threat as indispensable to their future security; they must join in a global endeavour to press Washington and Moscow to maintain the momentum given by the INF and START processes and to urge all nuclear states to move towards the goal of a world free of nuclear weapons.

More countries, new powers

It would indeed have required extraordinary prescience at the end of World War II even to divine that the map of the world would change so much in the decades following the war. Nothing that had happened in the inter-war period would have prepared anyone to expect imperial withdrawal after 1944 to be so rapid and sweeping. It led to a virtual quadrupling in the number of sovereign states – a measure of the hegemony that a handful of European powers had gained over the world between the sixteenth and nineteenth centuries. More recently, this process of proliferation has continued after the end of the Cold War, with the fragmentation of the Soviet Union and Yugoslavia, and the amicable divorce of the Czechs from the Slovaks.

Besides expansion in number, there have been changes in the relative standing of states, driven mainly by differential rates of economic growth. The economic vigour and political respectability of the former Axis powers – Germany, Italy and Japan – and the ascent of such countries as China, India and Brazil has its obverse in the relative decline of once powerful countries that have slipped down the ladder. But the world community has failed to reflect these important changes in its institutional arrangements for managing global affairs.

The global centre of gravity has long been located in the developed North-West segment of the world map, but it has been shifting east and south, and that shift is set to continue, no doubt with hiccups. One estimate – made in 1995 using World Bank figures of gross domestic product at purchasing-power parity – has China leading the table of the top ten world economies in the year 2020, way ahead of the USA and Japan. These are followed by India, Indonesia and South Korea, with Germany, Thailand, France and Brazil coming next.[3] Only four of today's Group of 7 leading industrial countries figure in the list. There are six Asian countries and only two from Europe. This estimate was made well before the 1997-98 economic debacle in Asia; but, while that reversal may slow the shift in the global centre of gravity, it is unlikely to change its direction.

The security of people

The world is set to move into the third millennium confronting a range of problems that did not exist, or existed in a much less acute form, at the time the present machinery of international governance was established, and which it is therefore not well-equipped to handle. To the problem of conflict within states must be added that of countries in which law and order break down and governmental authority disintegrates to the point where the state may be said to have failed. The world must also deal with countries in which state power is usurped and illegitimate regimes oppress their people, habitually and flagrantly violating their rights as citizens and human beings. No longer can the world community stay aloof from what happens in such countries on the ground that the long-hallowed principle of the sovereignty of states denies it the right to intervene in a country's domestic affairs. World public opinion – the opinion of citizens – is no longer always ready to subordinate the rights of people to the rights of the regimes that exercise state power over them, or to accept without demur the world community's acquiescence and inaction, when there are gross violations of human security. But our institutional arrangements, reflecting the earlier orthodoxy on non-intervention in internal affairs, are not geared to take effective action on behalf of victimised or endangered people.

These arrangements, influenced by the dominant perspectives of the mid-twentieth century, do not accord sufficient recognition to the rights of people as against states. But perceptions have changed in the last half century, and states have themselves taken legal cognisance of the rights of people. They have collaborated in creating a body of human rights law that recognises the need – indeed, an obligation – to protect the rights of people. The several international conventions on human rights, the global agencies set up to deal with human rights, the non-governmental bodies that function as human rights watchdogs, and the increase in public concern and activism on human rights issues have all been important advances. But large-scale abuses of human rights still occur: Bosnia, Kosovo, Rwanda and Somalia are deeply shameful chapters in our recent history.

These tragedies underline the need to strengthen further the world's capacity to protect people whose human rights are threatened or violated – particularly oppressed ethnic groups, who are perhaps the most vulnerable – and to deal effectively with transgressors. International action to protect people in danger needs to be subject to rational judgement and to follow broadly accepted criteria; it should not be dependent on the willingness in particular capitals to take part in such action, as such willingness is often influenced by domestic political considerations. The world community needs to put in place arrangements which will give it the capacity to take collective action when there is a clear need for intervention and a democratic consensus favouring intervention.

Refugees and displaced persons

The increase in internal conflict has greatly aggravated the world refugee problem. It is now customary to make a terminological distinction between those forced to leave their country – termed refugees – after 1944 and those forced to leave their homes but live in another part of their own country – displaced persons. The 1997 estimate of the total number of people so uprooted was as high as 50 million, or one person in every 120 of the world's population, of whom 20 million had sought safety outside their country. Women and children make up 80 percent of these people in distress, according to the United Nations High Commissioner for Refugees.

Some refugees are part of a massive exodus, as from the genocide in Rwanda in 1994, when more than two million people fled to neighbouring Tanzania and Zaire, or from Afghanistan in the 1980s, where the fighting drove an even larger number, mainly to Iran and Pakistan. In Europe, the Bosnian tragedy created 800,000 refugees. Internal displacement can also involve large numbers. Of the 1.85 million people forced to flee by the civil war in Sierra Leone, 1.6 million were displaced within the country; 1.2 million people were displaced and another 500,000 made refugees by the strife in Liberia. The past few decades have been punctuated by a series of man-made emergencies causing immense suffering and dislocation. These are a challenge not just to our humanitarian instincts but to global governance – to the world's capacity, not merely to cope with their consequences, but also to anticipate, prevent and control them.

International intervention

The proliferation of crises arising from situations internal to countries – armed conflicts, ethnic cleansing, repressive violence – has been imposing new demands on the United Nations Security Council, which was intended mainly to deal with conflict between states. As previously argued, it would be unwise to expect no interstate wars in the new millennium; but it is also prudent to expect that intrastate situations will continue to demand international attention. Enabling the UN better to respond to these demands is therefore a priority task.

One need is to revise the UN Charter so as to make it expressly legitimate for the Security Council to authorise intervention within a state in defined circumstances. The Charter now envisages intervention only when there is a threat to international peace and security. The Security Council has had to devise ways to get round the Charter's proscription of intervention in internal affairs, whenever it wishes to check large-scale violations of human security within states that do not necessarily pose an international threat. The global rule of law would be better served if the UN's rules were suitably rewritten to acknowledge the new reality of a world community – a community which now accepts the need to override the principle of national sovereignty in circumstances that require the interests of people to be placed above those of states and regimes.

Any change that regularises and enhances the Security Council's power to intervene should be accompanied by action to define as precisely as possible the situations in which the new powers may be exercised. While it should be within its power for the Security Council to authorise intervention, that power to intervene should be fenced in. It is important that any decisions to intervene should be taken according to principled criteria, so as to ensure consistency and minimise the risk of action being determined by the interests of particular powers. One very important justification for strengthening the UN's powers is the need to guard against unilateral action by global or regional powers, whose motives may or may not be above suspicion. If there has to be intervention in the domestic affairs of any country, it should be by the world community or authorised by it, and conducted on its behalf and under its control. Any other action, of which there have regrettably been instances in the not distant past, violates the global rule of law. In the new millennium, we should not countenance a return to the rule of the sheriff's posse on the global stage.

The UN's recent experience has also demonstrated that its enforcement capacity needs to be strengthened if it is to be more effective in preventing or curbing violence. Its ability to respond to a situation requiring troops to be deployed now depends entirely on the readiness of an assortment of governments to provide them, and its success in co-ordinating contributions from several sources and making the necessary complex logistical and other arrangements. A government's position can often be influenced by domestic political considerations. Even when there is willingness, valuable time can be lost. Large-scale military operations will continue to need contributions of troops from governments, but there are many other situations in which modest deployments very promptly made could be effective in preventing a disaster. A deeply tragic example of the cost of inaction was provided by Rwanda in 1994, when a force of 5,000 men could have saved half a million lives. This was the view of the Canadian General in charge of the UN assistance mission there – a view later supported by independent expert opinion, when an international panel of military leaders, convened by the Carnegie Commission on Preventing Deadly Conflict, examined the Rwandan experience and assessed the validity of the general's statement.[4]

It has long seemed clear that the UN could be more effective – by being able to act more promptly when the Security Council decides that action is appropriate – if it had at its disposal a modest body of well-trained troops ready for quick deployment. The knowledge that the UN would be able to move quickly to enforce Security Council decisions, which the availability of a UN rapid reaction force would signify, could itself be a deterrent against violence in some circumstances. To have such a force is by no means to give more power to the UN secretariat or to free it of governmental restraints. All enforcement decisions would continue to be taken by member states through the Security Council. It is the Council that would benefit by having its capacity to enforce its decisions improved.

Reforming the Security Council

Giving its interventionary role clearer definition and enabling it to act more promptly would help to make the Security Council more effective and more powerful. This would render all the more compelling the case to make it more representative of the United Nations of today. Reform of the Security Council has been on the global agenda for several years – certainly since 1995, when the United Nations marked its 50th anniversary. This would have been an appropriate occasion to make the Council more contemporary in character. Reform is patently overdue.

Only once since the Council was established in 1945 has its structure been changed. This happened in 1963, when decolonisation, far from complete, had lifted UN membership from 51 at its founding to 113. The Council was enlarged from eleven to fifteen members by raising the number of non-permanent members from six to ten. UN membership has since continued to increase, with further imperial dismantling and the post-Cold War dissolution of the Soviet Union. It rose to 185 in December 1994, nearly 40 percent up on the 1963 figure.

A mere enlargement of the Security Council, as in 1963, by adding a few more non-permanent seats cannot now be the answer. Power in the Security Council resides in its inner coterie – the five permanent members: Britain, China, France, Russia and the United States. That they should have written themselves into a privileged position when the UN was formed is understandable, but it is indefensible that this structure should be frozen and that five countries should form an élite group in perpetuity. These countries now seem to accept that it is time for a change, but the only change they seem ready to countenance is to take a few new countries into permanent membership, while holding on to their own seats; and, at the same time, upholding the principle of permanent membership and its particular attribute, the right to veto.

The principle of permanent privileged positions for some states is clearly unsatisfactory in a dynamic world in which changes in the relative positions of countries are the norm. To enlarge the ranks of permanent, veto-armed members would be a retrograde step. Having two classes of members in the Security Council is itself anomalous from a democratic perspective, the equality of member states being a founding principle of the United Nations. But it may be an anomaly that has to be accepted because these member states, though equal in sovereignty, are also so diverse, comprising countries as large as China (population 1,208 million) and as small as Palau (population 16,000), as rich as the US (GDP US$6,650 billion) and as poor as Guinea Bissau (GDP US$0.2 billion). While the United Nations consists of states as various as these extremes, the Security Council may need to have a class of members drawn from those states that are better able, by virtue of their financial, military and other resources, to contribute to 'the maintenance of international peace and security'. There is no justification, however, for such members to hold their seats in perpetuity. They should be elected for specific terms, say, of ten years, and be eligible for re-election. This would allow the Council to reflect the rise and fall of nations – and changes in their capacity to contribute to peace and security – which is a feature of the real world.

The reform of the Security Council should also address the issue of the veto. Each permanent member now has the capacity to block a decision supported by a majority or even all of the other members. Accepting this highly undemocratic provision may have been a price worth paying to ensure that the Soviet Union came on board when the UN was founded. Enlarging the upper level of membership without scrapping the veto would make matters worse than they are now, by adding to the number that could, individually, frustrate the will of the others. Now that the Cold War is over, the use of the veto has become rare and it is surely time for a more enlightened approach to the issue. It may be too naive to expect the countries with veto rights to relinquish them immediately, but persuading them to restrict the use of the veto to particular circumstances seems a realistic short-term aim. The Commission on Global Governance has urged permanent members to agree to use it only in exceptional circumstances related to their national security; and the Independent Working Group on the Future of the United Nations has suggested restricting the use of the veto to decisions on peacekeeping and enforcement, i.e. action involving the use of military personnel. In the longer term, however, the aim must be to ensure that no one member can exercise a veto and frustrate the will of the others.

In the informal discussions that have been going on for some time, Germany and Japan have emerged as front-runners among aspirants to permanent member-ship, with wide support also for the elevation of three developing countries – one each from the Asian, African and Latin America/Caribbean regions. Leaving aside the question of permanence, discussed earlier, it is to be noted that, if the present permanent five remain, bringing in Germany would raise Europe's representation to four countries, three of them members of one regional group, the European Union. Asia would have three seats; given its population numbers, this would be more equitable than the treatment of Europe. The balance between developed and devel-oping countries would also be improved if three developing countries were to be added.

Conflict and arms

The issues of armed conflict and large-scale abuse of human rights criss-cross those of militarisation, military spending and the arms trade. The nineties saw further reversals for men in uniform who had grabbed power, as more countries returned to democratic systems; global spending on arms also declined. But military expen-diture continues to absorb funds that could be used for productive and socially bene-ficial investments, and so to stunt development and deprive people of health and educational services in many countries. Meanwhile, the international arms trade, whose chief suppliers are the great powers with permanent seats in the Security Council, continues to make profits by the sale of lethal weapons, some of which find their way, directly or through clandestine channels, to zones of conflict.

It is small arms – portable weapons such as rifles, pistols, hand grenades and light mortars – that provide the firepower in today's intrastate conflicts. As the British military historian John Keegan has pointed out,[5] most of those who have died in war in the past half century have been killed by 'cheap, mass-produced weapons and small-calibre ammunition, costing little more than the transistor radios and dry-cell batteries which have flooded the world in the same period'. US shipments of small arms which averaged US$2-3 billion during the Cold War years climbed to more than US$25 billion in 1996, according to a *New York Times* report.[6] Unlike such heavy items as fighter planes and tanks in which transactions invari-ably involve governments, small arms are handled by the private trade, with few controls and little monitoring. It has become increasingly doubtful whether it will be possible to curb illegal trading in weapons without greater oversight of the legal trade. But arms trading has throughout been cloaked in secrecy, and governments have preferred to have it that way.

Terrorist groups now form part of the regular market for small arms.

International terrorism is another phenomenon that has confronted the world community only recently, and global governance has only just begun to take it on board. Its destructive reach is demonstrated by incidents as varied as the assassination of Rajiv Gandhi by a Tamil Tiger bomber from Sri Lanka in 1991, the bombing of the World Trade Centre in New York in 1993, the slaughter of American servicemen in Dhahran in 1995, and the bombing of US Embassies in Nairobi and Dar es Salaam in 1998 causing much loss of life, both of Africans and Americans. Connections between terrorist groups, gun running, criminal gangs, drug trafficking and money laundering make the task of combating each of them more formidable.

The war on drugs

The highly organised trade which feeds the market in narcotics is also a problem that has lately assumed worryingly international dimensions. Drug addiction, which blights the lives of increasing millions of people and which is a particularly vicious threat to young people, is a matter for national action by health, educational and law enforcement authorities. But even their most strenuous efforts have limited chances of success without robust action to break the supply chain; this necessarily means international action – active collaboration among consuming, producing and transit countries.

As *The Economist* has pointed out, the drug industry is now 'impeccably international' and 'drugs are as much part of international business as cars and hamburgers'. Its earnings are now counted in billions of dollars. The UN reckoned the world-wide drug business to be worth US$400 billion a year in 1997. This put it well ahead of the trade in medical drugs (around US$300 billion) and catching up on tourism (about US$450 billion).[7] The UN's estimate has been questioned, but even if it overstates the size of the trade, there can be no doubt that drugs are a very big global business. Monitoring financial movements has therefore become an important part of efforts to beat the drug barons.

This trade has focused attention, in turn, on the role of 'shell' companies and offshore banking facilities in hiding and laundering drug money. It has become far easier in recent years to buy ready-made companies, to open secret bank accounts, and to move funds round the world. One small group of islands was recently reported to have a bank for every 57 of its citizens. The number of countries offering easy facilities for setting up companies or tax-free banking has increased. Establishing companies and using tax-havens are legitimate activities, but where there are lax supervisory regimes, drug merchants exploit them. Employing clutches of 'respectable' lawyers and accountants, they use these facilities to wash money by passing it through various phoney companies and a succession of secret bank accounts.

Recent years have seen intensified efforts to combat the spread of drugs. A new UN strategy aimed at ending coca and opium cultivation was signalled by a drug summit in New York in June 1998. There have been several successes in tracking down large shipments, apprehending couriers and clipping the wings of drug cartels. But the war against drugs is not being won, and the global narco-economy is becoming even bigger. Human lives, families and communities are being wrecked, officials suborned and democratic governance threatened. The global reach and virulence of the drug menace presents a very formidable problem as the world moves into the third millennium. The primary effort must continue to be to curb demand for drugs, prevent the drug habit and tackle addiction. But such work, in each country with an addiction problem, must be complemented by more effective international action against the supply chain. International co-operation should become easier; for the distinction between consuming and producing countries is becoming less valid. Afghanistan and Pakistan, hitherto best known as sources of drugs, are now themselves grappling with addiction; they are said to have more heroin addicts than the United States.[8]

Poverty

Not all the challenges that face us as we move into the new millennium are of recent origin. Poverty certainly is not. That there should be so much poverty, such widespread deprivation, in the midst of great affluence is an indictment of humanity's performance during the twentieth century when, for the first time in history, it had the capacity to end acute poverty. The century has seen science and technology take breathtaking strides. Entrepreneurship and corporate ingenuity have made parallel advances. Space travel, nuclear energy, robotics, genetic engineering are only some of the century's achievements. Many people have enjoyed the benefits of supersonic aircraft, 200 m.p.h. trains, satellite telephones, global television, the Internet and 24-hour banking. But, in the same world and at the same time, there are many millions who have never ridden in a motor vehicle, used a telephone, switched on an electric light, used a cheque book or had pipe-borne water; millions wholly unfamiliar with such articles of everyday amenity elsewhere as refrigerators and television sets; millions who cannot read or write; millions of children who do not go to school and millions of infants who do not get enough food to survive.

A more vigorous effort to eliminate extreme deprivation should surely be a high priority for the new millennium. Economic progress in recent decades has lifted large numbers out of acute poverty, and success has been particularly notable in a populous country like China. But, globally, there has been no decline in those classified as 'absolute poor' – people whose income does not rise above the dollar a day threshold. These are reckoned to amount to 1.3 billion people, or one in every three persons in the developing world. Nearly three-fourths of these people live in Asia, but the highest proportion in relation to total population is in Africa. UNDP reported, in 1997, that the total figure appeared to be rising with growth in population, except in South East Asia and the Pacific; but this was written before the economic typhoon that battered South East Asia, impoverishing many millions.[9]

Although income is the handiest tape we have to measure poverty, other indicators can fill out the picture. Over 800 million people do not get enough to eat; more than 500 million are chronically malnourished. Over one billion do not have safe water to drink – a daily trek to fetch a pot of water is the lot of many women and girls. More than 800 million adults, nearly two-thirds women, are illiterate. Over 100 million children are not in school. Around 500 million are likely to die before they reach forty.[10]

Against the comfort enjoyed by the majority of the world's people – and the luxury enjoyed by many – the deprivation these figures document must surely be considered unacceptable. The hardship they reveal would certainly be considered too extreme – and the disparity between the richest and poorest too inequitable – to be tolerated within any industrially advanced society. These societies, irrespective of the political ideology that has been dominant in shaping them, have erected bulwarks against such hardship, assuring citizens sufficient income to meet their basic needs and access to such services as health and schools. They have accepted that the provision of this level of security to all citizens should be a common responsibility, where necessary financed from the public purse, i.e. by taxpayers. There may be debates about the precise level of provision and the criteria for eligibility for certain entitlements and services; there may also be concern about abuses; but the principle of a publicly guaranteed minimum level of social security and well-being is widely accepted across the political spectrum. It is regarded as a benchmark of a civilised society.

But the application of this principle has so far stopped at national borders. Nations that have responded to the domestic claims of social justice and equity by making statutory provision to protect the underprivileged have failed to acknowledge the transnational validity of such claims. They have treated appeals for social justice from outside their borders as calls to be met only from the charity box they have labelled 'aid'. And most countries have failed to lift their aid towards the level of 0.7 percent of their GDP – the internationally agreed target level set several decades ago. That this was a reasonable, attainable target and not a pie-in-the-sky fantasy is confirmed by the fact that some countries – the Scandinavians and the Dutch – have consistently surpassed it for many years. But others have been slipping back even from their much more meagre levels. And total aid contributions have declined for several years, at the same time as rich countries, while intoning the charms of the globalised market, have become both richer and meaner. Aid from the G7 countries fell in 1997 to just 0.19 of their collective GDP. 'America is the global Scrooge, with an aid budget of just 0.08 per cent of its GDP', as *The Economist* pointed out in June 1998.[11] Rich nations have appeared to use globalisation as an excuse for their withdrawal from active engagement in the battle against poverty.

An especially chilling aspect of rich countries' disengagement from the predicament of the abjectly poor is their prolonged foot-dragging over the debt burden of the poorest developing countries. There has been an inordinate delay in providing to these countries at the margin of the world economy any meaningful relief from the harshly debilitating effects of unpayable debts. After years of deliberation and ministerial confabulations in a string of capitals, the Heavily Indebted Poor Countries (HIPC) initiative was announced in 1996. But it will be at least some years into the new millennium before most of the eligible countries qualify for any assistance. The tardiness of rich country action on poor country debts contrasts sharply with the speed with which bail-out packages involving far vaster sums are put together for more economically significant developing or transition countries whose collapse would threaten western banks.

Globalisation

Globalisation is a reality, and virtually every country has accepted it. It has had a dynamic, integrative impact, knitting economies together and creating opportunities for trade expansion and economic growth. Greater interlocking into global systems of trade, production and finance has helped several developing countries make remarkable economic advances. But it has become clear that globalisation has a downside too. A UN report has observed: 'The greatest benefits of globalisation have been garnered by a fortunate few. A rising tide of wealth is supposed to lift all boats, but some are more seaworthy than others. The yachts and ocean liners are rising in response to new opportunities, but many rafts and rowboats are taking on water – and some are sinking'.[12]

The chief dynamic impulse of globalisation is believed to lie in stimulating trade. World trade has been rising faster than world production, so that countries sharing in this growth have seen their trade – exports and imports – become a larger proportion of their gross domestic product. But, for 44 developing countries with more than a billion people, the UNDP reported in 1997 that this ratio had been falling over the previous decade.[13] So globalisation had bypassed these people.

Another key advantage claimed for globalisation, and the liberalisation of financial markets which is one of its key components, is that it increases the availability of foreign capital. Financial flows, of which investment funds are a part, rose even faster than world trade up to 1996, with foreign direct investment showing a marked rise. But investment flows have been even more selective than trade. Nearly half the developing countries did not benefit, and just eight received two-thirds of all the direct investment going to developing countries.

The devastation of several East Asian 'tiger economies' in 1997, three years after the Mexican crash, has reinforced fears that globalisation has made the world economy more unstable. The countries that were laid low had been held up as models for the new age of global free markets. That even such star performers could be battered so severely suggested that vulnerability had increased significantly with globalisation. Foreign money could leave as quickly as it came in – and leave currencies and countries in deep trouble. Professor Jeffrey Sachs, who heads the Harvard Institute for International Development and has been a consultant to governments on economic modernisation, has written: 'The real meaning of the Mexican crash and the East Asian financial crisis is still far from clear, but both experiences have shown that unfettered financial flows from advanced to emerging markets can create profound destabilisation'.[14] And George Soros, no stranger to currency trading, was reported as saying: 'Instead of acting like a pendulum, financial markets have recently acted like a wrecking ball, knocking over one economy after another. There is an urgent need to rethink and reform the global capitalist system'.[15]

Rising inequality

An equally, if not more, worrying aspect of globalisation is the growing inequality experienced both globally and in individual countries. UN figures show that the share of global income received by the poorest 20 percent of the world's people fell from 2.3 percent in 1960 to 1.4 percent in 1991; it then fell sharply to 1.1 percent in 1994, suggesting that the decline had accelerated as globalisation advanced. The ratio of the income of the world's richest 20 percent to that of the poorest 20 percent, which had risen from 30 to 1 in 1960 to 61 to 1 in 1990, correspondingly jumped to 78 to 1 in 1994.[16] The gap had widened dramatically.

Developed countries have not been immune from this process of widening inequality, as evidenced by chronic high unemployment and a sizeable underclass in many of them. The process has been going on for some time but been quickened by globalisation. In the United States, the income of the poorest fifth of the population is reckoned to have been falling since the early 1970s while that of the richest fifth kept increasing, with the income of the richest hundredth rising by more than 100 percent.[17]

Economic governance

Issues of this magnitude and wide significance – persistent poverty, economic instability, increasing inequality – require truly global consideration, but there has been a lacuna in the structures of global economic governance. There is no democratically legitimate body at the highest level to address these issues on behalf of all the world's nations. There are, of course, a number of organisations engaged in aspects of economic oversight: the UN's Economic and Social Council (ECOSOC); intergovernmental agencies such as the International Monetary Fund, the World Bank and the World Trade Organisation; and other bodies such as the Bank for International Settlements and OECD. All these have their distinctive roles and, of them, ECOSOC has the broadest remit, but it has been drained of real authority and functions only at the margins. With ECOSOC weakened, the role of an apex body for considering major economic issues has been appropriated by the Group of 7 (G7).

This is not a healthy state of affairs, as the G7 is only a self-constituted body accountable to its members alone – 'a caucus of like-minded advanced democracies', in the words of one apologist.[18] There can be no objection to the G7 as such; for any set of countries may function as a group. What is wrong is that such a group should set itself up as a kind of directorate for the world and for the world economy in particular. It may not claim that title, but it certainly gives that impression – and it has been so described.

The G7 – Britain, Canada, France, Germany, Italy, Japan and the United States – cannot any longer even claim to comprise the seven leading industrial countries in the world. It is clearly not representative of the world in terms of either geography or economics. It represents a small proportion of the world population (12 percent) and this proportion is falling. Heavily overweighted in favour of Western Europe and North America, it ignores the emergence of new economic powers such as China and Brazil.

The G7 has, in its political incarnation, metamorphosed into the G8 by including Russia; but it is in its key *economic* role that its unrepresentative character matters most. In the political domain, the G7 is of less consequence, as crucial issues of peace and security are matters for the UN Security Council. As discussed elsewhere in this chapter, the reform of the Council to make it more representative of the United Nations as a whole, has been long overdue but, even in its present form, it has a greater claim to legitimacy than a privately constituted body such as the G7. In economic matters, however, the G7 has the field to itself, the UN's economic forum, ECOSOC, having been emasculated and marginalised.

In April 1998, the G7 transiently transformed itself into a G22, when Finance Ministers from 15 developing and transition economies responded to an invitation to join G7 ministers at a special session. Its agenda was confined to the Asian financial crisis and what action could be taken to strengthen the world financial system to avoid such crises. The meeting, carefully publicised as a one-time event, was not at the level of Prime Ministers; nor were the invited ministers asked to the regular discussions of their G7 colleagues.

The G7's action in co-opting some developing countries on an *ad hoc* basis in this manner is an acknowledgement of the limitations inherent in its narrow membership. But it is not only in considering a new architecture for the global financial system that these limitations are a handicap. Other economic issues of global importance also need to be addressed by a more inclusive body that represents not just one set of countries. The right approach to filling the gap that now exists is not, however, to invite the G7 to enlarge itself. If it wishes to open its doors in a temporary or permanent way, that is its business. But the resulting entity would not have legitimacy as a representative organ of the world community. What is necessary in the interests of the world as a whole is that a new body should be formed to represent the world community which draws legitimacy both from the way it is set up and from its composition.

An Economic Security Council

The Commission on Global Governance came to the conclusion that the needs of global economic governance would be best served by an Economic Security Council within the United Nations.[19] The Commission envisaged that, while it would respond to crises such as the Asian financial crisis, the Council's primary function would be continuously to assess the state of the world economy and evolve a long-term strategic policy framework; at the same time it would promote consensus, give leadership on international economic issues and secure consistency between the policy goals of the Bretton Woods institutions. While it would have the same standing as the Security Council, it would work by consensus and have only a deliberative role.

Within the Commission, thinking on this issue was strongly influenced by such members as Jacques Delors who, as President of the European Commission, had a seat at the G7 table, and Barber Conable, the former President of the World Bank, who has also been a Republican Congressman in the United States.

Calls for a new apex body for economic governance have also come from other quarters, notably the Independent Working Group on the Future of the United Nations and UNDP's Human Development Report. An important critic of the G7's inadequacies has been Peter Sutherland, the first head of the World Trade Organisation and a former European Union Commissioner, who has more recently been head of British Petroleum and Goldman Sachs International; he has repeatedly drawn attention to what he sees as 'a structural deficit' in the world's governance.

Consumption and resources

An Economic Security Council would be the appropriate kind of body to consider strategies to tackle such global issues as poverty, rising inequality and the volatility in financial movements of exchange rates associated with globalisation. It would also help the world address the issue of the balance between resources and consumption which is likely to emerge as a critical issue early in the new millennium. Much attention has been given to population pressure, but it is not population *per se* that is likely to be the most worrying factor but the pressure of people on resources. It has been fashionable to deprecate the pressure people in poor countries exert, for reasons of survival, on the land and to bemoan their clearing of forests to plant food crops, but there is little censure of the pressure that rich people, in both rich and poor countries, place on many resources through their high-consumption lifestyles.

Rich people consume much more per head of all manner of resources: cereals, meat, fish, land, water, energy, paper, plastics, timber, fabrics, metals, construction materials, rubber. What they eat, the houses they live in, the way they get about, the appliances they use, the clothes they wear, the holidays they take, the games they play – these all account for high levels of resource use. The American humorist, Art Buchwald, wrote that 'the biggest crisis the world faces in the next century is the shortage of land for golf courses'.[20] There may be more than a golf ball of sense in this light-hearted comment! Golf courses demand not just land but another resource in short supply, water; and they are perhaps a suitable metaphor for resource-intensive lifestyles that are increasingly at odds with environmental prudence.

The pressure a rich person exerts on resources is, on average, several times that exerted by a poor person. Poor people will certainly consume more as they become less poor, but it is wholly inconceivable that they would consume as much as rich people do today. This would entail a level of total consumption for which the resources would either just not be available, or which, in the case of greenhouse products such as petroleum, would have grievous consequences. As the living standards of today's poor improve and their consumption levels go up, there will come a stage when total world consumption threatens to exceed the resources available. These will, of course, expand as demand rises; but one does not have to be a doomsayer to doubt whether this process can go on indefinitely.

Arrangements to conserve resources and to bring about equitable access to them, which would involve some restraints on high consumption, will therefore become necessary sooner rather than later in the new millennium. This will be a complex task, and the strategies to accomplish it will need to be addressed at a high level before the problems become acute. An Economic Security Council commanding universal acceptance through its representative character, would be the most appropriate type of body to steer the world towards a rational accommodation between contending interests to the benefit of humanity as a whole.

Shared values

This is a century that has elevated 'management' – always an attribute of human survival – to a science and made it an icon of progress. Ironically, the judgement of history could be that in the 21st century humanity failed to manage the global society which its genius had created. The global neighbourhood is a reality but it has not yet come under the sway of neighbourhood values or of governance informed by them. Global governance needs to be supported by world-wide commitment to a set of shared values – a global neighbourhood ethic – to which the followers of all creeds – or none – can comfortably subscribe. The Commission on Global Governance suggested that people of all cultural, political, religious or philosophical backgrounds could unite in upholding the core values of respect for life, liberty, justice and equity, mutual respect, caring and integrity. As it argued, these could 'provide a foundation for transforming a global neighbourhood based on economic exchange and improved communications into a universal moral community in which people are bound together by more than proximity, interest, or identity. They all derive in one way or another from the principle, which is in accord with religious teachings around the world, that people should treat others as they would themselves wish to be treated'.[21] In the 21st century *Homo sapiens* may not live up to the evolutionary perfection we attribute to ourselves unless we become *Homo sentiens* as well.

Chapter 14

The dissolution of law?

N.E. Simmonds

Author

N.E. Simmonds teaches law and jurisprudence at Cambridge University and is the author of numerous books and articles on the philosophy of law. His most recent publication was *A Debate over Rights* (Oxford University Press 1998), written jointly with M. Kramer and H.Steiner.

Chapter 14

The dissolution of law?

N.E. Simmonds

Past and future in law

How can one predict the future of law? Within the tradition of thought called 'legal positivism', law is a product of authority and will; to seek to predict its future is therefore to seek to predict the variabilities of human *choice*. For another tradition (that of 'natural law') law is grounded in certain permanent requirements of reason, reflective of human nature and unalterable circumstance: here law's future is predictable only in the paradoxical sense that law lacks a temporal dimension. Many legal theorists, rejecting extreme versions of both alternatives, have sought to portray law as delicately poised between *voluntas* and *ratio*; but such middle positions, though no doubt nearer to the truth, do not help to answer our question: law remains arbitrary in so far as it has a history, and without a history in so far as it is grounded in reason.

Poised between reason and will, law exists also in the tension between past and future. More than any other aspect of human society, law forms the intersection between that which was done in the past and that which we prescribe for the future. To govern by law is to shape future action by reference to decisions made in the past, and to address new problems by deploying the resources of established juridical traditions and ideas. Yet, if law extends the governance of history into our future, it also renders that history a matter of contested and negotiable interpretation, existing in an uneasy relationship with the demands of the present. In no legal tradition is this more evident than within the common law, where judges claim to be bound by the very law that their decisions continuously re-shape; but in less evident ways it is true of every developed system of legal thought. Judges and legal scholars are bound to follow the law and interpret it faithfully; yet their interpretations are a source of change and adaptation, giving to the law an appearance of organic development rather than simple unchanging uniformity. Our efforts to understand what the law is are of one piece with our efforts to grasp the meaning of its history; and, since the law's nature seeks to determine our conduct prospectively, its history must be central to any prediction of our future.

Is not the law's ambivalent position between reason and will of one piece with its location between past and future? Law can embody and enforce political choices, and those choices are acts of authority; but such choices must be fitted into a complex system of legal principles and a long intellectual inheritance of legal ideas. At the same time, the steady incremental addition of new laws (new choices) forces upon the system of legal ideas a process of development and assimilation. The fact that law is an intellectual tradition, and not simply a collection of governmental decrees, shapes and constrains the laws that result from the political will. The tradition itself is not one of detached reflection upon the requirements of reason, but of interpretation and assimilation of the products of authority. Yet 'interpretation' and 'assimilation' here refer to an intellectual process which is, in part, one of the construction of reasons and the discovery of forms of moral coherence that may have been opaque to the authorities who enacted the law.

Law exists and has its being, therefore, within a field of complex intellectual tensions. Perhaps it is this very location that can give us a key to law's future, rendering intelligible much that is currently happening in law and that might otherwise seem merely bewildering. If law were simply the passive register of the political will, its future would be unpredictable and its nature would not merit independent study: law merits a place in our reflections precisely to the extent that this is not so. By focusing upon matters internal to the structure of legal ideas we can discover certain unresolved conflicts that struggle for resolution. It is this struggle that gives to law a dynamic historical trajectory of its own.

Being concerned chiefly with matters internal to the structure of legal ideas, this essay may well strike the layman as adopting a surprisingly arcane perspective; explaining its concerns by means of a historical narrative, it may well seem at times remote from the business of contemplating the future. It is hoped, however, that this perspective will help to render intelligible many current tendencies in law that are both obvious and troubling, while demonstrating (something that would not be obvious to most lawyers) that they are the manifestation of quite deep shifts in the very nature of legal thought.

The problem of law's nature

It might surprise many people to discover that there is a complex body of literature addressing the question 'what is law?'. One strand of this literature is concerned with the fact that law purports to impose *obligations* upon us (and to confer *rights*), yet also backs its demands with a threat of force. So should law be thought of as a body of orders backed by threats? Or as a body of prescriptions grounded in moral considerations of justice and the common good? Debates of this sort go back to ancient Greece, and pose a host of fundamental philosophical questions.

More recently, however, another set of questions has emerged as part of the general inquiry into law's nature. These questions raise their own philosophical difficulties, but they are also unusually indicative of some of the special problems faced by a modern legal order. For example, questions are raised about the legitimacy of legal doctrinal reasoning and of adjudication. When judges justify decisions they do not generally rest their judgment upon a direct appeal to moral values: they appeal to law, it being assumed that the law represents a set of objective standards that are independent of variable moral judgements. Yet, in the contentious cases likely to reach court on a point of law, there will probably be experienced lawyers on both sides who believe that the law supports their case. If the law is a set of *objective* standards, how is it possible for lawyers to disagree about its content in this way? Is legal argument simply a charade which employs technical mumbo-jumbo to conceal an arbitrary exercise of will by the judge, or a surreptitious reliance upon his or her personal morality?

These questions become particularly acute within common law systems where much of the law is made by the judges in the form of precedent, and where the law is continuously altered in the course of its application. For how can judges be *bound* by rules that they apparently have the power to *alter*? If I can alter a rule whenever I think that that is best, I cannot be bound by the rule, for the notion of being bound involves having to comply with the rule whether I think compliance desirable or not.

What matters for present purposes is not the substance of these questions but the fact that they have gained increasing prominence within modern society. This prominence is doubtless a function of the way in which modernity has rendered law *central* to its political thinking, whilst simultaneously rendering the very possibility of law *problematic*.

Ancient and medieval communities regarded law as an important element in the complex of human institutions that make possible a valuable and worthwhile life. The fundamental question for politics was one concerning the nature of a good life: having determined the answer to that question, one might set about designing or establishing institutions that would inculcate and encourage good lives. It should not be assumed that this perspective necessarily required the state to assume an oppressive or intrusive character. Human flourishing might well be thought to require some degree of diversity and some consequent scope for free choice; that something should be freely chosen might even be considered to be an essential precondition of a good life. The fundamental questions, however, were always questions of excellence. When questions of (say) 'freedom' or 'equality' arose they derived their significance and content solely from the deeper concern with excellence or 'the good'.

With the bloody sectarianism that followed the Reformation it became increasingly obvious that communities could be united around a shared conception of the good life only by extraordinary good fortune, or by unflinching repression. Political settlements were achieved that relied to a greater or lesser extent upon tolerance. Forms and procedures of governance evolved that did not rely for their legitimacy upon any claim to represent the 'good' for man: rather, the claim was that they made it possible for citizens to pursue diverse notions of the good, without undue mutual interference. Tolerance began as a pragmatically acceptable *modus vivendi* rather than as a significant moral ideal, but attempts to bolster the political settlement with an appropriate moral vision were not long in appearing. Thus it is that, in more recent times, the idea of tolerance as *modus vivendi* has metamorphosed into the idea of 'autonomy' as itself the central value of life. What matters now is not so much what one chooses, but that one chooses; and the dependence of all choice upon a background of taken-for-granted values (that must themselves be taught rather than chosen) is ignored.[1]

Within ancient and medieval communities, law occupied an important place as one facet of the arrangements of governance that were to foster good lives. Within modern communities founded upon the value of autonomy, however, law occupies *the central* place. For an orderly community of agents choosing and pursuing diverse goals can be made possible only if one can demarcate clear bounds within which choices may be made and options pursued. Societies that emphasise the importance of choice and diversity are unlikely to generate a high degree of moral consensus (or if they do possess such a consensus, they will be unwilling to acknowledge its existence, fearing that it denotes some failure of autonomy amongst the citizens). Clear bounds between conflicting domains of liberty can therefore be demarcated only by some device that is independent of varying moral judgement: only, that is, by law.

Yet what *is* law? The modern political imagination seems to postulate a community of great moral diversity held together by 'the rule of law'. Presumably, therefore, the content of the law must be identifiable independently of any individual's moral judgement. The law must be, as it were, a vast assemblage of written rules inscribed in some book or collection of books. We can discover the content of the rules simply by consulting the written text. In this way we can have shared rules even in a world of moral diversity.

A number of problems have haunted attempts to apply this 'rule-book' conception of law to the complex realities of legal traditions and bodies of doctrinal thought. Once again, the historical character of the law is a source of deep problems, for traditional features of legal thinking that seem to be incompatible with the rule-book conception do not conveniently disappear when the modern vision of political life pronounces them to be unwelcome or embarrassing. Suppose, however, that the 'rule-book' conception were correct. Would it really solve the problem that it purports to address? Even given shared *texts* there is no guarantee that we would have shared *rules*. For, approaching the texts from a great diversity of different perspectives, and on the basis of very different moral assumptions, would we not reach quite diverse conclusions about the *meaning* of the texts? Shared texts will yield convergent legal conclusions only when they are interpreted against the background of a shared inheritance of moral and cultural assumptions, as well as more specific traditions of legal interpretation and systematisation.

Early modern European legal systems secured a degree of convergence in the informed opinions concerning the content of law by reliance upon the tightness of an inherited tradition of legal thought. This tradition was centred upon the universities' teaching of Roman law in the case of continental Europe (and Scotland), and upon the Inns of Court and the common law in England. The virtue of a firmly rooted tradition of thought is the fact that it is invisible to its inheritors, and as seemingly natural as the air that they breathe. Thus, the business of legal reasoning could appear to be the unproblematic application of determinate rules possessed of transparent and uncontentious meanings.

However, the fluid and individualistic society that had created an intense *need* for the clear demarcation of legal rights was to undermine the stable inherited traditions that made such clear demarcations seem possible. The growth in the scale and complexity of legal systems made the assimilation of new law within old frameworks increasingly problematic. An ever more agonised debate began to emerge concerning the alleged 'objectivity' of law and legal judgment. The decline of univocal and unchallenged traditions of legal interpretation exposed the extent to which law consisted of an immensely complex assemblage of fragmentary materials (doctrines, statutes, cases, texts) that were capable of being construed in a great diversity of ways. What, in this context, could justify the claim of judge or legal scholar to possess a knowledge of the objectively correct interpretation?

One possibility was to seek an answer within the law itself, by endeavouring to reconstruct a coherent set of principles (a conception of justice) that was implicit within the fragmentary complexities of the legal rules and cases. Once unearthed and articulated, this implicit conception of justice could provide a guide for the interpretation of discrete legal doctrines and rules. Shared texts would now generate shared rules, because the proper interpretation of the texts would not be wholly dependent upon the varying moral judgement of individual interpreters. Nor would convergent interpretations simply be the product of an inherited tradition of unarticulated assumptions, dependent for its survival upon the limited diversity and complexity of a preindustrial society. Furthermore, if the implicit conception of justice could be shown to turn upon such values as 'freedom' and 'equality', it would avoid any reliance upon a particular conception of value or well-being, seeking instead to sustain equal freedom to pursue diverse notions of a good life: in this way, it would reinforce the modern polity's vision of itself as a realm of freedom, rather than an association committed to the service of distinct ideals of excellence.

Classical legal thought

The form of juridical thinking that resulted from this situation received its classical and most influential expression in the philosophy of right proposed by Kant. Kant's view focused upon two salient features of law: its internally systematic character (the lawyer's concern to relate each discrete rule or decision to a set of inter-related principles) and its coercive character (more accurately, the character of law as justificatory of coercion). These two features were linked by Kant to the notion of 'a right' and, through that notion, to a separation between the form of the will and its content.

When one asserts the existence of a right to act in a certain way, one does both more and less than assert that one's action is good or estimable. One does less than that in so far as one may have a right to perform actions that are without value, being unworthy or demeaning; yet one asserts more than that in so far as the existence of a right justifies the use of coercion in a way that the mere goodness or excellence of one's action never could. If I have a right to speak freely, my free speech should be protected from interference even if that involves the use of force (by myself, or by the state on my behalf) against those who would interfere with me; at the same time, my right to speak freely may be exercised by my saying that which is valueless, or even false or pernicious. To determine the circumstances in which one may possess a right, Kant believed that one should determine the circumstances in which coercion may be justified. Coercion is wrong in so far as it violates freedom; therefore, Kant concluded, coercion may be justified when it is deployed against those who would violate freedom: in other words, when it is used to maintain equal freedom.

If the connection between rights and coercion explained the way in which legal principles are invoked as a justification for ordering the use of organised force against miscreants, the notion of 'equal freedom' explained the law's concern with internal system. For Kant believed legality to consist in the set of conditions within which equal freedom obtained: the system of principles which made the exercise of independent wills mutually compatible. Such a system of principles would concern itself solely with the form of the relationships obtaining between wills, in the sense of the extent to which my pursuit of my will would impact upon the ability of others to pursue their own projects; it would abstract from the content of such projects, in the sense that it would be unconcerned with the value or otherwise of the goals pursued.

Lawyerly concerns with the internal coherence of the system of legal rights implied the possibility of some ideal model of pure coherence immanent within the practices of doctrinal reasoning, and it was easy to see the Kantian philosophy of right as articulating precisely such a model. Moreover, the separation between the form of the will and its content suggested a perspective for interpretation that might be shared even by those who held quite diverse conceptions of value; and the theory was also suitable for an increasingly secular age, in so far as it derived juridical principles not from a natural order established by God, but from the formal relationship between human freedoms. General juridical theories of this type may be expected to underpin and explain the autonomy of law from open-ended political debates. Here again the Kantian view proved to be a fruitful and congenial one, for the separation between form and content, upon which the theory was founded, created a distinct realm of juridical thought (concerned with the formal relationships between individual freedoms) which might be distinguished from more substantive questions of desirable goals and projects.

The Kantian theory of right exerted a powerful influence upon the development and systematisation of legal doctrine in the western legal tradition. This was most evidently so in the civil law tradition of continental Europe,[2] but the influence was felt within the common law as well.[3] Not only did the Kantian view suggest a way to resolve the problem of law's nature upon a philosophical plane; it also proposed a conception of an individual right that offered an immediate and appealing model for the systematisation of complex bodies of legal doctrine.

According to the Kantian view, an individual right protects an area of liberty. Consequently, a right possesses a number of distinguishable attributes: actions performed within the scope of the right are permissible (they do not constitute legal wrongs); they are protected from interference by others, in that they are correlative to certain duties incumbent upon others; appropriate actions performed within the scope of the right may (with certain exceptions for 'inalienable' rights) effect a transfer or limitation of the right; and such transfers and alterations are impossible without the right-holder's consent. The internal complexity of the notion of a right fitted perfectly with the Kantian notion of legality as consisting in 'the sum of conditions under which the choice of one can be united with the choice of another'.[4] for the domains of liberty constituted by rights could form nodal points for the systematisation of legal doctrine. A great multiplicity of discrete provisions could be inter-related by their varying relationships to notional domains of liberty or entitlement; and the resulting complex entitlements could be inter-related by the formal conditions for jointly possible freedoms. Questions concerning the legitimacy of such systematisation and interpretation of the law (for, we might ask, what gives the jurist's interpretations of law any greater authority than the interpretations of anyone else?) could be responded to by the claim that such systematisation represented the very condition of legitimacy or legality itself. Challenges based upon the idea that interpretations must rely upon variable and contestable moral judgements could be responded to by pointing out that the Kantian notion of right rests solely upon the formal relationship between freedoms, and not at all upon the ethical issue of how those freedoms ought to be exercised.

In speaking of the importance of this theory for the juristic systematisation of legal doctrine, it should not be imagined that we are addressing some esoteric issue of scholarly interest but of little real importance. For the task of doctrinal systematisation is one of reducing the necessity for legally unregulated choice by the judge. In place of a long list of discrete rules peppered with gaps (the latter forcing upon the judge a choice unregulated by law), the jurist hopes to offer an orderly system of doctrines, the internal coherence of which can provide a basis for the proper interpretation of discrete provisions and the proper resolution of cases not explicitly covered by any enacted provision.

A good illustration of this is provided by the decision of Lord Lindley in *Quinn v. Leathem*.[5] Quinn, a trade unionist, had tried to force Leathem, a butcher, to sack his non-union workers and employ only union members. Quinn threatened a strike at the shop of one of Leathem's customers if the customer would not stop doing business with Leathem. Leathem sued Quinn. There was no case or recognised legal doctrine that appeared to give Leathem a cause of action. Nevertheless Lord Lindley decided in favour of Leathem. He reasoned that Leathem had a right to earn his living in his own way, and this right included the right to deal with others who were willing to deal with him; therefore, everyone (including Quinn) was under a duty not to interfere with this right, and Quinn had violated that duty. Lord Lindley here employs a Kantian notion of right[6] as entailing (amongst other consequences) both permissibility and inviolability. From the permissibility of Leathem's trade he infers the existence of an entitlement, and from that entitlement he deduces a duty of non-interference upon others. No evaluation of the merits of union organisation or of butchers' shops is required, since the situation is taken to be one where Quinn is encroaching upon Leathem's liberty in a way that violates the conditions for equal freedom. The decision is seen by Lord Lindley, not as requiring a legislative decision on grounds of social policy, but as dictated by the existing structure of legal entitlements.

The fragmentation of rights

The doctrinal systematisation of law stemming from the Kantian tradition relied upon the idea that a right is essentially complex. This internal complexity enabled a judge such as Lord Lindley to pass from one attribute of a right to another by a seemingly deductive process of reasoning. Thus, assuming that rights entail both permissibility and inviolability, Lord Lindley reasoned from the permissibility of Leathem's actions to the impermissibility of Quinn's.

The erroneous nature of the basic assumptions upon which this theory was founded was gradually exposed by a number of jurists, but the exposure found its clearest and most thoroughly convincing expression in the work of the American law professor, W.N. Hohfeld.[7] Hohfeld demonstrated that the notion of a 'right' is not *complex* but *ambiguous*. Whereas the Kantian conception of right was thought to entail a diversity of legal consequences (permissibility, inviolability, power of alteration, and immunity against alteration by others), Hohfeld demonstrated that there was no single legal concept that had all of these consequences. Rather, the single word 'right' was employed by lawyers to refer to four quite different concepts, each of which entailed *one* of the complex consequences attributed to the notion of right in general by the Kantian view.

Seemingly deductive passages of legal reasoning were exposed by Hohfeld as containing a logical equivocation. Thus Lord Lindley's argument in *Quinn v. Leathem* proceeds from the premise that Leathem possesses a right to the conclusion that Quinn was under a correlative duty. Yet the only uncontested right that Leathem possessed consisted in the mere permissibility of his actions, and this type of right (which Hohfeld called a 'privilege') did not entail any duties incumbent upon others. Such duties were entailed only in the type of right that Hohfeld called a 'claim-right'. The question raised by *Quinn v. Leathem* was really one of whether

Leathem's privilege (of operating a butcher's business) ought to be protected by a claim-right against interference, and the answer to this question was not logically dictated by any uncontentiously accepted legal doctrines. The equivocation in Lord Lindley's argument thereby served to conceal the element of creative choice in his decision. Moreover, it enabled him to avoid some of the hard questions posed by the creation of a new claim-right against interference. What, for example, were to be the limits of the new claim-right? Since my opening a butcher's shop next door to Leathem might also interfere with his business, the need for some such limits is evident: but in trying to articulate the limits of the claim-right we would be trying to spell out in full the limits on fair competition and the rules within which a market should operate. This daunting task is avoided by the employment of an obfuscating terminology.

Hohfeld's analysis showed that the logic of entitlements would not in itself provide a basis for the systematisation and interpretation of legal doctrine. Moreover, Hohfeld's analysis suggested the need for a disciplined and precise deployment of the language of 'rights'. Any assertion of a right could be met with a challenge to specify the type of right that was being asserted, and (in appropriate cases) this might require one to specify the persons on whom the correlative duty lay, and the precise content of the duty. It is hardly surprising, therefore, that Hohfeld's analysis has proved to be uncongenial to the numerous ardent enthusiasts for rights who have become such a feature of the late twentieth century. They have been anxious to preserve a usage whereby rights can be asserted without any serious effort being made to assign responsibilities and delineate the precise implications of the right. Unable to find a flaw in Hohfeld's argument of a kind that would invite direct assault, they have chosen to *dilute* the notion of a right in order to preserve its comforting pliability.

Their argument invites us to distinguish between the rights that are being protected and the legal instrumentalities whereby the rights are protected. Hohfeld's analysis, it is then suggested, is really an analysis of the latter instrumentalities. The rights being protected by these Hohfeldian atomic elements are very weighty interests. To say that someone has a right (in free speech, for example) is simply to say that they have a very weighty interest (in speaking freely) which would be sufficient to justify the imposition of a duty on some (unspecified) person, other things being equal. This weighty interest might be protected in a variety of ways, as appropriate. It might be protected by the imposition of duties on others, by the conferment of privileges upon the right-holder himself, by recognising certain powers of transfer or immunities, and so on. Hohfeld succeeds in demonstrating that a right cannot *entail* all of these different consequences, but it can (in appropriate circumstances) provide a good reason for conferring them.

For those who insist on debating every issue in terms of 'rights', this anti-Hohfeldian view is a Godsend. Notice, however, that unlike gifts from God it bears a price ticket. The price is the acknowledgement that a right does not *entail* any of its consequences. The Kantian view saw the assertion of a right as possessing peremptory force, in the sense that, if I have a right against you, you are under a correlative duty. All debate about the wisdom or decency of my action is cut short (the exclusion of such issues being a part of the object of the theory) and I can simply insist upon your compliance. Although a critic of the Kantian view in other respects, the idea that rights entail duties is shared by Hohfeld, at least in relation to claim-rights. But in the newer anti-Hohfeldian theory by contrast, the peremptory force of rights is lost. A right is now an interest which may justify the imposition of a duty if circumstances are judged appropriate. To assert the existence of a right is *not* (on this theory) to assert a proposition that, once accepted as true, will conclude the matter in dispute; it is simply to point to a consideration that the court must take into account.

The emptiness of equality

The flawed conception of right that was integral to Kantian jurisprudence was not the most substantial deficiency in that theory. The very notion of a set of principles constituting the conditions of equal freedom had been denounced by both Hegel and Schopenhauer[8] as devoid of content, and such denunciations have never been silenced. To understand this point we must first remind ourselves that the Kantian theory was supposed to pronounce upon the legitimate bounds of freedom whilst setting on one side all questions about the desirability or worthiness of the projects pursued in the exercise of that liberty; the theory purported to make the 'right' independent of the 'good'. Given the absence of any metric for the measurement of freedom, however, the injunction to protect equal freedom yields no guidance whatever. Suppose that your freedom to practise the trumpet interferes with my freedom to study. To be told that the law should aim to enforce *equal* freedom here is unhelpful; for we have no way of saying whether your trumpet playing is a greater or lesser interference with liberty (the liberty to study) than my preventing your trumpet playing (by, say, an injunction) would be. The only sense in which we can measure interferences with freedom is by reference to the judged *importance* of the respective exercises of freedom. Scholars who take study to be more important than martial music will reach a different conclusion from jazz musicians or army bandsmen; but to judge the issue in this way is to reintroduce the very considerations (of the 'good') that the theory seeks to exclude.

If the basic idea of equal freedom was truly empty, how was a form of legal analysis based upon that idea influential for so long? The answer is easily found as soon as we realise that pliability is a consequence of emptiness, and the extreme pliability of a theory frequently explains its popularity, provided that the fact of emptiness can be successfully concealed from view. The emptiness of Kantian jurisprudence could be so concealed in so far as the given institutions and practices of civil society were simply taken for granted as a natural landscape for human activity. Viewed in this light, Quinn's actions in the case quoted above constitute an interference with liberty in a way that Leathem's actions do not. We begin to question this understanding of the situation only once we unearth by analysis the way in which the situation of both parties is structured by the law's coercive constraints upon freedom. Thus Leathem can pursue his business as a butcher only in so far as he enjoys exclusive control over certain resources, an exclusivity which is guaranteed by the organised force of the state; similarly, he depends upon the legal enforceability of contracts. The location of the parties' actions within a social situation structured by the law's constraints makes the analysis of the situation in terms of liberty and encroachments upon liberty far less straightforward than might at first appear to be the case.

In his essay *On the Jewish Question* Karl Marx observed that the liberal market polities of post-feudal Europe took civil society[9] as a natural basis which needed no further grounding; the bourgeois political revolutions did not extend to the institutions of civil society, which were not subjected to criticism.[10] We have seen the element of truth in Marx's claim, in so far as juristic thinking relied upon the established structures of civil society to give content to its basic conception of equal freedom. Once, however, those structures were identified as a subject for political evaluation and possible change, the notion of 'justice' took on an extended significance. Ceasing to be primarily a matter of conduct between specific individuals, concerned with the justice or injustice of particular transactions (contracts, accidents, etc.) justice came to be a matter of the overall structure of benefits and burdens in social life: 'civil' justice became 'social justice'. Whereas the justice of a particular transaction might be assessed by reference to accepted norms for conducting the daily business of life, such norms ceased to be appropriate as a basis for evaluating 'social justice', for it was precisely the rectitude of a state of affairs structured by such norms that the latter conception called into question.

Aristotle distinguished between 'corrective' and 'distributive' justice in Book 5 of his *Nichomachean Ethics*. He tells us that distributive justice 'is shown in the distribution of honour or money or such other possessions of the community as can be divided among its members', while corrective justice is shown 'in private transactions' which may be 'voluntary or involuntary'.[11] Although Aristotle's meaning is far from clear, it is quite significant that modern philosophers and jurists have tended to construe the notion of 'distributive' justice very broadly, and to subordinate 'corrective' justice to the distributive variety. Thus, rather than taking distributive justice to arise only in certain specific contexts where a fund is being deliberately distributed (as when I consider how to share my property amongst my children on my death), the notion is taken to apply to the global distribution of resources, benefits and burdens. The objection may be raised that (at least in market societies) such benefits are not distributed intentionally but are the unintended outcome of uncoordinated actions directed to other ends. To such an objection one receives the reply that the process of uncoordinated distribution is nevertheless structured by an institutional framework that is alterable, and for the results of which we therefore bear collective responsibility. Since distributive justice has, in this way, arrogated so much to its sphere, one might well wonder what is left for Aristotle's other category, of corrective justice. In fact this tends to be treated as a subordinate and essentially procedural form of justice concerned with the restoration of an originally just distributive situation that has been upset by some particular transaction (such as an unjust contract or an accident).

Having thus floated free from the specific contexts that might give it a meaning, from where does the extended notion of distributive justice derive its content? On the level of popular rhetoric, no content is required, for 'justice' comes to be the utopian condition in which all conflicting claims are reconciled and all goals costlessly achieved. Within more reflective juristic and philosophical theories, however, the ever available notion of 'equality' offers an apparent solution. Equality is closely linked to justice since justice requires that like cases be treated alike and different cases differently.[12] This principle (sometimes called 'the principle of formal justice') lacks any content until we have spelt out our criteria for similarity and dissimilarity, but it has proved to be easy to overlook that fact and to assume that some more substantive connection exists between justice and equality. One can treat people equally only with regard to this or that criterion, and in treating them equally in one respect you treat them unequally in all others. Thus, to pay my workers equally with regard to their needs is to treat them unequally with regard to their merits, and so on. Any determinate criterion of equality will therefore ignore the complex features of real situations, and will treat people unequally in relation to those features. Yet far from this fact alerting the modern world to the emptiness of one of its central ideals, it supplies the motor behind many corrosive social critiques. By pointing to the gap between the narrow respect in which people are treated as equals and the great multiplicity of respects in which they are treated as unequals, it is all too easy to suggest that the equality that obtains is merely 'abstract' or 'formal', and to propose its replacement with a 'rich' and 'substantive' form of equality.

It was in fact Karl Marx who pointed out the fallacy at work here. In his *Critique of the Gotha Programme*,[13] in the course of denouncing any reliance upon juridical notions such as justice, he observed that every right of equality is also a right of *inequality*; for 'unequal individuals (and they would not be different individuals if they were not unequal) can only be measured by the same standard if they are looked at from the same aspect.' Marx's own response to this discovery relied upon his belief in the possibility of a realm of direct and transparent social relations unmediated by rules or juridical categories. Even if we reject this response as incoherent, however, we can acknowledge the accuracy of Marx's critique of conceptions of social justice and equality. For Marx saw clearly that these ideas, being in themselves empty and formal, would always derive whatever content they might possess from a taken-for-granted background of social institutions. Detached from such a background, justice and equality become hopelessly pliable ideas:

> 'What is "just" distribution? Does not the bourgeoisie claim that the present system of distribution is "just"? And given the present mode of production is it not, in fact, the only "just" system of distribution? Do not the socialist sectarians themselves have the most varied notions of "just" distribution?'[14]

The modern polity therefore finds itself in a hopeless quandary. The central ideas to which the modern world pays homage are themselves but empty structures that must derive their content from elsewhere. Such a content might be derived (as in the past) *either* from the taken-for-granted background of an established social structure *or* from a deep moral consensus upon the features of humanity that are of distinct moral significance. Either of these alternatives would provide appropriate criteria for judging what was a 'like' or 'unlike' case, and for defining the dimension by reference to which equality was to be secured. The first alternative, however, is ruled out by the modern polity's desire to ground itself not in tradition or history (which are seen as mere perpetuators of illegitimate oppression), but in abstract 'justice' itself. Similarly, the second alternative is ruled out by the commitment to moral pluralism and diversity.

Fundamental rights and the stability of law

Law must always depend upon a background of informal moral and cultural understandings, settled institutions and the like. Determinate meanings cannot be ascribed to the texts of the law except by reliance upon such a background. It is true that these informal understandings would, in many cases, be themselves too imprecise to yield a definite and unambiguous account of the requirements of justice in this or that situation. Furthermore, it is this lack of determinacy in background understandings that would make law necessary even in a community of persons of good will where problems of deviance and predatory conduct were absent. Yet this is to say no more than that law and the background of moral understandings exist in a symbiotic relationship, where each lends greater determinacy to the other.

Many of the most puzzling features of modern political communities stem from their tendency to adopt self-conceptions that are radically at odds with this situation of symbiotic dependency. The popular ideology of many developed societies is one that views morality as a matter of individual commitment or autonomous choice rather than being an essentially shared part of a communal fabric. This not only tends to blind people to the large areas of moral consensus that continue to exist,[15] but also encourages them to view such consensus as they do acknowledge as a failure of individual autonomy which should not exist in a society of truly free citizens. The modern vision of the ideal polity as radically pluralistic makes law of vital importance, for it is law that must demarcate the bounds of liberty in a world where people lack other shared standards; but this vision simultaneously makes law immensely problematic, for it is hard to see how law could possess the necessary determinacy of meaning once detached from the shared background of understandings upon which it has always relied.

This basic paradox receives a further epicycle with the rise of 'human rights' or 'fundamental rights' in domestic and international law. In the first place, it is in the highest degree surprising that these notions should have become a part of accepted conventional wisdom in a world that generally rejects all notions of natural law or of binding universal moral standards. If moral conceptions are matters for the autonomous choice of individuals, how can there be universal fundamental rights that override the autonomous and democratic choices of entire nations? One suspects that, at the heart of this vaporous fragment of popular political culture, lies a confusion that associates moral subjectivism with tolerance, and moral universalism with intolerance. Such an equation would explain how people who view morality as 'subjective' can also espouse universal human rights: they are assuming that subjectivism *entails* tolerance and human rights *enforce* tolerance. This is, of course, nothing more than a muddle. If morality is subjective, a commitment to tolerance is no more (and no less) well-grounded than a commitment to intolerance; a commitment to human rights and a commitment to racist supremacy are, for the subjectivist, exactly on a par, since neither can claim to be 'true' or even more reasonable than the other.

The assumption that subjectivism is linked to tolerance is, therefore, an error. No less erroneous, however, is the assumption that human rights necessarily make for tolerance. In many instances the value of tolerance comes into conflict with the law of human rights. Suppose, for example, that the customs of a minority cultural or religious group are regarded by the majority as involving the subordination of women by a variety of discriminatory practices in relation to education, marriage and the family. Here *tolerance* might seem to dictate a 'live and let live' attitude towards different communities, while human rights may be taken to require interference with the minority group's customs. The fact that some enthusiasts now urge upon us the idea of 'group rights' for minority cultures does not overcome this tension (although it certainly demonstrates the relentless determination of some people to discuss *every* issue in terms of 'rights'), but simply builds it into the notion of human rights.

Pragmatically speaking, there can be no doubt that great potential gains could flow from a wider concern for human rights. The international community might be more likely to take action against flagrant cases of oppression, and domestic judiciaries in tyrannical regimes might feel more empowered to resist the dangerous political projects of fanatics and cynical demagogues. Yet these gains are themselves an indirect consequence of the current refusal to take seriously any moral issue that is not framed in terms of rights. When confronted with outright cases of brutal tyranny and oppression, why do we need the language of 'rights' in which to address the issue? The 'pragmatic' argument for rights therefore feeds the very conditions that give it purchase; it is our failure to be mobilised by any language other than that of 'rights' that makes it necessary to speak of 'rights' rather than 'welfare' or 'oppression'; but our very willingness to act upon this pragmatic imperative fosters the dangerous confusions from which it arises.

The danger with the notion of rights in this context is that it encourages us to confuse distinct issues. On the one hand is our desire to mobilise the international community in such a way as to address fundamental problems of human welfare and of barbaric cruelty; on the other hand is the desire of some to remove issues from the agenda of democratic debate by imposing the judicial scrutiny of legislation upon basically just and humane regimes.[16] Insistence upon reducing every question to an issue of fundamental human rights confuses these quite distinct problems; it appropriates our intense desire to address outright atrocities in oppressive regimes, and deploys it on behalf of an assault by élite groups of lawyers upon the power of democratically elected representatives within basically just and tolerant regimes.

The general tendency of the domestic and international law of human rights is to transfer power from democratically elected legislatures to the judiciary. Domestic Bills of Rights, and regional[17] and international conventions on human rights, are invariably drafted with great brevity and in terms of immense generality that leave enormous leeway for interpretative dispute. In its international and regional dimensions, human rights law can operate to transfer issues away from the domestic institutions of a state altogether. Each of these characteristics is likely to aggravate a widespread sense of popular impotence which is, in any case, an increasingly prominent and dangerous feature of modern democracies. Should that angry sense of impotence be seized upon by unscrupulous politicians, it could easily provoke a repressive backlash against which the constraint of human rights law would prove ineffective. The record of judges in standing up to majorities bent on oppression is not impressive; the record of the international community in dealing with even the most gross atrocities is still less impressive.

Thus, the protection of allegedly 'fundamental' rights may well prove to be effective in taking power away from broadly just democratic bodies, and quite ineffective in dealing with the flagrant injustice of outright tyrannies.

The problem of responsibility.

We noted above Marx's observation that the classical liberal political communities regarded civil society (the world of property, the market, and the family) as a natural basis that needed no further grounding. Once the structures of civil society were seen as themselves suitable objects of evaluation, and as inevitably constituted by and alterable by law, the notion of 'justice' itself began to change. From being a matter of the propriety of individual transactions between specific individuals (corrective justice, or civil justice), it came to be a matter of the overall distribution of benefits and burdens in society (social justice, or distributive justice). This (we suggested) detached the notion of 'justice' from the accepted norms of daily life that otherwise gave it content. Justice became essentially pliable and irresolvably contested; genuine conceptual connections between 'justice' and 'equality' merely served to conceal the emptiness of each notion when detached from a content-conferring context.

Modern legal systems are structured by a division between public and private law. The nature of this distinction is not always clear, and lawyers themselves disagree about its significance. We may see in it, however, a reflection of the distinction between distributive and corrective justice, or between social and civil justice. That is to say, public law (e.g. tax and welfare law, planning law, administrative law) is concerned with securing some favoured overall distribution of benefits and burdens, while private law is concerned with the justice or injustice of particular transactions between specific individuals. This would explain the distinct *form* of private law, which tends to reflect bilateral litigation between individuals, whereas public law is more typically instrumentally related to the attainment of some overall social goal.[18]

Lawyers tend to assume that private law (unlike public law) exhibits a certain independence from politics. Certainly the historical continuity of some of the principal concepts typical of private law is remarkable, and it has been suggested that this reflects 'the close connection between private law and certain moral ideas which have remained relatively static over long periods.'[19] This picture of continuity could, however, be misleading. It is, in the first place, hard to square with the widespread perception that modern societies have become more *litigious* than hitherto, and that the law is used by litigants in increasingly predatory and unpredictable ways. Optimists will explain such perceptions by invoking their own perception of history as the onward march of individual 'rights': no longer downtrodden and subordinated, ordinary people are now prepared to litigate where once they would have accepted wrongful loss and injustice as their unalterable fate. Pessimists, however, can easily offer a different explanation.

The full realisation that the given structures of civil society were alterable brought about a metamorphosis in the notion of justice: 'civil justice' (or 'corrective' justice) was marginalised by 'social justice' (or 'distributive' justice); private law began to play a part subordinate to that of public law. These were not, however, the only changes; there were also changes *internal* to the notion of civil or 'corrective' justice itself.

Corrective justice between individuals, of the kind with which private law is concerned, revolves around the causation of harm. Whether my complaint is of an intentional or unintentional injury, or of a breach of contract or breach of trust, I will be alleging that, had it not been for your wrongful act, I would not have suffered some loss, be that a physical injury or a loss of possible profit. Especially in the context of the law of tort,[20] the requirement of causation is an important constraint upon the proliferation of liability. Once causation is established, liability depends upon other factors, such as the 'foreseeability' of the injury, and the failure to take 'reasonable' care; but such notions as 'foreseeability' and 'reasonable care' are inherently vague, and they invite open-ended discussions of a kind that would render legal liability unduly uncertain and unpredictable, were liability to depend upon such issues alone.

The question of causation is, however, not a purely factual issue. When a scientist speaks of the 'cause' of an event, he may mean the set of preconditions without which the event would not have occurred: this is what lawyers call *conditio sine qua non* or 'but for' causation. The lawyer, however, (like the historian) must make an *attributive* decision about which of these various preconditions should be regarded as the cause for legal purposes (the 'proximate' cause, as lawyers say).

Let us suppose that a factory is destroyed by fire. The loss may have resulted from the conjunction of many different events and circumstances: the dropping of a match; the presence of combustible material; the presence of oxygen in the atmosphere; the springing up of a light breeze which fanned the flames; the night-watchman's lack of attention; the late arrival of the fire-brigade. Not all of the essential preconditions for the occurrence of an event are regarded as 'causes' of the event. In attributing responsibility for harm, the courts must select from the range of preconditions those factors that can appropriately be regarded as having caused the harm.

In the first half of this century, jurists associated with the American 'legal realist' movement adopted a position which has come to be known as 'causal minimalism'. Causal minimalists argued that the distinction between the causes of an event and its essential (but non-causal) preconditions was simply a cover for policy judgments by the courts. According to such theorists, the only genuinely causal issues addressed by courts were issues of 'but for' causation: i.e. the identification of the factors but for which the event would not have occurred. So far as the factual questions go, they argued, all such preconditions were equally entitled to be treated as the 'cause' of the event: the presence of oxygen in the atmosphere has as good a claim to be the 'cause' of the fire as has the dropped match. In deciding that some preconditions are causes and others are not, the courts are making a policy judgment about where it is desirable to locate responsibility and impose liability. Legal principles proclaiming the existence of liability for the causation of harm are not (it was argued) the stable juridical principles that they appear superficially to be: they might better be regarded as empty frameworks offering ample scope for flexible policy judgment.

If soundly based, the claims of causal minimalism would have dramatic implications. Liability in private law is to a very large extent based on the causation of harm. Private law demarcates certain rights or protected interests, and imposes liability upon those who are (intentionally, negligently, or on the basis of strict liability) causally responsible for damage to the relevant interests. It is assumed that, while questions of general social policy may enter into the conferment of rights and the choice of general rules, adjudicative decisions about the infringement of those rights and the violation of those rules are manifestations of the intellectual integrity of private law doctrines, and are not simply matters of pliable social policy.

If, however, issues of causal responsibility are in fact just policy issues in disguise, the integrity of private law is an illusion. Adjudication (on this view) loses its character as an application of coherent doctrines, and individual rights are revealed as having no genuine stability or tolerably clear parameters.

Hart and Honoré, in their classic work *Causation in the Law*[21], mounted a powerful attack upon this line of argument. The causal minimalists had moved too swiftly from the observation that there were no simple principles regulating the courts' causal judgments to the conclusion that there were no principles at all: merely an open-ended contest of policies. In fact, Hart and Honoré demonstrated, the causal principles employed by the courts are complex but orderly, subtle but intelligible. They reflect similar causal judgments made in the course of everyday life, embodying principles which are fundamental to our sense of possessing a distinct identity. In particular, the central model of causation which underpins many of our judgments is one of 'intervention on a stage already set'. Factors which are constant and normal will not be treated as causes, even when they are essential preconditions for an occurrence. Only those factors which can be regarded as interventions or departures from the normal course of events will be judged to be causes.

Hart and Honoré fully accept that this model shows causal judgements to be, in a sense, relative to perspective and focus of interest. Thus a man's stomach pains may be attributed by his doctor to ulcers; but the man's wife may attribute them to the eating of parsnips. The difference is that the doctor treats the eating of parsnips as a part of the normal course of events, and regards the ulcers as the abnormal feature that explains the incidence of pain in this case. The wife regards her husband's ulcers as a part of the normal background to life, and attributes his pain on this occasion to the parsnips. The wife and the doctor are in effect asking different questions. The doctor is asking 'Why does this man suffer pain (on eating parsnips) when most men do not?' ; the wife is asking 'Why (given the existence of his ulcers) is my husband suffering pain *now* when usually he does not?' This seeming relativity of causal judgments invites the conclusion that only the question of 'but for' causation is a factual issue, with all other supposedly 'causal' issues being in fact a matter of policy. But, Hart and Honoré urge, such a conclusion would be overly simplistic and inherently misleading. Our values and interests may shape the questions that we consider it relevant to ask: but they do not determine the answers that we give to those questions. Moreover, even if our common sense causal judgements exhibit a degree of contextual relativity, they are not mere functions of social policy judgements. Causal judgements in law reflect a background of common sense understandings: they are not a cover for decisions about desirable social goals.

While exposing the naiveté of the causal-minimalist attack, however, Hart and Honoré's own analysis unintentionally points us in the direction of a different kind of attack. For the conclusion of the Hart and Honoré analysis is that causal judgements in law are relative to a taken-for-granted background that constitutes the 'stage already set', upon which cause 'intervenes'. We may therefore conclude that such causal judgements will seem most secure and determinate in contexts where the background may indeed be taken-for-granted as an established background like a natural landscape. The more that we come to appreciate the alterability of human arrangements, the less secure such causal judgements will seem.

We are now in a position to see how the *concepts* of private law may have exhibited considerable continuity, while the law has nevertheless grown more pliable, and more available for predatory or speculative litigation. Where once a famine might have been attributed to the unexpected absence of rain (for example) we now have public bodies that might be expected to anticipate such events and do something to avert their possible consequences. Their failure to act may then be viewed as the cause of the famine. When once the car crash might have been attributed to the icy road (or to the driver's negligence) it might now be attributed to the Highway Authority's failure to grit the road, or perhaps even to the Automobile Association's failure to warn of low temperatures. The more extensive the paternalistic scope of the state's authority, the less we will be inclined to regard tragedies as the result of ill-fortune, and the more we will seek to ascribe blame; the more extensive that paternalistic authority, the less we will be willing to accept the idea of uncompensated loss. This will extend the potential liability of both public and private bodies.

Consider the connections between this development and the changes (described earlier) in the notion of 'a right'. The Kantian conception of a right shared with the Hohfeldian conception the assumption that a right is a peremptory demand which is correlative to a duty incumbent upon some other specific individual. The most popular account of rights at the present day, however, abandons that idea. A right is simply a weighty interest which must be taken into account; it need not be correlative to any duty on someone else; I can, on this view, truly claim the existence of a right even though I am unable to truly attribute a duty to anyone. When allied to the erosion of notions of causal responsibility, we have a situation where legal rights are simply important interests, and liabilities are freely assignable according to open-ended judgements concerning the proper distribution of responsibility.

Conclusion

The modern political community aspires to be a peaceful and orderly realm of rights founded upon no shared conception of excellence or value. This pluralistic vision necessitates the ascription of a central role to law, for it is law that must demarcate the bounds of otherwise conflicting liberties. Yet, if liberal pluralism makes law central and necessary, it also makes its existence problematic. For how can we sustain a shared set of laws when we share no other moral conceptions?

The problem is deepened by the modern unwillingness to accept current institutions and inherited traditions as a workable *modus vivendi* without deeper grounding or significance. The radical desire to ground all institutions in justice rather than tradition actually detaches the notion of justice from the contexts that could give it content. The resulting pliability of law follows inevitably, even though it percolates through a number of distinct channels. These channels include the dilution of the notion of 'right' to the point where it no longer entails assignable duties but simply marks out an important interest or demand; the subordination of civil or corrective justice to 'social' or distributive justice; and the erosion of settled baselines from which judgements of causal responsibility might otherwise be made.

If the law is to demarcate the bounds of liberty for autonomous agents bent upon the advancement of their own projects, it must possess a determinacy that is independent of the will of those same agents. In the absence of some shared set of values or base-lines, however, it would seem that the law can enjoy no such determinacy. Rather than being the framework of rules within which social life is conducted, and which sets the limits to competitive self-seeking between individuals, the law becomes itself a resource over which the contending factions struggle. The unrestricted assertion of rights, the invocation of unspecified standards of justice or equality, and the litigious willingness to exploit the lee-ways of causal attribution, are becoming common features of modern society. What is less common is any willingness to face up to the political landscape that these features present.

Chapter 15

The practice of conservation by religions

Martin Palmer

The Author

Martin Palmer has pioneered religious involvement with ecology, economics and education for 20 years.He is Religious Advisor to the World Wide Fund for Nature (WWF) International and Secretary General of the Alliance of Religions and Conservation. He heads the International Consultancy on Religion, Education and Culture (ICOREC), is author of over 30 books, and broadcasts regularly on religious issues.

Chapter 15

The practice of conservation by religions

Martin Palmer

One day in 1953 two men stood on the summit of Mount Everest. Sir Edmund Hillary, a Western scientist, and Sherpa Tenzing, a Himalayan Buddhist. Separated as they were by culture and beliefs, they had together scaled the highest mountain in the world and had, for perhaps the first time in history, reached its summit. What the press said of the significance of this event and what the press did not report speaks volumes for the different cultures from which they came. Edmund Hillary was depicted sticking the Union Jack flag into the snow on the summit and the Western press claimed he had 'conquered' the mountain. Of Sherpa Tenzing's actions on the summit we heard nothing.

The absurdity of the belief of the press that by climbing a mountain a human being could claim to have conquered it seems like nothing more than hubris today. In this context, it is interesting to hear what Sherpa Tenzing actually did when he reached the summit. This devout Nepalese Buddhist sank to his knees and asked forgiveness of the gods of the mountain for having disturbed them. When the newspapers of the West reported the ascent, they all took up the 'conquered' theme. Not one saw anything to report in the prayers which were offered.

Some thirty years on Edmund Hillary commented that perhaps Sherpa Tenzing had more fully understood what was happening than he did. Everest, like the rest of the Himalayas, is in a desperate state. Invasion of its peaks by tourist climbers and others in the last few decades has caused environmental degradation exacerbated by massive pollution, destruction of fragile habitats and logging in the foothills. The UK government regularly sends a team to remove the 10 tons or so of rubbish left on the higher sections of the mountain by climbers.

Yet this is a sacred mountain to the local people, who for millennia have lived in a more balanced relationship with that mountain. And the truth is that Hillary did not feel he had 'conquered' the mountain. Instead he knelt in the snow too and gently laid a crucifix on the snow, given him and blessed for the purpose by a Catholic monk, as an act of similar humility to that of his Buddhist colleague. Of this virtually nothing was known for decades. To the Western world's press this was conquest and that was all.

A material view and spiritual views. One proclaims that a man who might live for 90 maybe 100 years, can 'conquer' a mountain some 50-60 million years old – a mountain that will still be here when *Homo sapiens* has become extinct. The other, views of human beings as just part of something much, much greater. In trying to understand the practice of conservation by religions, it is insights such as these which we need to notice, as much as the actual, 'hands on' projects which so many religious organisations now run on conservation.

We all inhabit worlds created by beliefs. This frightens many people. They believe – note how we still use that word – that they inhabit the 'real world'. But of course every world view is a construct. We create a manageable world around us, bracketing out certain issues which, were we to face up to them, would crush us.

As Carl Jung said, without creation myths which give us a mythic significance, we would be crushed by the sheer awe-full-ness of the universe. So we create stories which help us feel a place, a home, a meaning in the midst of vastness and in the face of awe. This is entirely human and understandable. We simply cannot function any other way. We have to have a place from which we start; and we leave unanswered the unanswerable questions. For example, the Big Bang theory is a classic case of belief and of story. As Terry Pratchett, the wonderfully insightful science fiction writer, says: the theory is that there was nothing; then the nothing exploded! I know this will cause apoplexy in some but essentially it is a story which gets us started and helps us feel we understand the impossible to understand.

What religions do is admit that they have stories, beliefs and teachings from which they construct the vision of the world that they inhabit. I wish science, and in particular the different environmental movements, were so honest. For the modern environmental movement is a belief structure which tells stories and creates beliefs by which it asks us to live; yet it often presents itself as being something more removed, more analytical than that.

I want to stress again that this is what human beings do. We live mythologically and by beliefs. So rather than being surprised by this let us look at how and why this has particular impact on the environmental movement, and especially why the massive involvement of the faiths in ecology in the last fifteen years is so significant.

If I believe that human beings are the only important species on earth, then I will use, or even abuse, the environment in a particular way. This ideology is found, amongst others, in those industries that profit from the use of non-renewable resources for immediate needs. But it is also found amongst development agencies which, for wholly laudable reasons of charity and compassion, drill waterholes to meet human needs now, but which destroy water tables, leading to salinisation and ultimately to a serious decline in flora and fauna – including human beings – in the region. In other words, rapacious greed or human-centred development can both operate on a 'human use only' model and, despite different motives, can end up with the same disastrous environmental results.

If, however, I believe that human beings are but part of something vaster, deeper and more profound – call that Creation, Nature, Buddha-Nature, the Tao or God – then I will look to understand and locate human activity within this wider framework. I will, as for instance Islamic Shariah law does, consider the water needs of animals and plants, birds and insects as being as important as the needs of the human beings or their domestic animals in an area. For example, this world view led to the development of a bill of legal rights for animals in the 13th century AD by the classic Muslim jurist Izz ad-Din Abd as-Salam. It leads to the world view of a St. Francis who, when 'rescued from the birdbath' of popular piety, is in fact a revolutionary thinker. It finds expression in the reincarnational beliefs of Hinduism, Jainism, Buddhism and Sikhism which explicitly state that all life is interconnected and that, as the Bhagavagita (chapter 5, verse 18) says:

'The humble sage sees with equal vision a learned brahmana, a cow, an elephant, a dog and an outcast.'

Let me give a very simple example. What is a fox?

To some people it is a reddish brown harmless creature hunted brutally by cruel hunters. To the farmer it is a cruel hunter who needs to be killed. To a fox hunter on horseback it is part of tradition and really very important to social life. To story tellers in many lands, the fox is a cunning and witty character as in the French story of Reynard the Fox. To many Hindus the fox is a soul, a spark of the divine as they are themselves, and may even contain the soul of someone they knew as a human being in an earlier life. To traditional Japanese, the fox is the harbinger of evil, capable of capturing your soul and turning you into the living dead. To others, the fox is a species which would be very useful in experimentation for scientific products. To some, the fox is a fashion accessory.

So I return to the question: What is a fox? The only answer we can really give, other than a Mr Gradgrind answer which simply describes its physical shape, is that it depends on what we see, experience and believe the fox to be.

What we do is conditioned by what we believe. It is as basic as that. Put quite simply, the environmental crisis is a crisis of the mind and the imagination and of beliefs, not a crisis of nature itself. Therefore, no matter how much information or data we supply, if we have a world view which views the world as something there just for us, our action and interpretation will reflect that world view. Let me give an example.

The first really major international gathering on the environment under UN auspices was the Stockholm Conference of 1972. At that conference, vast amounts of data were provided by scientific organisations and conservation groups, hoping to persuade the leaders of the world's nations to take the decline of the environment seriously.

One particular area of concern was the rise in the destruction of the tropical rainforests. Scientists and environmentalists presented a terrible picture of the loss of these vital ecosystems. They told of nations selling off their forests for cash rather than preserving them for posterity. The intention in the minds of those presenting the information was to shock others by this catalogue of destruction. But what happened at the receiving end was rather different. Some leaders of delegations, especially from countries struggling with debt burdens and massive poverty amongst their people, as well as some leaders of delegations from corrupt governments, heard something different. They heard that people out there in the commercial world would pay good, hard currency for the useless forests in the badlands of their countries.

The result? The rate of destruction of the rainforests leapt upwards after Stockholm. It is not information that will change the world. It is a world view which can use that information within a wider ethical, moral, and spiritual ethos. A fusion of heart and mind.

Such a fusion is what the great faiths of the world have always sought to maintain. Thankfully, the days have gone when 'secularism' opposed 'religion' as rationalism versus superstition. Today, we know and understand that, for example, physics is about sifting through value statements which help us for a time to express an understanding of our perceived notion of what might actually be at work or exist. The frank openness of contemporary astrophysics or scientific cosmology finds an echo in a similar and increasing openess in the faiths. Of course there are fundamentalists who dig in on both sides: Dawkins on one side and Pat Robinson on the other. But they are, despite the press attention and noise they make, the minority.

It was this realisation, that the faiths could engage open-handedly with a secular concern such as the environment, which led the World Wide Fund for Nature International (WWF), to bring together in 1986 in Assisi, Italy, the leaders of certain major religions. The intention was to see if either together, or perhaps more significantly, working through each of their own unique networks, the faiths could bring their own understanding of the message and practice of care for the environment to people who otherwise would never hear of it or take it seriously.

WWF had been in existence for 25 years at this point. 25 years of pumping out information, data, analysis and facts. Yet little had improved. It was recognised that the message simply was not going home. Nor was it being heard, except in certain circles, with any sense of authority behind it. This is what led WWF to invite Buddhism, Christianity, Hinduism, Islam and Judaism to come together with leading environmentalists in 1986.

At Assisi, five of the world's major faiths, for the first time ever, clearly and unambiguously stated their position on the environment. The four faiths which have joined since Assisi – Baha'is, Sikhs, Jains and Taoists – have done likewise. The Declarations made by these faiths show that all of them have fundamental teachings about why we should care for the environment. Each faith brings a distinct and powerful understanding of what reality is, how it is experienced, where humanity fits within creation or nature, and thus what it is that can and should be expected of believers in terms of environmental protection.

For example, Buddhism stresses the interconnectedness of all life. It is beautifully expressed by Thich Nhat Hanh:

'Buddhists believe that the reality of the interconnectedness of human beings, society and Nature will reveal itself more and more to us as we gradually recover – as we gradually cease to be possessed by anxiety, fear and the dispersion of the mind. Among the three – human beings, society and Nature – it is us who begin to effect change. But in order to effect change we must recover ourselves, one must be whole. Since this requires the kind of environment favourable to one's healing, one must seek the kind of lifestyle that is free from the destruction of one's humanness. Efforts to change the environment and to change oneself are both necessary. But we know how difficult it is to change the environment if individuals themselves are not in a state of equilibrium.' [1]

The idea that the environment is both inner as well as outer was news to most of those environmentalists who were famous for the fastest rate of burn-out in any campaigning organisation!

Jainism, also arising from the Vedic culture of ancient India, takes the ideas expressed above further:

'The Jain ecological philosophy is virtually synonymous with the principle of *ahimsa* (non-violence) which runs through the Jain tradition like a golden thread. *Ahimsa* is a principle that Jains teach and practise, not only towards human beings, but towards all nature. It is an unequivocal teaching that is at once ancient and contemporary.

There is nothing so small and subtle as the atom nor any element so vast as space. Similarly, there is no quality of soul more subtle than non-violence and no virtue of spirit greater than reverence for life.

Jain cosmology recognises the fundamental natural phenomenon of symbiosis or mutual dependence. All aspects of nature belong together and are bound in a physical as well as a metaphysical relationship. Life is viewed as a gift of togetherness, accommodation and assistance in a universe teeming with interdependent constituents.'[2]

The Abrahamic faiths come at the issue from a somewhat different slant. The interconnectedness which is such a feature of the Jain, Buddhist and indeed Hindu perspective is expressed more hierarchically and gives rise to the notion of duty and responsibility.

The Abrahamic traditions of Judaism, Christianity and Islam preserve the attitude that the meaning and significance of nature – creation – is not exclusively instrumental. In these traditions, the world is a creation of God. The use of the world by human beings constitutes a pragmatic relationship between humanity and God, because God gives and humanity receives the riches of nature as an offering of God's divine love for the sake of all creation, the whole world.

Central to the Abrahamic understanding of creation is belief in one God, creator and sustainer of all that has been, is and will be. Nothing exists but for the Will of God. Genesis 1 in the Torah captures this:

'In the beginning God created the heavens and the earth.'

The Quran likewise:

'To Him is due the primal origin
of the heavens and the earth:
When he decrees a matter He says to it "Be", And it is.'
(Surah *2:117*)

The consequences of such a belief are that all creation has meaning and purpose in and of itself before God. It is not humanity which gives meaning or significance to any living thing. As the Quran says:

'There is not an animal that lives on the earth, nor a being that flies on its wings, but forms part of communities like you. Nothing have We omitted from the Book, and they shall be gathered to their Lord at the end.' (Surah *6:38*).

To the central question, fundamental to any vision and understanding of our place in creation: What is the purpose and reason for Humanity? The Bible answers in poetry:

"I look up at the heavens, made by your fingers,
at the moon and the stars you set in place —
ah, what is humanity that you spare a thought for us,
the son of man that you should care for him?
Yet you have made us little less than an angel,
You have crowned us with your glory and splendour,
made us lord over the work of your hands,
set all things under his feet."
(Psalm *8:3-7*)

All three Abrahamic faiths see humanity as central to the loving purpose of God's creation, even if we do not deserve it under our terms and within our limited understanding. The Abrahamic faiths do not see humanity as just another species. They emphasise the immense power of humanity to do good and evil. For some in the environmental movement, such a recognition of power is to be scorned. There has been a tendency, not supported by any of the faiths, to de-emphasise our responsibility – to seek some form of Golden Age when humanity just lived alongside all creation. While interconnectedness is central to many faiths, none regard this as meaning that humanity doesn't have a special role, a special place in all creation.

The Abrahamic faiths express this openly. Whether the dominance of humanity was created by the Biblical texts or, more likely, the Biblical texts reflect the reality of human power, the Bible comes down squarely in favour of the idea that humanity is different and carries different responsibilities from the rest of creation.

Islam confronts this boldly. The creation of humanity is even opposed by the angels in the Quranic version of the Creation story:

> 'Behold Thy Lord said to the angels: "I will create a vice-regent on earth". They said, "Will Thou place there one who will make mischief and shed blood? We celebrate Thy praises and glorify Thy holy name." He said, "I know what you know not". (Surah 2:30)

As the Islamic Declaration at Assisi made clear, this gives us tremendous responsibility:

> 'For the Muslim, humanity's role on earth is that of a *khalifa*, vice-regent or trustee of God. We are God's stewards and agents on Earth. We are not masters of this Earth: it does not belong to us to do what we wish. It belongs to God and He has entrusted us with its safekeeping. Our function as vice-regents, *khalifas* of God, is only to oversee the trust. The *khalifa* is answerable for his/her actions, for the way in which he/she uses or abuses the trust of God.'

In Christianity, many traditions would follow closely the Muslim statement above. In particular, the Western Churches – Catholic as well as Protestant – have had a managerial or stewardship model of creation. It is true, some forms of Protestant belief developed into an outright exploitative model. But the bulk of Western Christianity has sought to be stewards of God's creation, taking seriously the words of the psalm, 'The Earth is the Lord's, and the fullness thereof.'

But in Orthodox Christianity a very different model has arisen. There is an inherent risk in a model which says we are little less than angels and have God's authority to rule the world. Sin leads to destruction of that role and to corruption of the authority we have received.

Orthodox Christianity resolves this power nexus issue by turning it on its head. St Paul says that God emptied himself of power and came to earth as a child, weak and defenceless. He took upon himself the role of a servant, and the Gospels often show Jesus in this light and in his own words as a servant. Orthodoxy takes the Christ-Lord-made-servant model seriously. It sees humanity's role to be the servant of the rest of Creation. Only the most powerful can take the risk and have the authority to humble themselves. This is the model Orthodoxy has developed. It is likened to the priest in an Orthodox community. Traditionally the priest was someone nominated by his community to take a special role in the life of the Church and the local community. As the Ecumenical Patriarchate has so clearly stated it:

> 'Just as the priest at the Eucharist offers the fullness of creation and receives it back as the blessing of Grace in the form of the consecrated bread and wine, to share with others, so we must be the channel through which God's grace and deliverance are shared with all creation. The human being is simply yet gloriously the means for the expression of creation in its fullness and the coming of God's deliverance for all creation.' [3]

Put simply we are called to be a blessing to creation. Look around you. Do you see much of this in our world today? Could such a radical view of our power and place in creation help?

There is a serious danger that we become trapped in seeing the world too narrowly, in too short term a view. For example, it has become common in environmental thinking to try and put a price on every living thing. The argument is that if we 'value' a whale or a rain forest for what it 'contributes' to the environment – usually meaning here human welfare – then we will treat it with more respect. But such a vision runs dangers of equating what a living creature or eco-system is with its cash value. We need constantly to be shifted away from a vision of the world in which meaning is given through human use of creation. It is what lies at the heart of the Abrahamic vision above with its emphasis on the meaning and significance of creation not lying in its instrumentalism.

One of the values of working with the systems of different faiths is that we encounter views which come at our vision of reality from the side and can, if we allow them, jolt us out of a comfortable complacency of world view. Hinduism and Buddhism both have this power in the West because of the highly different philosophical and cosmological premises from which they operate. Stephen Batchelor puts it succinctly:

> 'Buddhists [see] the universe as teeming with diverse living beings, longing not just to be, but to be something, to have something, to feel something. This collective craving is seen as the reason for the existence and constant renewal of the universe. It translates itself into external environments – complexes of thoughts, feelings and impulses with the tragic habit of grasping themselves as separate, solid and permanent selves.

> 'Human existence is just one of the six forms of life spread throughout the universe. Some are visible to the human eye, as with animals, but most are invisible. . . The presence of human beings is certainly not restricted to the planet Earth nor is the human species regarded as the best effort so far in evolutionary unfolding of matter. Nonetheless, existence as a human is seen as an exceptional opportunity; for it is the kind of life most suited to finding out what is going on here.'[4]

One issue which the environmental movement finds difficult to handle is what the Abrahamic traditions have called sin. Yet the desire to find 'sinners' runs very deep in environmental propaganda. I recall being very taken aback when I visited a very secular colleague in the International Union for the Conservation of Nature – the main international scientific body monitoring environmental problems and species loss. On his wall he had a picture of the world from space and underneath it words from the Biblical Book of Revelation: 'The time has come to destroy those who are destroying the world.' (Chapter 11).

Whether it is multi-nationals, governments, consumerism, neo-colonialism, the European Community or Communism, every one has their block of sinners who have brought us to this present state. What is missing so often is the insight which Thich Nhat Hanh spelt out in the quotation earlier: that the environmental problems are a personal problem; that the environmental crisis is a crisis of our own making; in other words, that we are all caught up in sin and the result is destruction. What is needed is a way out. Stephen Batchelor expresses this very well from a Buddhist perspective:

> 'If the Buddhist analysis of the ecological crisis is correct, then we are going to have to

do more than just switch to recycled envelopes and ozone-friendly hairspray to prevent the potential environmental catastrophe that a growing number of responsible voices are predicting. Yet to be realistic, we also have to accept that selfishness and greed are not going to vanish overnight.[5]

Buddhism has a tradition, the Four Noble Truthes and the Eight Fold Path, which offers a way out of the sins of selfishness and greed. Other traditions offer their experiences, garnered over centuries, even millennia, of working with human folly, pride and greatness, of how to tackle the issue of the wrongs we do, sometimes even for the best of reasons. It is perhaps time for the environmental movement to listen and watch these ancient wisdoms to find out more about how to work with and even redeem human beings.

For ultimately, this is what the environmental crisis confronts us with. Is it possible to change?

All the great faiths believe that the normal state of human affairs is less than the glory, the wonder and the capacities to which we can be called. I can think of no clearer statement of this than the heart of the Taoist statement on nature, issued in 1995 by the China Taoists Association. It has already begun to transform the relationship of ordinary Taoists to their environment in China through its dissemination throughout China. Let me quote it in translation at some length:

'With the deepening world environmental crisis, more and more people have come to realise that the problem of the environment is not only brought about by modern industry and technology, but it also has a deep connection with people's world outlook, with their sense of value and with the way they structure knowledge. Some people's ways of thinking has, in certain ways, unbalanced the harmonious relationship between human beings and nature, and over-stressed the power and influence of the human will. People think that nature can be rapaciously exploited.

'This philosophy is the ideological root of the current serious environmental and ecological crisis. On the one hand, it brings about high productivity; on the other hand, it brings about an exaggerated sense of one's own importance. Confronted with the destruction of the Earth, we have to conduct a thorough self-examination on this way of thinking.

'We believe that Taoism has teachings which can be used to counteract the shortcomings of currently prevailing values. Taoism looks upon humanity as the most intelligent and creative entity in the universe (which is seen as encompassing humanity, Heaven, Earth within the Tao).

'There are four main principles which should guide the relationship between humanity and nature:

'1 In the Tao Te Ching, the basic classic of Taoism, there is this verse: "Humanity follows the Earth, the Earth follows Heaven, Heaven follows the Tao, and the Tao follows what is natural." This means that the whole of humanity should attach great importance to the Earth and should obey its rule of movement. The Earth has to respect the changes of Heaven, and Heaven must abide by the Tao. And the Tao follows the natural course of development of everything. So we can see that what human beings can do with nature is to help everything grow according to its own way. We should cultivate in people's minds the way of no-action in relation to nature, and let nature be itself.

'2 In Taoism, everything is composed of two opposite forces known as Yin and Yang. Yin represents the female, the cold, the soft and so forth: Yang represents the male, the hot, the hard and so on. The two forces are in constant struggle within everything.

'When they reach harmony, the energy of life is created. From this we can see how important harmony is to nature. Someone who understands this point will see and act intelligently. Otherwise, people will probably violate the law of nature and destroy the harmony of nature.

'There are generally two kinds of attitude towards the treatment of nature, as is said in another classic of Taoism, Bao Pu Zi (written in the 4th century). One attitude is to make full use of nature, the other is to observe and follow nature's way. Those who have only a superficial understanding of the relationship between humanity and nature will recklessly exploit nature. Those who have a deep understanding of the relationship will treat nature well and learn from it. For example, some Taoists have studied the way of the crane and the turtle, and have imitated their methods of exercise to build up their own constitutions. It is obvious that in the long run, the excessive use of nature will bring about disaster, even the extinction of humanity.

'3 People should take into full consideration the limits of nature's sustaining power, so that when they pursue their own development, they have a correct standard of success. If anything runs counter to the harmony and balance of nature, even if it is of great immediate interest and profit, people should restrain themselves from doing it, so as to prevent nature's punishment. Furthermore, insatiable human desire will lead to the over-exploitation of natural resources. So people should remember that to be too successful is to be on the path to defeat.

'4 Taoism has a unique sense of value in that it judges affluence by the number of different species. If all things in the universe grow well, then a society is a community of affluence. If not, this kingdom is on the decline. This view encourages both government and people to take good care of nature. This thought is a very special contribution by Taoism to the conservation of nature.'

Encapsulated in this text is a radical new slant for many on what the world is and how to value it. I particularly like the final statement of how to judge affluence. If nothing else, this should be read by those who believe that nothing has meaning unless we either use it or 'value' it.

The rising involvement of the religions in the environmental movement has had dramatic effects not just on the scale and depth of ecology but also on the faiths themselves. To engage with other traditions is to put yourself at risk. From such risk can come changes of perception which help each faith and each strand within each faith to explore perhaps more deeply and more honestly the riches of their traditions as well as the flaws in their traditions. For example, while each faith has powerful teachings and insights about our role in creation/nature, few had done anything to awaken the social dimension of these teachings in the face of environmental decline. It took a secular body such as WWF to challenge the faiths to put into practice their teachings. I offer some case studies later on.

The changes within religions have been sometimes subtle, sometimes dramatic. One of the most astonishing has been in Sikhism. Sikhism has a model of cycles of time which span three centuries. In 1999 it celebrated the end of the first 300 year cycle from the founding of the Khalsa, the Brotherhood of Sikhism which marked the start of formal Sikhism as we know it today. For 300 years Sikhism has had to struggle against oppression and invasion – first from the Moguls, then the British and in recent decades, in its relationships with the State of India. The last 300 year cycle has thus been described as the Cycle of the Sword. In April 1999 the Sikhs inaugurated a new 300 year cycle. Through involvement in environmental work, and through the calling of an interfaith meeting, the Sikhs declared that the next cycle was to be that of Protection of Nature. To see such a major commitment from a faith which five years before had not a single environmental project to its name is to realise the power of releasing deep traditions and teachings to play a contemporary role.

In Christianity, the struggle to understand our special place in creation has led to many books, conferences and projects. But it took some strength of courage for the senior representative of the Orthodox Church to say in 1995 that what they had learned from Hinduism about the interconnectedness of all life was now informing their own understanding of traditional Orthodox Christian theology.

Meanwhile the Taoists, freed from decades of persecution, are discovering that ancient ways are now suddenly very contemporary; while Buddhism, which has often been a passive force, has now become active.

In Hinduism, old stories and legends of Krishna have been given new significance. For example, there is a tale that Krishna once fought and overcame a terrible serpent which dwelt in the sacred river Yamuna. Today that legend has been given relevance because the Yamuna is now terribly polluted. Once again Krishna needs to grapple with the serpent in the water – only this time it is human and industrial waste pollution. But, by using the legend to express a sense of shock at the desecration of the sacred river Yamuna, ordinary people can be motivated to action, for whom formal reports of the UN or by environmental specialists would be irrelevant. And one of the main geneses of the Yamuna project was interaction between Christian and Hindu religious ecologists.

When WWF launched its programme of work with the faiths in 1986, little was being done by the faiths to give life and meaning to these teachings. To some extent this was because they had been forgotten or glossed over. Sometimes it was because the faith itself had become caught up in other modes of thinking which blunted the message of care for creation. Or social and economic forces had so penetrated their culture with a vision of the consumerist lifestyle, that the old teachings were ignored by many as being out of date or irrelevant.

When WWF issued its challenge to the faiths to make their teachings real in the struggle to save the natural world, few could have foreseen the scale of the response. Some undertook a fundamental assessment of what their faith should be doing officially on conservation – such as the Baha'is and some of the branches of the Christian faith such as the Orthodox Church. Others began to realise that they had been practising passive conservation for centuries, through the influence of holy sites, mountains and temples. For example, Buddhism and Hinduism offer sanctuaries for endangered species by the simple fact that people do not log, hunt or trap on sacred land.

Classic examples of this are the Buddhist monasteries set within forests in Thailand. Because of the presence of the monasteries, the forests are sacred and thus less hunted in and damaged than elsewhere. This passive conservation input has now been made active. Some Buddhist communities in Thailand are now building small monasteries in endangered forest zones, thus making them sacred!

Another example of this passive conservation made active, from Islam, is that, in the city of Istanbul, which has grown from one million to eleven million inhabitants in twenty or so years, the only remaining breeding site for storks is in the grounds of the Mosque of Eyup. Here, gravestones occupy the hillside and the trees, which once covered all this area, survive only in the graveyard and mosque precincts. Without them, the storks of Istanbul would be history.

Yet others have begun to rediscover their teachings and to look afresh at their own lifestyles. The Jain programme to apply the principle of *ahimsa* – non-violence – with renewed insight to Jain business practices, has led to shifts in behaviour and in investment in line with traditional Jain practices.

In Zambia, all the main faiths – Baha'i, Christian, Hindu and Muslim plus people from traditional African Religion – have worked, together with the State Education Department, to create a multi-faith environmental education curriculum and provide resources for the state and religious schools of the country. This exploration of many teachings also includes many ideas for practical action.

On Mt. Athos, the Orthodox Christian peninsula of Greece, monastic ways of life have led to the preservation of the largest native Mediterranean forests left in the region. These are now protected by edict of the monks who own the whole peninsula and will be protected as Orthodox Christian Nature Reserves in perpetuity.

There are literally tens of thousands of other religious conservation projects that I could cite. World-wide, it has been estimated that over 120,000 religious based conservation projects have been created in the last fifteen years. Much has happened. Much still remains to be done.

But perhaps the most important way in which religions practise conservation is one which it is almost impossible to quantify. It does not manifest itself in grand schemes, nor in long papers. It is not presented to international conferences and does not figure in the assessments of political or economic leaders. It is presented through teaching, music, dance, festival and fasting, through sacred texts and prayers, liturgy and meditation, through song and silence. It is the vision of our place within a greater order; a sense of being part of a Divine purpose, however that is understood. It is the feeling of community with each other and with the rest of creation; the inspiration that we gain from an understanding of the wonder and glory of all creation.

These may seem to the hard-nosed scientist, the over-worked economist or the stressed-out conservationist, to be vague even sanctimonious ideals. Yet for millions upon millions of people, throughout history and to this very day, it is by way of the songs of their faith, the festivals and fasts of their calendar, and the rituals of their life-cycle that an understanding and relationship with the Divine, with themselves and with nature has been communicated.

Practical programmes are important. But they are fundamentally flawed unless there is the will, the vision and the belief in them to make them not only work, but become part of one's being. The most exciting thing about the rising tide of religious involvement in conservation is that, increasingly, it is becoming impossible to call oneself a person of faith without this affirmation having consequences for one's attitude towards nature. It is this which in the long term offers more hope than any number of reports and surveys of what is happening to nature.

We need to keep the dynamic balance between the practical and the inspirational. We need to keep the tension between the information which science can offer and the meaning which religions can provide. We need to strive to push for more and more work by the faiths, but not at the cost of losing the reflective and educational aspects of the religions.

When we started this work in 1986, we saw it as a mutual challenge. We saw the conservationists challenging the religions to make real their profound teachings. But we also saw the faiths questioning whether the secular conservation world had really understood what motivates people. That dialogue continues to this day. It is this challenge that the Alliance of Religions and Conservation (ARC) has been established to respond to – the creation of a new fund to assist the faiths in becoming more involved in conservation.

As a Christian, I believe in hope. I believe that people can change. If I did not, then I could not work on conservation. If I believed we were doomed by our greed and selfishness, I think I would curl up in a corner and die. It is because I believe in redemption, liberation, change – call it what you will – that I can continue to work in this area.

Perhaps this is the greatest thing that religions bring to the practice of conservation — hope. Hope that we can change, rediscover our truths and then journey on together, each in our own ways, but side by side, towards a world in which the whole of life is loved, respected and appreciated. For this is a truth that we all can find within our own traditions and even within our own hearts.

Chapter 16

And after all that?

Rom Harré

The Author

Rom Harré took a first degree in mathematics and physics and a second in philosophy and anthropology. He has had an abiding interest in the philosophy of science – in particular the role of language in the development of the sciences and in human interaction in general; and has pioneered the theory and practice of discursive psychology. He is an Emeritus Fellow of Linacre College, Oxford and Professor of Psychology at Georgetown University (Washington DC); also adjunct Professor Philosophy at American Universtiy (Washington DC). He is the author of many books.

Chapter 16

And after all that?

Rom Harré

'Philosophy is not the underlabourer of the sciences, but rather their tribunal; it adjudicates not the truth of scientific theorizing, but the sense of scientific propositions. Its aim is neither to engage in nor abjure science, but restrain it within the bounds of sense.'[1]

Introduction

The second millennium presents a panorama of one thousand years of philosophical discussion that looks immensely diverse if studied closely. By standing back a little we can see that it can be roughly divided into two periods of about equal length. In the first five hundred years of our millennium we could say that Western philosophy was driven by religion, and, in particular, the Christian religion. Philosophers struggled to make the Christian world view into a coherent and defensible account of human life and the material universe. The last five hundred years of philosophical reflections on the great themes of human nature, the universe at large and our place in it, have been driven by two new demands, that began to emerge clearly in the fifteenth century.

Science began to make claims to knowledge, challenging and displacing those based on the sacred books and other authoritative texts. How can the scientific world view and the methods of enquiry associated with it, be defended? Science takes the place of theology as the focus of philosophical reflections. However, this did not mean that the relation between the claims of science and those of religion was not taken by many to be problematic.

In matters of ethics the focus shifts from reflecting on the nature of Divine authority, either as it is manifested directly in sacred texts, or indirectly in such philosophical constructions as natural law theory, to the search for a source of moral authority within the confines of human life and the conditions for its flourishing.

After this journey of one thousand years, where have we arrived? There have been all these clever people reflecting on the nature of the concepts that we need for understanding the Universe and our place in it. There has been all this passion, these disappointments, these triumphs. Mistakes have been made. Errors have been rectified but they continue to plague us. This seems to be endemic to philosophy. Insights have been achieved, and have guided people for a while. Then they have ceased to satisfy. The cycle of intellectual attitudes, from the austere to the lush, from the atomistic to the holistic, from the sceptical to the naive, from the active to the passive, has been trodden again and again. It is visible in the orient as clearly as it is in the occident, though the points of equilibrium have been different.

One thing is sure: the broader issues have remained pretty much the same throughout the millennium. What is the nature of a person? What sort of universe do we inhabit? What sort of lives should we lead? The philosophical temperament which focuses on the concepts that we might employ for answering these questions, rather than trying to answer the questions themselves, is reawakened in every generation.

As Wittgenstein remarked: 'One keeps hearing the remark that philosophy makes no progress, that the same philosophical problems that had occupied the Greeks are still with us.'[2] The wheel turns on its own axle, but the juggernaut does seem, however slowly, to move on.

The homogenisation of language

Despite the arbitrariness of the celebration of a millennium which has neither mathematical, historical nor social legitimacy, nor indeed any great necessity, it does seem as if there is a definite character to the last few decades before the calendar changes. I will examine some that seem to me significant, auguring a future a little less confused intellectually than the recent past.

This discussion has to be seen against the background of the creation of a huge and quite pervasive system of information promulgation and exchange, and, with it the appearance of a nearly universally understood language, for the first time since the demise of Latin as the common language of literate Europe. At the same time we have learned how fine differences in the linguistic resources of a culture make for sometimes quite striking differences in cognitive techniques, emotional repertoires and moral sensibilities. If International English is not only the language of commerce and science, but also of adjoining domains such as psychology and medicine, we can hardly escape a degree of homogenisation of what not so long ago would have been our culturally distinctive casts of mind and ways of life. To take a simple example, the absence of any distinction in the second person forms of address in English allows one to postpone assessment of one's social relations to another person, sometimes indefinitely. But, if one must choose between 'tu' and 'vous', one must pay some attention to social 'rank'.

In this chapter I want to examine a newly hatched threat to the celebration of the achievements of human reason – post-modern relativism. I want also to look at some of the ways that scientists, adopting outmoded and naive conceptions of science, have made themselves vulnerable to anti-science criticism.

The most visible philosophical movement in the last years of the millennium has been postmodernism. Beginning in literary studies it has spread even into the philosophy of the natural sciences. It expresses a particular kind of scepticism or even nihilism about the prospects of progress in knowledge, in morality and so on. But to understand its origins and its characteristic claims we must first examine briefly, the tradition its advocates claim to succeed.

Modernism

Despite very great differences in detail the philosophical consensus that has dominated European thought since the sixteenth and seventeenth centuries has been that the rational analysis of observational and experimental studies of the natural world and, more recently, of people in interaction is the only legitimate method of arriving at knowledge of anything. This methodology is grounded in the deep assumption that in studying natural phenomena we confront, albeit through lenses of our own devising, a fixed world, of which we are thereby creating more and more accurate representations. This is how and ultimately why we do science. We tend to generalise this way of working as a tool for most human enterprises, be they conducted by philosophers such as Russell, almost all natural scientists, administrators in contemporary bureaucracies, or practical people digging tunnels or nurturing their kitchen gardens. This movement of thought has been called 'modernism'.

The philosophical content of Modernism can be summed up in four main theses:

1. *Foundationalism.* Every human practice, be it practical or intellectual, rests on some taken-for-granted foundation. For example scientific theorising rests on a foundation of observations and experiments. Legal processes rest on a foundation of promulgated law and the evidence of witnesses, and so on. There is something in every domain that cannot be queried. In the strongest versions of Modernism the foundations of a practice are to be found in the unmediated presence of some objective – that is culture and cognition free – reality.

2. *Universalism.* By the progressive application of scientific method it is possible to ascertain some principles that apply everywhere in nature. There is such a thing as a common human nature which is exemplified among all human beings. This might be found in genetics or in the deep structures of cognition, but it sets a limit to what is acceptable as a form of human life.

3. *Objectivism.* There are matters of fact which are independent of human thoughts and wishes, and we can find out at least some of them.

4. *Determinacy.* Though different aspects of the world show up from different points of view, the things and events in the world are determinate, that is have a definite character independent of the versions of them we create in our attempts at representation.

Modernism has been under attack from various sides since the eighteenth century. The Romantics of the late eighteenth and early nineteenth centuries tended to identify scientific and economic rationality with the Industrial Revolution, and the blight that cast on the quality of human life. William Blake, for example, gave us the image of industrial England as 'these dark Satanic mills'. The philosophers Nietzsche and Kierkegaard remarked on the demise of the old religion-driven consensus. Both were ready to contemplate the moral character of a world no longer dominated by religion with its ready-made answers to human problems. Nietzsche offered the 'will-to-power' as an alternative source of social order and secular authority. Hierarchical social organisations would replace the Church as the source of moral authority.

Much of the character of philosophy at the turn of millennium takes its flavour from a more systematic and vigorous rejection of established values, than these gadflies of the past, like Blake, could mount. I refer to the loose movement we now call post-modernism. I now turn to a brief description of this point of view. The main characteristics of this movement have been nicely set out by Cahoone.[3] He sees four themes in the post-modern interpretation of and criticism of modernist thought. Three are negative, more or less defined by those aspects of modernism they each reject. Complementary to these, there is a positive theme, from which characteristic post-modernist theses concerning specific topics are derived.

Post-modernism: the negative themes

Presence. Modernists, at least some such as David Hume, assumed that there were 'items' which were immediately given by the senses. Hume called these 'impressions' . The legitimacy of every concept was referred to them. If one wanted to know the meaning of an idea, a mental content, one sought the impression from which it was derived. This has been called the 'doctrine of presence'.

> '*Presence* refers to the quality of immediate experience and to the objects thereby immediately "presented". Post-modernism . . . denies that anything is "immediately present", hence independent of signs, language, interpretation, disagreement etc.'

The defects of the doctrine can be illustrated from the case of simple perception. Even to see an apple as an apple one must have some kind of cultural and linguistic preparation. There is no such thing as a pure perception, or even a pure sensation, one into which one's culture has no input at all.

Origins. Modernists, perhaps influenced by the success of the hypothetico-deductive method in science, or perhaps in the long run by Descartes' enthusiasm for a deductive presentation of the system of the world, tended to try to construct regresses from problematic situations to less problematic foundations. Post-modernists saw this as an illegitimate search for 'origins' in every field of human endeavour, and a characteristic modernist illusion.

> '*Origin* is the notion of the source of whatever is under consideration . . . Inquiry into origins is an attempt to see behind or beyond phenomena to their ultimate foundation.'

For example, it is futile to try to ground our perception of things more securely by analysing them into elementary sensations, in which they might have had their origin. The claim to be seeing a horse is no less secure than the claim to be seeing a brown patch, so it is futile to try to base one's perceptual claims on allegedly more secure claims about sensations. 'Brown' is no less a culturally embedded word than 'horse'.

Unity. Modernists can, with some justice, be accused of trying to find, in any complex field of experience, the essence of whatever there was. What is justice, really? What is the one authentic basis for moral judgements? What are the necessary and sufficient conditions for the use of the word 'beauty'? And so on. Modernists, for the most part, thought there must be such unities.

> '[*Unity*] . . . post-modernism tries to show that what others have regarded as unity, a single, integral existent or concept is plural.'

For example, enlightenment rationality presumed there was only one kind of truth, correspondence with the facts. But, according to post-modernists, there are many kinds of truth, relative to persons, purposes and projects. Thus feminists of a post-modern persuasion claim that there are truths for women, and even a feminist kind of science, with its own domain of problems and its own standards of good work, which differ from those of the science created by men.

Transcendence. It is true that modernists have taken for granted that it makes sense to try to identify and then to argue for norms and principles that stand above their moments and contexts of application.

> 'Norms such as truth, goodness, beauty, rationality are no longer regarded as independent of the processes they serve to govern or judge but are rather products of and immanent in those processes.'

For example, most people presume, even when they are not professionally engaged with investigating the standards of sound reasoning, that there are principles of logic that are independent of particular processes of reasoning in everyday contexts. So, for instance, the denial of some claim, if denied, affirms the original claim. We tend to think of the principle that something is either A or not-A, an apple ripe or not-ripe, as a self-evident basis for all logical thinking. But, post-modernists have argued that whether it is to be made use of depends on the context and the situation, and the kind of reasoning we are employing.

These quotations are taken from Cahoone.[4] Each identifies a 'modernist' concept or leading theme which is denied, rejected or drastically revised by various authors and to different degrees. The authority of immediate experience, the existence of unshakeable foundations, the universal character of concepts like 'truth' and their independence of context, the determinacy of meanings, are all denied by post-modernists.

Post-modernism: the positive doctrine

The techniques of post-modern analysis can be summed up in the simple principle of *constitutive otherness*. The strategy is to argue that 'what appear to be [free-standing] cultural units – human beings, words, meanings, ideas, philosophical systems, social organisations – are maintained in their apparent unity only through an active process of exclusion, opposition and hierarchisation'.[5] The upshot is a stance characterised by the rejection of any attempt to find foundations for beliefs, doctrines, practices or social orders. Applied to the natural and human sciences the principle leads to a denial of the pre-eminent place of scientific method as the source of the best accounts of nature and humankind. Every scientific story, it is argued, excludes all sorts of other stories which, in the absence of foundations, have as much claim on our attention as the best that chemistry, physics and biology can do. Every moral insight can be matched by another, in which the moral quality of an act can be seen differently. Every claim to authority can be resisted by demonstrating its dependence on mere convention.

Post-modernism and analytical philosophy

Reflection on these facets of post-modernism soon reveals that, in a sense, there is nothing new here and little that most philosophers of the analytical tradition, particularly those for whom language use has been the main focus of attention, would disagree with. The rejection of 'presence' is no more than an oblique way of denying the long outdated sense datum theory in general philosophy, and hardly differs from the principle of the theory-laden-ness of observations that supported the rejection of the main tenets of logical positivism in the philosophy of science half a century ago.

As regards 'origin', scarcely anyone would seriously support naive forms of realism, either in philosophy of language or philosophy of science. Contemporary realisms are subtle and refined interpretations of how our knowledge is related to a world that exists independently of ourselves.

The denial of 'unity' is not so different from Wittgenstein's warnings against simplistic interpretations of the uses of words, ignoring the way that language use displays fields of family resemblances rather than hidden semantic essences.

The denial of universal standards is another matter. There is a gap between the patterns of reasoning to be seen in the natural sciences and the authorised standards of rigour expressed in formal logic. There is also a gap between the multitude of ways of life that history reveals, and anthropology has discovered, and unduly simplistic moral theories. Is there then no authority in scientific method over any other source of knowledge, and must we accept that there can be no entitlement to a universal moral standard to which all regimes, whatever their surface variety must conform? Many contemporary philosophers, I believe, would conclude, instead, that logic and moral philosophy were incomplete. There are other standards of rationality and modes of reasoning in use in the sciences than those captured by logic, and there are universal moral intuitions yet to find a soundly argued grounding in some universal feature of a specifically human form of life.

Finally the principle of constitutive otherness is a commonplace technique in analytical philosophy. Austin,[6] echoing observations that go back at least to Thomas Hobbes, pointed out that it is a mistake to search for the positive characteristics of such important concepts as 'reality', 'freedom' and so on. Their meanings in use come from what they are being used to exclude, context by context.[7] However, this quick dismissal of constitutive otherness is not quite fair to the insight that post-modernists are trying to convey.[8] The excluded includes the 'voices' of those who have been unheard, either by tradition or by deliberate exclusion from a domain of discourse, such as psychology. Here post-modernism shows its political aspects most clearly. Modernism favours accounts of things based on empirical evidence and organised according to the principles of traditional logic. But there are other voices offering accounts of things, sometimes more or less the same things, drawn from other sources and ordered according to other principles.

A critical assessment of post-modernism

Post-modernism strikes me, as I believe it has struck others, as the product of a recurring pattern of unwarranted exaggerations of commonplace observations, analyses and arguments that few would disagree with. The post-modernist literature is shot through with sloppy and careless uses of words and with ill-understood borrowings from the advanced sciences and mathematics.[9] Dramatic claims such as 'that there [has] been a radical dissolution of the self in the post-modern era', can be made plausibly only if central concepts like 'person', 'self' and 'identity' are used without due care. These concepts are not univocal. They encompass subtly various fields of family resemblances. For example, the word 'self' is used both for personal identity, the formal singularity of a person, and also for the set of attributes that a person possesses. The fact that, in each of us, the totality of our personal attributes seem to fall into groups, some salient in some circumstances and some in others, does not show that each individual does not have a robust sense of individuality. The same can be said of the related concept of 'personal identity'. The multiplicity of 'selves' in the second sense is a mundane fact of life, but the thought that I may be multiple in the first sense is thrilling. If these and other vital distinctions are ignored, the case for post-modernist declarations of the 'death of the self' are dissolved into a sort of philosophical sludge.

The problems that are thrown up by this movement show up in their discussions of the sciences and in their relativistic moral philosophy. Post-modernism is one facet of a broader slide into relativism that characterises a good deal of the philosophy of the recent past. While the nineteenth century saw the decline of religion as the source of moral relativism, in the twentieth century the expansion and dissemination of anthropological studies of distant cultures has fuelled some of the enthusiasm for both scientific and moral relativism.

Moral philosophy at the turn of the millennium

In academic circles, moral philosophers have been turning from the project of analysing the logical grammar of moral assertions, such as can be found in the writings of Hare[10] and Nowell-Smith,[11] to defences of moral 'realism'. The analysts tended to leave open the question of which ways of life to advocate, since they were concerned with the nature of the *processes* of advocacy. This opened the way to moral relativism. The content of moral judgements as a criterion for picking them out as moral, gave way to the revival of formal criteria which can be found in Kant's moral philosophy of two hundred years ago. In particular, the principle of universalisability was central. That simply required that, if I advocate a certain moral principle for others, I must also subject myself to it. If I am not willing to do so, the principle is not a moral principle.

Post-modernists deny that there is any universal content to be found in the moral intuitions and judgements of all human societies. Even if some feature of human associations were to be found to be ubiquitous, it could not serve to ground any moral universals. Hume[12] famously denied that one could validly derive an 'ought' from an 'is' – a moral principle from a matter of fact. How could Hume's principle be breached? There might be a general notion of well-being that is such a pervasive feature of human life, that it would irrational not to take it as a goal towards which all human beings should direct their actions. Griffin has explored some of the consequences of taking up this idea.[13] It has echoes of Aristotle's idea of the flourishing life, which he took to be a universal source of moral and political concepts. Moral relativism is not just a consequence of post-modern criticisms of the heritage of the enlightenment. It has been a recurring theme in this century. It seems to have received a strong boost from anthropology.

Anthropology and moral relativism

During the 1940s the realisation that there were many seemingly viable moral systems, other than the Christianity that the colonial regimes had taken for granted, began to filter through into Western philosophy as a ground for moral relativism. The beginnings of this interpretation of the discoveries of anthropologists are usually credited to Franz Boas. It is worth remarking that Boas did not argue from cultural diversity to relativism, moral or otherwise. His argument is much more subtle, and deserves to be set out in some detail.

He begins by pointing out that cultural diversity can be explained in various ways; there are 'external conditions or internal causes which influence [people's] minds'.[14] But if we are to treat these explanations as showing merely superficial variations on universal aspects of the human organism or the human mind, we must establish that there are relevant universals. The assumption of universalists is that, 'if an ethnological phenomenon has developed independently in a number of places, its development has been the same everywhere. . . This leads to the generalisation that the sameness of ethnological phenomena found in diverse regions is proof that the human mind is always the same everywhere'.[15] Thus, diversity would be superficial variation on a fixed foundation. But, he argues, there is a flaw in the reasoning, since 'the same phenomenon may develop in a multitude of ways'. There is no logical basis for the 'conclusion that there is one grand system according to which mankind has developed everywhere'.[16] Boas emphasises the importance of specific historical connections in the development of specific cultures. We must study ethnological phenomena, not in isolation, but in the whole specific cultural setting. The best presumption, when we find 'an analogy of single traits of culture among distant peoples. . . is that they have arisen independently.'[17]

This is 'Boasian relativism'. It is a very different way of arriving at relativism from those which come from outright rejections of either universal biological determinism such as that proposed by Tiger and Shefer[18] or the psychological universals determinism such as those proposed by the followers of 'Chomskian' searches for cognitive universals in the deep structures of patterns of thought.[19]

A resolution of the inconclusive debate between universalists and relativists in the human sciences, inspired or at least initiated by anthropological explorations, has been proposed, with his usual panache and subtlety by Geertz.[20] He draws a distinction between two conclusions that might be drawn from the anthropological literature, which describes so many diverse human customs and practices. He cites, at random, 'ghost marriage, ritual destruction of property, initiatory fellatio, royal immolation and . . . nonchalant adolescent sex' as examples of ways of acting alien to our way of life.

What should we say? The wrong thing is to conclude that there are no cultural universals. It should never have been any anthropologist's intention to prove such a thing. Indeed Boas had already shown that any such proof would be invalid. The correct thing to conclude is that we should not uncritically take it for granted that our local cultural forms and practices are the universal ones, to be looked for beneath the 'strange versions' that have appeared elsewhere. Wittgenstein made the same point in his critical remarks against Frazer's treatment of alien cultural practices as bad or primitive science. They were not science at all, so they could not good or bad science.[21]

Anthropological field work establishes neither radical relativism nor radical universalism. It should be read rather as a 'repositioning of horizons and the decentering of perspectives'[22] comparable to the Copernican revolution in astronomy.

The post-modernist attack on the possibility of moral universals

The same relativist argument that purports to debunk empirical criteria of truth-to-nature as the basis of confidence in science-generated knowledge, has been applied to political and moral issues as well. Arguments against the viability of the idea of an essential human nature are treated by post-modernists as arguments for moral relativism. They are supposed to dispose of any claims for trans-situational, trans-cultural moral criteria.

> 'The more basic problem these [Aristotelian and Kantian ethics] and similar attempts to ground ethics in an account of intrinsic or essential human nature is our strong post-modernist suspicion that there is no such thing. . . There is no historical essence that is both universally found and yet is also determinate and substantial enough to generate or justify. . . a definite ethical theory.[23]

To do justice to the post-modernist localisation, and hence relativisation, of morality I shall summarise the best short argument in the literature on this matter – an argument developed in various places by Richard Rorty, but elegantly expressed in a recent essay.[24] It has all the characteristics of the best post-modernist advocacy, an air of sweet reasonableness that should lead to a conclusion that anyone would say was morally outrageous. But Rorty escapes our moral censure only by abandoning a central principle of the post-modern position, and, indeed, subverting his own argument.

His discussion begins with a distinction between two ways of making sense of one's life. One is 'by telling the story of their contribution to a community', the other 'is to describe themselves as standing in an immediate relation to some non-human reality'.[25] The one finds a basis in *solidarity*, the other in *objectivity*. Realists wish to reduce solidarity to objectivity, by developing procedures by which human beings can form representations of non-human reality, that are not merely social. For example, I would add, by using the causal powers of material things in experimental apparatus. Pragmatists wish to reduce objectivity to solidarity. For pragmatists, 'the desire for objectivity is not the desire to escape the limitations of one's community but simply the desire for as much inter-subjective agreement as possible'[26].

Now comes the fatal compromise. Rorty agrees with the inheritor of the values of the enlightenment that we should both adopt and seek to justify 'toleration, free enquiry and the quest for undistorted communication. . . [for the pragmatist this] can only take the form of a comparison between societies which exemplify these habits and those which do not, leading up to the suggestion that nobody who has experienced both would prefer the latter'.[27] There are two glaring flaws in this important move, indeed the vital move of the whole defence. How would one be able to pick out exemplifications of tolerance, free enquiry and relatively undistorted communication unless one had 'objective', enlightenment standards for so doing? Rorty admits to falling across the dilemma between ethnocentrism and relativism, and opts for ethnocentrism – that is for societies which exemplify *his* three good things. But this option is just what leads to the vicious consequences of moral relativism. Now we have no criterion for preferring his three good things to other triads of social goods, such as 'Duty to God', 'Respect for one's betters' and 'Woman's place is in the home'; or why not 'One Folk, One Land, One Führer'?

Geertz[28] rightly castigates those who would find cultural universals in genetics and the biological endowment, since this is surely too coarse grained to touch the parts of life that only philosophy can reach. And he also castigates the Chomskians, with their claims for the existence of species-wide cognitive structures, a revival of the doctrine of innate ideas once roundly criticised by John Locke. It seems to me that we are ending the millennium with a problem that was thrown up at the time when there arose an alternative vision of the Cosmos to the theological. Where can we find a moral grounding that will be binding on all human beings?

Some disquieting liaisons

Reflection on the state of our moral sensibilities, at least in the West, cannot ignore the phenomenon of 'Political Correctness', 'PC'. The basic principle of argument by PC criteria is that we should assent to or reject a narrative, be it scientific, literary, moral or even journalistic, only on the grounds of its likely efficacy in some political engagement. The basis of this stance to moral issues is paradoxically authoritarian. While the larger motivation is to redress injustices that come from the fact that certain groups of people are treated as defectively different with respect to the goods taken for granted by the majority, PC works, in practice, by certain moral entrepreneurs claiming to have privileged access to what is morally right. This claim to privileged access is based on a post-modernist attitude to knowledge. In Rorty's words, knowledge is generated by solidarity rather than objectivity. The claims of knowledge concerning the natures of men and women, the characteristics of different cultures, the abilities of the disabled and so on are to be related to practical programmes which promote the interests of those deemed to be disadvantaged. Factual correctness (objectivity) is to be sacrificed to political correctness (solidarity). According to post-modernist views of epistemology, objectivity in knowledge creation is an illusion.

Political Correctness and post-modernism are logically linked . A major argument for privileging political correctness over factual correctness is the post-modern doctrine that factual correctness is the result of an arbitrary choice of discursive conventions, favouring those who are already privileged. Political Correctness becomes the privileging of some non-arbitrary moral intuitions, the moral values inherent in the lives of some hitherto unprivileged group of people. Paradoxically, these intuitions are treated as if they were absolutes, and not themselves subject to a post-modernist critique. But they, too, are the product of social and linguistic conventions, as we outsiders can easily see.

The recovery of morality

We need to establish some universal moral principles, which override the extreme relativist interpretation of local variations that would allow a defence of the killing fields, or of the holocaust, on the grounds that they were ways of living that seemed right to the people who carried them out. The post-modernist plea that the voices of the underprivileged or otherwise silenced members of our society should be heard, and their moral intuitions respected, can, if generalised, turn into as monstrous a doctrine of moral permissiveness as one dreads to find. The principle of constitutive otherness and the denial of 'origins', central principles of post-modernism, lead straight to the conclusion that the moralities of the proponents of the Final Solution or of the methods of social transformation by massacre, adopted by the Khmer Rouge, are 'other' than the advocates of liberal, Western Enlightenment moral and political principles and so are voices that must be heard. According to the denial of 'Origin', our disgust and abhorrence of such moralities has no force beyond the confines of our culture. The denial of 'Origin' can lead someone of sensibility and moral decency like Rorty to such a disturbing statement as this:

'. . . what counts as being a decent human being is relative to historical circumstances, a matter of transient consensus about what attitudes are normal and what practices are just and unjust'.[29]

Rorty proposes that 'something behind history' such an 'Origin' as a general trans-historical conception of human decency, is to be replaced by a 'sense of human solidarity'. On what might this be based? Fellow feeling? And how would such a sense distinguish the fellow feeling of the guilty thrill of a sadistic solidarity with those who organised the vanishing of the *desaparacidos* in Argentina or Northern Ireland, from the solidarity one might feel with the victims of the Kosovo massacres? This is thin moral gruel indeed. We need something more robust than that. Fellow feeling is a poor criterion for attaining moral value.

Holiday and the conditions for the possibility of language

Offering a somewhat less tortured and abstract argument than that made famous by Habermas,[30] Holiday has made a good case for the necessity of certain moral universals.[31] His argument is based on an analysis of the conditions for the possibility of language. This makes a powerful case for some over-arching moral principles, since all human cultures are language based. The argument has the added advantage of forcing post-modernists up against yet another paradox. The conditions under which their arguments are advanced in favour of moral relativism – that is of the claim that moralities are strictly local and tribal – must include the conditions for the possibility of whatever language they are using to make the point. And if those conditions include moral conditions, then these they must tacitly accept. To *argue* against, or even to speak against, there being any universal moral principles is to presuppose at least some.

If language is a social phenomenon, existing in the actual and potential communicative relations between people, then there must be certain relations that are necessary for meaningful linguistic acts to be possible. Here are Holiday's three necessary conditions for the possibility of language. There is no suggestion that they are jointly sufficient.

1. There must be a general acceptance of the power of ritual to fix the conventions on which the possibility of linguistic communication depends. Meanings cannot be established by force. Each person in the community of speakers and listeners must be willing to be bound by the conventions in force in that community. It cannot be that some are and some are not. In this way it is unlike sexual mores or the structures of social hierarchies. Variations in any of these would not undercut the possibility of there being a social order of some sort.

2. There must be a mutuality of respect in giving all persons the right to make meaningful statements. As Wittgenstein would say, this is a remark about the grammar of 'person'. For example, if members of one group issue orders which they expect another group to understand, but ignore the speaking of the dependent group as having no meaning for them, then they are denying to the dependents the very linguistic capacities which their giving of intelligible commands presupposes must exist.

3. Holiday's third condition concerns the institution of truth telling. Even if there were a great deal of lying and deception, in the absence of the institution of truth-telling the distinction 'true/false' would lapse. It would have no application. All statements would then have the same status as efforts at communication, namely none. Language, as the cement of society, would not exist. People might vocalise but they could not speak.

The point is that these are not merely pragmatic conditions for the efficacy of communication, such as clarity and loudness. They concern relations between people, which boil down to mutual respect and sincerity. In other words they incorporate minimal conditions for the preservation of persons.

Summary

Take away the authority of the Church as the mundane expression of God's moral authority, and the diversity of moral orders as the foundations of the normative systems of human forms of association seems very striking. The problem for moral (and political) philosophy is to find a way of acknowledging this, yet of resisting the slide into moral relativism. We need something that all societies must have, in order to provide a source for some minimal moral practices that all societies must be based on. Language offers itself as the most plausible universal source of morality, in that, as Holiday and Habermas have argued, among the conditions for the possibility of language there are some which are moral.

Science at the turn of the millennium

The prevailing tendency to scepticism, illustrated by the post-modernist attitude to the possibility of universal moral principles, is also to be found in attitudes to science. We can catalogue four major positions in contemporary thought about the nature and status of scientific knowledge. There are many who are committed to the defence of scientific realism. This is the thesis that the sciences, thanks to the evolution of a cluster of special methods of enquiry, are able to give human beings better and better insights into and symbolic representations of the world, as it exists independently of human beings, their thoughts and practices.[32] There are many who reflect on the aesthetic quality of scientific accounts of nature and humankind. Some, such as Appleyard,[33] find the scientific picture to be bleak and soulless, somehow missing the point of what it is to be a human being in a material world. Others, such as Dawkins[34] and Wilson[35] thrill to the power and pleasure of 'doing science' and delight in the depth and intricacy of its insights. The third group are the sociologists of science, who have asked themselves how the actual beliefs of the members of scientific communities can be explained, when the evidence for these beliefs falls far short of what might be demanded by strict logic. The answer, they argue, is to be found in the social relations between the members of scientific communities and the general social conditions of the time. The fourth position is post-modernism. If there are no secure foundations for beliefs and no sustainable universal truths, if social solidarity is to replace the creation of knowledge in engendering relations with the material world, the very bottom is knocked out of the pretensions of the scientific community to have a view of the world and of its inhabitants which is superior to that of any lay person.

Anti-science as a mix of myth and insight

The roots of anti-science in some pro-science attitudes

Purchase for anti-science comes in part from mistaken accounts of scientific work itself, promoted not only by philosophers but also by distinguished scientists, for example E.O. Wilson.[36] Ironically these mistaken accounts are survivals of long-since refuted philosophical analyses of science, particularly positivism and conventionalism – the most robust and long surviving alternatives to realism. If *that* is what science is, then we are not impressed. Sociologists of science seem to take, as the best shot at displaying the rationality of science, the problematic analyses of scientific method and the content of scientific claims to knowledge offered by positivists and/or conventionalists.

Positivism, as we may be reminded, is the doctrine that the sciences can offer us no more than detailed catalogues of observable phenomena, which, under certain circumstances, can be generalised into probabilistic laws of nature. We collect innumerable instances of the phenomena of falling bodies, tidy these up by rigorous measurements, and generalise them into the Galilean laws, which summarise what we know and allow us to predict new phenomena of the same kind. What of theories? Positivists, confining the content of the sciences to what can be observed, deny that theories are about an independent and unobserved domain of reality in which we might find the explanations of the phenomena we observe. Theories are, it is claimed, merely logically organised hierarchies of laws, which serve as devices for the ordering and prediction of descriptions of observable phenomena. Thus positivism is the very exemplar of anti-realism in the interpretation of science. Realists hold that theories can generally safely be read as referring to real entities and processes, beyond the bounds of what can be observed, be this in nature or in the reactions of instruments. Theories are the very means by which we can catch glimpses of a reality beyond the bounds of the perceptible. To deny this is to shake the very roots of science. Scientists, intent on censuring unwarranted speculations and excessive claims to insights into reality, are attracted to positivism's austere restrictions of the scope of scientific knowledge. Yet, in their everyday work, these same scientists make use of theories, and are usually to be found enthusiastically defending the reality of what was yesterday's speculative leap beyond the frontiers of the observable. Neuro-psychologists, while carefully eschewing talk of mental phenomena, are quite ready to talk about, refer to and explain observable phenomena by molecular processes. But molecules are no more publicly observable than thoughts. The positivism of most scientists is selective. They reject only those theoretical entities that they find obnoxious for some extra-scientific reason. Those they like, they keep.

Conventionalism, the point of view made popular a century ago by Poincaré[37] – and very much with us still – involves the claim that what is observable is a function of the conceptual system employed to identity and describe the phenomena of the domain in question. A chemist could not pick out examples of *oxidation* unless he or she was familiar with the concept. Conceptual systems are fixed by convention not observation. We do not first observe a phenomenon and then find out that it is a case of oxidation. We define oxidation and then we can identify the phenomena. So concepts are *a priori* and their applications *a posteriori*. However, concepts can be readjusted to fit whatever comes up as the results of applying them to what human beings can perceive. So thought and experience always match. It is easy to slip from the insight that we can perceive only what we can conceive, into the extravagant and obviously false but exciting claim, that what we perceive is created by our concepts. Now comes the irony of these anti-realist, pro-science positions.

Take these two supposedly pro-science insights together, and we get the postmodern position that the discourses of the sciences are just one genre – among an open set of possible kinds of stories that could be told – about the plants, animals, minerals, the astronomical environment of the earth, the nature of human cognition, and so on. Science is privileged neither as a royal road to reality, nor as a uniquely powerful way of categorising, recognising and ordering phenomena. Add to this the text metaphor, drawn from literary theory, that the world is like a text (and thus capable of indefinitely many alternative readings) or indeed, as some have said, *is* a text, and the work of the sciences would then be a branch of hermeneutics – the techniques of revealing the hidden meaning of texts. But since the text itself cannot determine how it is to be read, there are indefinitely many possible readings of the world – that is scientific claims about the nature of phenomena and their explanations.

The realist response to conventionalism is that the world does not admit of the use of any possible system of concepts, since the world is not created by us; some fit better than others. This is not by comparing the world and the concepts to see if they fit, but rather it is that, in trying to use conceptual systems, some are more practically effective and more enhancing of the logical coherence of our discourses about the world than others. Yes, we do, as Goodman says,[38] compare versions with one another, but on the assumption that there is world to which they are oriented.

Wittgenstein pointed out that though conceptual systems ('grammar' in his terminology) are not abstracted from the world or brought into being by it, nevertheless they are intimately related to the world.

> ' "So you are saying that human agreement decides what is true and what is false?" – It is what human beings *say* that is true and false; and they agree in the *language* they use. That is not an agreement in opinions but in form of life.'[39]

Whether what we say is true or false is a matter of how the world presents itself to us, while the conceptual systems we use to describe it are our creations. In so far as the way the world presents itself to us is a product of our uses of conceptual systems, then every view we could ever have would be selective. But it is one thing to say that, and quite another to say that the world as it is presented to us is wholly our own creation.

The post-modernist attack on science.

We can see this as yet another aspect of the rejection of the Enlightenment privilege of reason over tradition and feeling. This is tied in with the rejection of 'Origins', with the denial that there are secure and unchanging foundations for knowledge and for morality. If there were a fundamental text, it could serve as an 'Origin' for readings. Once this had been done, it would be possible that the readings could be ranked. If there a world beyond the text as an 'Origin' for theories, these could be ranked.

In practice post-modernism seems to move from the claim that there are a multiplicity of legitimate readings to the thesis that there is no constraining 'Origin'. Every kind of narrative can be infected by 'multiplicity' in two main dimensions. For each narrative as document or text, there are a multiplicity of readings. For every event as a topic for a narrative, there are a multiplicity of narratives. Suppose these dimensions are collapsed on to one another, simply by taking topics as themselves texts. Then the multiplicity of narratives for every topic can be treated as a special case of the multiplicity of readings of narratives as texts.

So far so good. But, now, let us ask whether there is any way of ranking readings as more or less authentic or true-to-the text? Literary scholars have argued that there is no such way. No reading has any better claim to authenticity than any other since the original author's intentions are either unknown or irrelevant.

The genesis of post-modernism in philosophy of science can be understood as no more than the extension of this claim to the double multiplicity of possible readings of any text. What is true of readings of literary texts is also true, so it is claimed, of narratives as readings of what are metaphorically also texts. According to the post-modernists no narrative has priority over any other as a better or worse account of its putative topic.

Science is not read off the world by a passive but attentive observer, by 'immaculate perception'. It is the result of active intervention by well-informed people using all sorts of probes to force the world to manifest itself in various ways. Where do the concepts come from? They are drawn from some reservoir in the culture. Concepts are prior to percepts, language is prior to that which it can be used to describe, and both are prior to acts of recognition of the type of events and things that constitute the material and the social world for those who deploy some *particular* conceptual resources. Here is one of the points at which radical relativism gets its purchase. Nature has be interpreted. So Nature is like a text, and texts can have many readings.

Post-modern radical relativism has come from borrowing from literary theory to analyse the writings of scientists. If no reading of a text is privileged, and the world as it is presented to us is also a text, and scientific theories are readings of it, how do we decide which story we prefer? It cannot be by consulting nature, because that consultation too can yield only another story. At best we have intertextuality, the attempt to judge one text in terms of another, when both are capable of indefinitely many readings.

Of course, this line of attack irritates professional scientists immensely and with good reason. It exploits superficial analogies between science and literature. While crying out in favour of contextuality, it abstracts from the very contexts which would serve to ground a ranking of stories in just the way that scientists informally rank the results of their work. How to rank a story depends on the project in which one is engaged. If one wants to find out about the origins of the sexual impulse, biology and genetics are a better guides for that task than Botticelli. But if one wants a superb display of the force of the female body to engage the male attention, *The Birth of Venus* would be a splendid choice.

Why is it thought that no distinctions of authenticity or truth can be drawn between a multiplicity of narratives by reference to how adequately they present the topics they address? This radical claim is based on the thesis that a new and complete world-for-us is created by the acceptance and use of each distinct story-line. So, any attempt to use the world as a resource in deciding between story-lines is pointless, since that would require that there to be an access to the world independent of the current story-line. There is, it is claimed, no such access. The world no more constrains scientific narratives as to their authenticity than the printed texts of a poem or novel constrains the readings of those who engage with it. Each narrative generates a world that is in general accordance with it.

> 'The word doesn't reflect or represent the world; the word contains the world and not the other way round. Therefore, texts are self-referential – they refer only to themselves, not to anything outside themselves. There is no such thing as the real world . . . All that's left is a succession of misleading signs, a parade of words beyond the power humanity to control them.'[40]

If we add one further twist to this tale, we have the whole story. What determines which reading any particular person makes, of the literal text or its metaphorical counterpart – which would be, to a scientist, some material state of affairs? Their interests, of course. And here we get the tie to the hitherto covert political dimension. If there are no distinctions to be drawn between narratives with respect to the authenticity of their accounts of the topics they address – whether they purport to be about nature or about human life – then, of all the criteria by which they had been assessed in the modernist era – the era of faith in science and the hegemony of logic – only two kinds of criteria have survived, aesthetic and political/moral.

What is right in anti-science? That scientific knowledge is not arrived at inductively, and that for any given domain there are alternative stories. But these stories are not on the same footing the moment we introduce the question of 'what task?' The sociological account is not only paradoxical but incomplete, since, though I may explain why Darwin believed that fossils were survivals from the remote past, and Fitzroy did not, by reference to upbringing, class affiliations and so on, none of these considerations explain whether each ought to have believed what he did. However, when standards of correct reasoning are introduced into the debate, the question of their origins has been raised.

The recovery of science.

Some of the impetus to anti-science has come from misunderstandings about scientific method, drawing a pattern from the reworked and rhetorically elaborated texts to be found in journals and monographs. These systematically conceal the route from the world to the text.[41]

Model making and model matching

Since the beginnings in antiquity of that human project that we now retrospectively recognise as science, its practitioners have set about explaining the observed in terms of the unobserved. The relative positions of the heavenly lights was explained in terms of the cosmic architecture of planetary motions, the state of the human body in terms of the balance between the four humours, and so on. The power to penetrate beyond the deliverances of the unaided senses is one of the glories of the scientific method, but it also offers a foothold for sceptics. Guesses about the unobservable substrate of the phenomena of the observable world have often turned out to be wrong. Nor can any hypothesis about the unobservable be held to be true with absolute security. None is immune from the possibility of revision.

This aspect of scientific practice raises two deep problems that have persisted in the philosophy of science for millennia: how does anyone know what is the nature of the relevant unobserved processes, structures and so on? And how does anyone know whether anything corresponds to what seems to be being described in these explanatory hypotheses? These questions must be answered to defend the authority of science in matters of the gathering of knowledge. A start can be made in looking very closely at the source of nature-penetrating ideas. We get our ideas of the character of the unobserved part of a domain of phenomena – for example what is happening in a plant that has been watered – by the use of metaphors and models. These devices allow thought to penetrate beyond what has already been given. Since both draw upon what we already know, this kind of thinking beyond the limits of observation is meaningful. But what exactly are models and metaphors? And why do they have this magic power?

Models are representations, but they are not arbitrary in the way that words are. ☺ represents a facial expression in a quite different way from the way the words 'smiling face' represent it. A famous scientific model is Darwin's Natural Selection. Whatever the process is that brings about the transformations of species, it was not observable to Darwin, nor to Capt. Fitzroy. While both saw the beaks of Galapagos finches and the bones of dinosaurs, neither could show the other the act of creation. Darwin's model is the epitome of superb scientific reasoning. He tells it in the first part of the *Origin of Species*. If we think of Nature as something like a farm, and of new species as something like new breeds and varieties, then we can look for something in Nature that is analogous to the selective breeding that we see on farms and in gardens and orchards. Thus, parallel to Domestic Selection, there must be an unobservable process of Natural Selection. Darwin spends the rest of his great work explaining what that process might be. The whole work is an interlocking pattern of analogies and metaphors, through which we apprehend something we shall probably never observe. Natural Selection is a working model of what causes speciation. How closely does it resemble the actual process going on in nature? That is the question that prompts scientific advances.

Summary

Post-modern relativism and sociological epistemology both contain a smidgen of truth, but we must be careful not to be tempted to generalise this into a global anti-realism. It is true that scientific knowledge is accomplished by using pre-existing instruments, material and conceptual. It is true that deductive logic cannot account for the beliefs that scientists have in the results of their knowledge-creating activities. But to claim that scientific knowledge is merely a local set of just-so stories, and that people come to believe them by virtue of the structural forces that operate on them from the social milieu in which they have been living, neglects the role of the world. An apparatus has to interact with something other than itself, and a conceptual system must have something to be applied to. But how is this 'something' to be represented? Model-making is the key, when what we need to imagine is something which, with certain improvements in our situation, we could perhaps perceive. But we need to adjust our metaphysics when the necessary aspect of the world is beyond all possible perception, however our powers of observation are enhanced. There is a long tradition in the use of dispositional concepts – beginning with Gilbert's *orbis virtutis* (the magnetic field) to the affordances that are at the heart of Niels Bohr's interpretation of physics – as the alleged royal route to objective knowledge. I think it is true to say that the sciences find themselves in less trouble from the rise of relativism than do our moral intuitions. We are surrounded by the products of that science which is about the material world. We look in vain for the signs of a universal morality of decency and respect for persons.

Chapter 17

Some thoughts on education

Sir Hermann Bondi FRS

The Author

Hermann Bondi was born in Vienna. A graduate of Cambridge, he was a Fellow of Trinity College from 1943 to 1949 and from 1952 to 1954; Professor of Mathematics, Kings College London from 1954 to 1967; Master of Churchill College, Cambridge from 1983 to 1990 and subsequently Fellow. Elected FRS in 1959 and knighted in 1973. Between 1967 and 1984 he had a career in the public service concerned with the European space programme, defence, energy and environmental research.

Chapter 17

Some thoughts on education

Sir Hermann Bondi FRS

Without in any way claiming to be an 'expert' (a term I dislike in most contexts) my varied career has given me some experience and views on education that it may be worth setting down. Perhaps my somewhat critical outlook can be characterised by a rather cruel quip: From the time it learns to talk, a child makes a constant nuisance of itself by constantly asking 'Why?'. To stop this source of irritation, society has invented an effective system named 'education'. The few failures of the system are called 'scientists'.

I am fully aware that professional educators in fact do their very best to stimulate the natural curiosity of children and young people, yet I think it would be widely agreed that in the later stages of secondary education there is usually far less curiosity (especially of the all-embracing kind) than there is at the beginning of the primary level. Is this an inevitable consequence of growing up or are we doing something disadvantageous but possibly avoidable in schools (and in tertiary education)? The discussion of the issues raised by this problem is the main theme of the first part of this essay.

The initial question must be: what is the aim of education? What are we trying to achieve? Surely our purpose is to help the individual to develop and to become a useful and appreciated member of society – that is somebody who can strongly and purposefully interact with others, and whose talents and interests are continually developing; somebody who contributes to the well-being of the whole of society through useful work (at least some of which should be paid for). Here I would like to insert my understanding of the contrast between education and training. Education enables us to work with others towards a large variety of ends and helps us to understand and appreciate other people. Training enables one to perform a specific task. There is some tension between education and training. A subject (e.g. French) can be taught either 'educationally' – to widen horizons through acquainting the student with a different culture and literature, or as 'training' – to enable the student to live in France. The two teaching tasks have much in common, but there are also some marked differences; so it is desirable to be clear about the aim of the course.

Since we live in a rapidly changing world, the direct value of training, particularly in some technical fields, will often have only a limited period of utility. But, each time we have to undergo novel training, we benefit from having learnt to learn and strengthen this ability. So, cumulatively, training becomes educational.

As we do not educate people to become hermits, communication skills are of the utmost and universal importance and their every aspect should be fostered energetically. Children do indeed have an intense desire to communicate, but perhaps this tends to be mainly by word of mouth and with people they already know.

It is important to widen this urge through introducing reading and writing, not just to acquire the technical ability, but to learn to enjoy and appreciate these activities. Moreover, the ability to communicate with all and sundry needs fostering. Throughout, one must aim for precision and general intelligibility. (There is no phrase I dislike more than the lazy: 'You know what I mean.') These considerations lead on to the question of how much, in the way of resources of time and money, should be devoted to the learning of foreign languages.

First, a point that may be encapsulated in the following observation: More foreign languages are taught to a large proportion of the population in Denmark and in the Netherlands than perhaps in any other country of Western Europe. But, far from this effort pre-empting what else can be taught, it is arguable that these two countries also achieve as high – if not a higher – level of general education than the others. Indeed, bilingual persons are not noted for being backward in other respects. It looks probable to me that fluency in more than one language is a powerful aid to learning in all other subjects. What I want to infer from this is that our educational efforts come nowhere near to filling up the human brain; on the contrary, I suspect that the more one learns, the easier it becomes to learn more. Underestimating the limits to what in principle can be achieved is an attitude easily acquired by hardened professionals. The boundaries of what in fact can be achieved are set by motivation far more than by brain capacity; yet our understanding of how to stimulate motivation is still rather limited. Certainly, it would be unreasonable to count on the same interest in acquiring proficiency in foreign languages in England as among our Dutch or Danish friends whose mother tongues are spoken by only modest-sized populations. But I suspect that our expectations of what can be learnt are quite generally too low. Milton is reputed to have mastered five languages by the time he was five years old; who expects this of a five-year old now?

Acquiring a foreign language is one way – but by no means the only one – to learn about societies other than our own, the understanding of which is deepened both by similarities and by contrasts with others. But we must not confine such a study to the present; the evolution of cultures in their setting is a splendid topic. Geography and History are therefore essential subjects. It is my impression that it is easier to excite interest in the remote rather than the near-Neolithic times rather than the repeal of the Corn Laws, life in isolated valleys of Ethiopia rather than in German towns. Do our choices of topics take sufficient note of students' inclinations?

If my attitude is not wholly mistaken, motivation is the central issue in all fields. Certainly with young children – and I suspect with all those older ones whose inquisitiveness has not dulled – interest comes naturally, but, later on, exciting them becomes a difficult and arduous task. I do not want to blame parents who, busy with their own worries, tell an inquisitive child to be quiet. Yet I wonder whether quite sufficient priority is given to encouraging the 'pull' of curiosity, at home or in school.

So we resort to the 'push', the fear of failing examinations, to encourage learning. This may well be unavoidable, but I suspect it is not always appreciated how high a price, in educational terms, has to be paid to make a subject examinable at all – let alone the extra price to make the marking demonstrably fair. Frequently, particularly in subjects like mathematics, history and foreign languages, what is readily examinable – accuracy, dates, irregular verbs – is indeed a necessary part of understanding the topic, but wholly insufficient to convey its spirit or to arouse curiosity and interest. Yet, because this 'brittle' material – an answer is either right or wrong – is testable, it is liable to loom large and to obscure the essence of the subject. The best teachers can get over this hurdle with many of their students and many teachers with their best students, but there are numerous young people who gain little enthusiasm or understanding of the beauty or spirit of the topics taught because of this emphasis on what is examinable.

Nevertheless, I abhor too 'soft' an approach. Accuracy, dates, irregular verbs are indeed, as I have already said, necessary and essential. Underemphasising the need for precision gives a poor education and breeds slovenly attitudes.

Yet, perhaps too little time is given to exciting the interest and enthusiasm of pupils. Maybe one often underestimates the benefit not only to understanding, but even to examination performance, of devoting time to stimulating motivation. I feel impelled to relate the comment made to me by one of my students: 'It is marvellous to have somebody lecture to us as you do. It would be dreadful if everyone lectured to us as you do.' (What he meant, I am sure, is that while my lectures were enjoyable and stimulating, they were of no direct help in preparing our students for their examinations. Fortunately for them, most of my colleagues' lectures aimed more directly to strengthen their prospective examination performance.) The balance between inspirational teaching and teaching to pass exams is difficult to strike. I suspect that, in allegedly 'soft' fields, it tilts often to the inspirational; in the supposedly 'hard' subjects, too much to the 'right or wrong' topics.

I trust it is appreciated that I am not attempting to belittle the achievements of professional educators, but rather to stress the difficulties of their work. Yet, we must not close our eyes to palpable failures of present-day education. Much has been said about the poor communication skills of young people – accuracy, intelligibility, spelling – and their deplorable performance in arithmetic, and especially their inability to make rough estimates. Without in any way wishing to distance myself from these criticisms, I want to concentrate in the second part of this essay on my attitude to present-day science education.

The worst of its results is the anti-science attitude so widespread through much of the western world. The failure to convey successfully either the excitement and spirit of science, or its role in our way of life, is sad and bodes ill for our future. In my view, a major reform of the aims and means of science education is called for.

When somebody expresses anti-science views to me, my standard response is to say: 'So you do not mind if children die?' To my usually rather startled interlocutor I explain that, 150 years ago, half of all children died before they were 14 years old and that the transformation to the present situation is essentially due to scientific insights and their application. (Better medical care accounts for only part of this; clean water is probably the most important component; but safer food, better housing and better recreational facilities have all made significant contributions.) Parents now expect their children to reach adulthood, contrary to what used to be the case. This is a great improvement in our lives which could not have happened without advances in many areas of science. To give an apparently humdrum example, the synthesis and cheap method of manufacture of polythene bags has done much to improve the cleanliness and therefore safety of food. One may well question whether enough time and effort is spent at school to avoid children acquiring an anti-science attitude and equipping them to counter it when they meet it.

But, perhaps even more important, is for the questioning, stumbling, provisional nature of science to be conveyed in schools. This is probably best done by teaching science through the history of a few carefully selected topics, with a little of its philosophy thrown in. Perhaps an example will make my ideas clearer.

We teach that the Earth goes round the Sun – usually 'on authority', rather than by describing successive tests of the Copernican hypothesis. In fact these constitute a very fascinating piece of history. By the late seventeenth century, astronomers generally accepted that the Earth went round the Sun and started efforts to measure the parallax of stars, that is the apparent annual movement of near stars, relative to the background of distant stars, due to the Earth changing *its position in space* in its yearly voyage round the Sun. Determining even only one parallax would be doubly significant: as a decisive test of the Copernican hypothesis and in giving an idea of the distances of the stars. This search was very difficult as it was not known which stars were near or how big their parallax would be. Over the years, numerous claims were made of having observed such a parallax, but none could be sustained until, in 1838, Bessel made an unambiguous and repeatable measurement of the parallax of a star. However, much earlier, in 1725, James Bradley had observed changes, week by week, of the angles between widely separated stars, and had also noted that this displacement of stars was at right angles to that expected for parallax. With astonishing brilliance he inferred that this effect – later named the aberration of star light – was due not to changes in the position of the Earth in its orbit about the Sun, but to the changing *direction of its velocity. This* was then the first clear demonstration that the Earth was moving – surely a thrilling piece of the history of science that should be presented at school.

In my opinion, science should be taught through a series of such topics, carefully selected for being exciting and for illuminating the fitful, stumbling evolution of scientific ideas. (Another good field might be to follow through how, under the influence of improving technologies, the notion of whole number atomic weights came to grief, was rescued by the idea of isotopes, and then had to be modified for mass defects.) I am sure that, in every field of science, a number of such stimulating and instructive topics could be identified.

Above all, such an education in science would show the innumerable errors made by scientists, and how sometimes more is learnt from mistakes than from successes. Such a course would present science as a thoroughly *human endeavour*. It would banish to the dustbin the stupid, but common, notion of scientists as cold, dispassionate and objective – whatever that word may mean.

I am fully aware that such a manner of teaching would take a lot of time per topic and that the allocation of time to science is unlikely to be increased. Therefore only a very limited number of topics could be included at school. Large chunks of science now taught would have to be left out. This would be a big sacrifice, particularly for the universities. But the alternative of proceeding with the present methods looks to me disastrous. It turns a large, and perhaps growing, part of the population against science. It so discourages young people from choosing science for study at university that whole science departments have had to be closed.

If the market for the product now on offer is shrinking, then the product has to be changed. The change I suggest would, I believe, result in a general population more friendly towards science and in more and better motivated, even though less comprehensively prepared, students choosing to read a science subject at university. Do we really believe that the universities are unable to handle such a change? Should not what I advocate be elaborated and tried out as an approach to teaching science in schools in collaboration with some universities? To go on as we are looks to me to be a recipe for disaster. So surely something else has to be tried out ? Just because a thin stream of well prepared and interested young students fills some of the science departments of some universities, we should not neglect the need to make changes so that more people become science friendly.

I am also bothered by the feeling that school science, as now presented, conveys a misleading picture of what a life in science is like. This, I suspect, results in young people who could have become excellent scientists and would have enjoyed such a career, opting for something else. Can teaching for readily examinable topics bring in the adventurous spirits and imaginative minds science needs? I fear not. Conversely, our present teaching of scientific subjects may make people opt for science at university, who are uncomfortable with the unknown, and prefer cut-and-dried answers to the fog of uncertainty that always exists on the frontiers of science. They will be unhappy with their choice – a very bad advertisement – and, if they persist with their career, bring dullness into science.

To sum up my worries: I suspect that the way we teach science is essentially a misleading prospectus for the subject. This has many ill effects of which I have listed only a few. Is not correcting the prospectus worth a major sacrifice in the volume of science taught and in the pain that a major change necessarily brings?

As a third and final section of this essay I want to set down my views both on the need to teach statistics at school and what features of the subject should be emphasised. Democracy is not just the only morally acceptable form of government, it is also the only one that, in the longer term, has proved to be effective. But one of the essential needs of a working democracy is a population that is sufficiently well educated to make adequate judgements on issues of public concern. Where at present this ability is perhaps weakest is in appreciating the significance and value – if any – of the statistics conveyed by the media so enthusiastically and often with little discrimination. What is needed is a suitably sceptical attitude about the meaning of the figures and, in addition, an appreciation of orders of magnitude, so that minor issues can be separated from major ones.

To be able to distinguish between meaningful and meaningless statistical statements can be taught by analysis and by working through realistic examples, without going to difficult levels of abstractions or through lengthy mathematical manipulation. It should not be too difficult to make the population aware of the nonsense of statements like: 'It has been revealed that nearly half of all school children are below average in ability'; or, 'The weakest firms must be brought up to average efficiency'. Similarly it is important, even in the thorny field of risk assessment, to bring in an ability to sort the important risks from the less significant ones. At present it seems to be all too easy to concentrate public interest on dangers of quite minor relevance while ignoring some major ones. Nor is it widely understood that the death rate per person is one.

Somewhat more difficult ideas to get across are notions of error bars and statistical significance; but orders of magnitude of uncertainty should not be too hard to convey. It would be real progress if people became suspicious of a claim that 35.4 percent was a major advance on 35.1 percent and queried whether the sample size was sufficient to support such an interpretation.

Much of this can be taught relatively easily as part of general education by examples and stories, though it may be more difficult to examine them effectively. Such lessons on 'elementary sceptical statistics' could greatly raise the discrimination of the general public and so improve the working of democracy.

What is emphatically not needed in such a course is the teaching of refined statistical procedures. So often this only leads to an examinable knowledge of the mechanics of, say, the chi-squared test with no understanding of its meaning and significance. It would be a big advance if the general electorate could appreciate major statistical differences. Sophisticated analyses of less obvious matters are for specialists.

The aim of all education is to improve our capacity for living together peacefully and effectively. I hope my discussion of points that particularly worry me is a step towards this end.

Chapter 18

Eddies in the flow: towards a universal ecology

P. J. Stewart

The Author

P.J. Stewart graduated from Oxford in Arabic and Hebrew, and translated a novel by Naguib Mahfouz into English, before taking a degree in forestry and working in the Algerian Forest Service for 7 years. Since 1975 he has taught economics to forestry students and ecology to human sciences students at Oxford. His introductory course on Islam for students of Arabic has been published as *Unfolding Islam*. He and his French wife have raised five bilingual children.

Chapter 18

Eddies in the flow: towards a universal ecology

P. J. Stewart

The shrinking of ecology.

It is time to re-launch ecology. That which seemed to many in the 1960s and 1970s to be a whole new way of seeing the world ends the century as a shrinking and under-funded branch of biology. What has brought about its decline is its failure to include most human behaviour in its subject matter, when it is evident that our agriculture and our industry are the dominant factors in the biosphere.

A narrow view of ecology was taken at the very outset, when Ernst Haeckel in1866 proposed *Ökologie* as the name for a new science.[1] He conceived it as part of what he called 'relational physiology', which was one of the four branches into which he divided zoology (sic!), the other three being anatomy, morphogenesis and 'conservational physiology' (vol. I, p. 238). He envisaged human ecology and geography as parts of one of the four corresponding branches of anthropology (vol. II, p. 433).

In the event, the ecology of plants developed well before that of animals, so much so that A. G. Tansley, in his 1914 presidential address to the British Ecological Society, could say: 'We claim for ecology that it is before all things a way of regarding the plant world, that it is *par excellence* the study of plants for their own sakes as living beings in their natural surroundings, of their vital relations to these surroundings and to one another, of their social life as well as their individual life, and that is why we put it in the forefront of botanical science.'[2]

Animal ecology eventually caught up, but in the process it divided its potential subject matter with another new science – 'ethology'. This division involved splitting an animal's environment into two parts – its fellow species members and the rest – difficult to do in cases like that of the termite, almost the whole of whose immediate environment consists of termites and their works. By implication, the ecology of our species should leave aside our human environment.

Putting together plant and animal ecology, Eugene Odum, in 1953, envisaged something much bigger than Haeckel's subdivision of zoology: a scientific framework for studying the whole of life on earth. However, one species was almost absent: our own.[3] The human ecology envisaged by Haeckel had failed to mature, though the term had been in use since the 1920s for a variety of approaches – seven according to Quinn.[4] It was – and still is – most commonly seen as a branch of human biology, with only timid references, if any, to culture. This makes a tidy little subject, but it is very incomplete, for it leaves out our most characteristic feature. The efforts of Gregory Bateson, 1972, failed to establish an 'ecology of mind'.[5]

It is clear why ecologists ruled us out of their subject matter: we cannot be subjected to their simplifying procedures. They try to study ecosystems that are as far as possible self-contained, and they divide each one into discrete compartments ('strata'), such as the tree canopy, dead wood and soil. They describe successions of plant communities of increasing total biomass, and they group organisms into trophic levels such as herbivores, carnivores and parasites. Energy provides the guiding thread as it passes up the food chain.

Modern transport and the growth of the world economy have destroyed the relative isolation of most ecosystems. The splitting of them into different strata breaks down for people even with quite simple technology, for we can climb, dig and swim, our influence reaching up to the treetops and down into soils and rocks, ponds and rivers. With our farms and plantations and gardens we disrupt the succession of plant communities. We do not fit into any trophic level either, for we belong to every level of the food chain, being not only plant-eaters, meat-eaters and sometimes eaters of meat-eaters, but also decomposers of a sort. Some day we may even learn to mimic the role of plants, using sunlight to make the molecules of life from carbon dioxide and water. Even the unifying flow of energy through the system becomes difficult to apply in the presence of humans, for we inject into living systems great quantities of energy from a wide variety of sources, using non-renewable fuels to promote human survival and multiplication where they would otherwise be limited or impossible.

As if all these difficulties were not enough, there is the central human peculiarity: that our adaptation is primarily not genetic but cultural. With any other species it is possible to take behaviour as something given and constant (though there is such a thing as animal culture), and to proceed with studying how that species interacts with the rest of the ecosystem. In the human case, a few tens of thousands of years of cultural change produced as much diversity as tens of millions of years of mammal evolution, so that the ecology of the Bushman differed as much from that of the Inuit as, say, that of the spotted hyena from that of the polar bear.

Today, such self-sufficient cultures as survive are rapidly being transformed into outliers of a world-wide cultural network, which changes too fast for description and analysis to keep up. Resources of soil and wild plants and animals are shrinking, human numbers are expanding, and technology is changing, so that it is increasingly difficult to assemble data that all relate to the same point in time. It is even hard to know what data are going to be needed for completeness; for example, which of the thousands of chemicals that we produce will prove to have the unforeseen importance of the CFCs that have damaged the world's ozone layer, and where do the viruses of future epidemics lurk?

The almost universal pressure of people on other living things has long since forced ecologists to give up trying to ignore us. Virtually all ecological studies now have to take account of 'anthropogenic factors'. However, these are still usually regarded as being outside the system. The object of study is still plants and animals, and human actions are treated as external influences comparable with, say, volcanos.

In spite of these difficulties, an ecology movement did develop. Aldous Huxley was perhaps the first to articulate its ideas clearly, in his 1959 lectures at Santa Barbara,[6] and in 1963 he entitled an essay 'The Politics of Ecology'.[7] In 1962, Rachel Carson attracted enormous public interest with her warning of the widespread damage done by over-use of pesticides.[8] People became concerned about the effects of growing numbers of increasingly affluent humans on the rest of the planet. By the end of the 1970s, ecology parties had appeared in most western democracies, and some people were forecasting that all political parties would have to become ecological if they were to survive.

There was a multiplication of 'eco-' words, mostly of questionable usefulness: 'ecocentric', 'ecocide', 'ecofriendly', 'eco-feminism', 'eco-tourism', 'eco-terrorism', 'ecodynamics', 'eco-ethics', 'eco-logic', 'eco-mystic', 'ecopsychology' 'eco-philosophy' or 'ecosophy', even 'EcoNomics' . . . Enemies very understandably retaliated with 'eco-nutter' and 'eco-freak'.

Two decades later the word 'ecological' has practically disappeared from politics, replaced in popular discourse by 'green' or 'environmental'. What went wrong? A first approximation to the answer is that leadership of the ecology movement fell naturally to plant and animal ecologists, most of whom saw people as a negative factor, interfering with the functioning of natural ecosystems. Many of them adopted an anti-human political stance, describing human population growth as 'popollution', advocating the introduction of breeding licences to stop people having too many children, and even welcoming famines as 'Nature's way of restoring the balance'. These were not ideas around which to unite voters. At the same time that some professional ecologists were busy giving the ecology parties a bad name, others were barricading themselves into their academic fastnesses, gruffly rejecting any suggestion that their pure science might be relevant to the messy world of politics.

At a deeper level the problem was intellectual. By failing to address human culture and mentality, ecologists had renounced the study of most of the activity of the dominant organism on the planet. The emergence of *Homo sapiens* is the biggest event in the biosphere for many millions of years, comparable in importance with the development of an oxygen-rich atmosphere, the appearance of multi-cellular organisms and the colonisation of land. The current wave of extinctions can be compared with only two or three previous ones. A science that cannot fully integrate humans is not a true science of the biosphere. No wonder ecology has contracted into a minor academic discipline.

The synthesis of plant and animal ecology is often inappropriately termed 'general ecology'; it will be referred to below as 'natural ecology', without any intention to imply that humans are not part of 'Nature'. The truly general system yet to be developed, in which humans will be completely integrated, will here be called 'universal ecology'.

The environmentalist takeover

Frustration with the narrowness of academic ecology helped to promote the popularity of concern with 'the environment'. Since the 1960s the words 'environmentalism', 'environmentalist' and 'environmental' have increasingly replaced 'ecology', 'ecologist' and 'ecological' in popular usage. Given that 'environment' is just one concept in ecology, this is an extraordinary shift. It is as if people tried to talk about marriage by referring only to husbands.

The term usually translated as 'environment' in Haeckel's definition of ecology, was '*die umgebende Aussenwelt* (the surrounding outside world)', by which he conceived a set of conditions, both organic and inorganic, existing independently of the organism. Jakob von Uexküll, in 1909, introduced the word *Umwelt* to express the fact that each animal species lives in its 'own world' (not the literal meaning of *Umwelt*, but never mind), defined by those factors that its nervous and motor systems enable it to interact with.[9] The concept can be extended to plants and microorganisms. Thus, for example, a cow, an owl and an earthworm find totally different own-worlds in the same field, as do an oak, a moss and an amoeba. Each in this sense can be said to 'internalise' a different environment and is largely blind to that of the others.

Human culture has enabled us to create a huge variety of social and individual own-worlds, to the extent that no two people interact with exactly the same range of factors. We have also introduced into the own-worlds of other organisms all sorts of factors with which they are not adapted to cope, for example antibiotics, oestrogen-like chemicals, and new radio-active materials.

No sharp frontier can be drawn between an organism and its environment or own-world. It is customary to take the surface of an organism as its boundary, but for most multicellular organisms this is largely internal: the lining of guts and breathing organs is several magnitudes greater in extent than the external skin or cuticle of an animal, and the same is true of the intercellular spaces on to which plant stomata open. On land, the only major exchange surfaces that are external are root hairs, and they are often encased in or replaced by mycorrhizae. The content of the gut is not strictly part either of 'the organism' or of 'its environment', but rather is a transition between them, and it is itself the environment of the gut microflora.

Outside of organisms, in the conventional sense of 'outside', there are equally important transitions in the form of solids, liquids and gases that they have produced or modified. A spider's web or a beaver's dam is part of the unit that survives as well as being part of the world in which the animal survives. On a larger scale, Earth's atmosphere, water and soil owe their composition to the activity of living things, and even the climates and rocks would not be what they are, if life had not constantly released carbon dioxide and oxygen. The biosphere is an intimate mixture of the living and the non-living, and there is no sharp boundary between the physical systems that are modified by life and those that would be the same on a dead Earth – least of all for our species, with the enormous ramifications of its products, by-products and waste products.

Universities are supposed to be the guardians of intellectual rigour, but the vagueness of things 'environmental' has not deterred them from following fashion. For example, in the United Kingdom in 1999, 115 university-level institutions offer some 417 honours degree courses with 'environment(al)' in the title, and these can also be taken as part of literally thousands of combined degrees. There is even 'environmental biology', which is in effect another name for ecology. Only 45 institutions offer honours courses in ecology, and only 3 of these allow it to be taken as part of a variety of combined degrees. Three institutions offer degrees in human ecology.

People may ask: does it matter whether we call things 'ecological' or 'environmental' as long as we pay attention to them? Do not the two words mean virtually the same thing? Alas, they do not. Ecology is the study of multiple *relationships*. Our environment is the sum total of *things* to which we can relate. One word suggests a reality of which we are part and of which the essence is our taking part; the other a reality that is, at least potentially, independent of our existence.

Once a class of problems has been labelled 'environmental', their solution becomes a specialist matter, just one of the various types of problem competing for attention and resources. Most countries now have a 'Ministry of the Environment', with a small share of the national budget to deal with a rag-bag of matters that do not fit easily into the brief of other ministries. This gives the impression that the problems are being dealt with on our behalf, and that there are not too many of them.

If it became generally accepted that our ecological relationships had gone wrong, it would be evident that this affected every aspect of life, that it concerned everyone and that changes were needed in the life of each of us. Appropriate responses would be required from all branches of government, and if any agency were to be created to coordinate them it would have to be above all ministries – at prime ministerial level.

One would not have expected a four-syllabled technical term like 'environment' to become popular. What seems to have happened is that 'the Environment' (with at least a notional capital E) has largely replaced 'Nature' as that with which westerners contrast humanity. For the general public, it is essentially the countryside – 'the Great Big Green Thing Out There' – as witness the adoption of 'green' as a virtual synonym of 'environmental'. 'Humans *versus* the Environment' has taken over from 'Man against Nature', itself the successor of 'Spirit over Flesh'. In this history there has been a reversal of signs, from the 'good' Spirit and the 'bad' Flesh, through the ambivalent Man and Nature to 'bad' Humans and the 'good' Environment, but essentially it is the same old battle.

This opposition between humans and everything else has its roots deep in western cultural history. Plato taught, 24 centuries ago, that humans alone have an immortal *psyche*, the seat of reason, which belongs to an eternal world of perfect forms or ideas. This he contrasted with the mortal parts of *psyche* concerned with emotion and animal drives. His pupil Aristotle endorsed the uniqueness of the human rational soul, though he stressed the embodiment of *psyche* and the presence of simpler forms throughout the living world.

Plato's doctrine eventually became the popular dualism of Christianity, according to which the immortal human soul is exiled or imprisoned in the fleshly body and will return to a heavenly union with the divine – or to separation from it in Hell. It is true that the creeds say a resurrected body is required for the afterlife, but something without needs, desires, ageing or suffering can hardly be called a body. It was only when Aristotle was rediscovered in Western Europe and made acceptable to Christians by Aquinas (1224-74), that the Renaissance became possible, with its exuberant development of physical interests and achievements.

In popular dualism, the spirit world can interfere in the material world, and many physical events are seen as acts of God or the work of the Devil, often mediated by angels or demons. The special contribution of René Descartes (1596-1650) was not that he invented the idea of two kinds of reality, but that he separated them rigorously, accepting only material causes for material events and mental causes for mental events, the two worlds meeting only in humans.

Knowledge came to be divided along Cartesian lines into the sciences, which dealt with the physical world, and the arts or humanities, which were concerned with the psychic or spiritual. The former were objective, the latter subjective. Knowledge about inanimate objects all fell on one side of the divide, and the West has become very good at it; but the study of humans was split apart, with knowledge about our bodies on one side of the divide and knowledge of our minds on the other, which gravely damaged our ability to understand either.[10]

Eventually, to bridge the gap, there was created a third category: the social sciences, which natural scientists regard as being little more objective than the humanities, and which arts people regard as little less remote from subjective experience than the hard sciences. Knowledge about humans is now split between three domains, with, for example, medicine and human biology in one, sociology, psychology and economics in the second, and the study of culture, history and religion in the third. This threefold division will here be called 'neoCartesian'.

More than anyone else, Descartes may be said to have brought about the demise of 'Renaissance Man', who combined arts and sciences, and to have ushered in the age of the specialist. There have been many gains, but at great cost. One of the most serious consequences has been the divorce of philosophy from science, which for two thousand years had been known as 'natural philosophy'. Since the Cartesian revolution, neither scientists nor philosophers have been expected to know anything of each other's work. For the reputation of Descartes, who was both scientist and philosopher, this had the happy effect that for a long time few people felt qualified to question his wisdom. New sciences could be slotted into his framework without any thought about the nature of their knowledge or about their means of reaching it.

The popular concept of knowledge and the structure of our institutions of teaching and research have been divided along these lines for more than three hundred years. Universities and research funds classify faculties and institutes into the three neo-Cartesian categories and treat them quite differently. Most universities do not allow students – still less teachers and researchers – to combine subjects across the frontiers. Within each category, knowledge has been further fragmented into ever more specializations, with boundaries that are only a little less rigid.

Attempts have been made to overcome the division of knowledge by assimilating one kind to the other. Materialists claim that mind can be explained in terms of the workings of neurons and molecules, and that the social sciences and the humanities should be absorbed into the physical sciences. Some even suggest – though they do not behave as if they believe it – that consciousness is an illusion. Defenders of the autonomy of mind go to the opposite extreme, claiming that science is just a 'social construct', though they seem happy to enjoy its benefits. Meanwhile, for want of rigorous ways to view the relations of mental and physical, the general public seems increasingly tempted to blur the boundaries by reverting to popular dualism, whether in the religious form of fundamentalism or in the secular form of belief in ghosts, telepathy, clairvoyance, astrology and so on. This is a hostile intellectual environment for a universal ecology.

Philosophical foundations for a universal ecology

The assumption that underlies most aspects of western culture in all its branches, whether Jewish, Christian, Muslim or secular, is that the world consists of discrete entities (from late Latin ens, 'being'). These may be simple or composite and short- or long-lived and they may or may not interact. Examples are atoms and molecules, organisms and species, individuals and nations, souls and spirits, self and environment.

Belief in entities seems like common sense in any culture since it reflects the universal nature of human language, in which there are 'things' (substantives) that have 'qualities' (adjectives) and 'enter into processes' (verbs). This in turn reflects the nature of human perception, which selects out that with which we can interact from that with which we cannot, and which sees that which changes rapidly as the figure standing out against the ground of that which changes more slowly. The naive view is that language represents the world and that the structure of the world is therefore like the structure of language.

Western science has been a long and largely successful endeavour to analyze complex objects and processes into simple constituents and to discover the enduring properties of these and the way they interact. It worked remarkably well for a long time, but difficulties arose when the supposed entities were obviously without clear boundaries, for example anticyclones or schools of fish or ecosystems.

Western religion has traditionally made the same fundamental assumption: there are discrete entities such as souls, angels and devils. Indeed, there are religions. There is one soul per person, and a creature either is human and has a soul or it is not and does not. In the popular view, a soul is bound up with an identity, a personality, a bundle of memories, and it will carry these into eternal life.

An entity-based view of reality is concerned with delimitation, definition, classification, isolation and analysis.[11] Attention is focused on entities rather than on the larger wholes to which they belong. Knowledge is sought in the form of essences, norms, constants and laws of combination. Contradiction is seen as conflict between truth and untruth. This approach has been enormously successful in the management of non-living things, much less so in the organization of human society. Concern centres on the individual, on the nuclear family, on the firm, on the social class and on the nation, at the expense of wider solidarity with humanity and with life on earth.

There is an alternative view: reality is a single vast flow in which eddies and cascades of eddies form and re-form, never quite the same, yet some lasting so long as to seem permanent. They have no existence separate from the overall flux, which gives birth to them and which swallows them up, and they acquire their character from their interactions. Effects feed back into their own causes, and causation loops and spirals and tangles. The whole explains the parts as much as the parts explain the whole.

This view may be said to be counter-intuitive in that it does not arise naturally from the structure of language or perception, and it has been arrived at and maintained relatively rarely in the history of ideas. It was anticipated two and a half thousand years ago by the Greek philosopher Heraclitus, but it has only lately entered the mainstream of western culture. It has spread out from two focal areas at opposite ends of the sciences: from physics and from ecology, aided by developments in mathematics. Modern physics has produced some shockingly counter-intuitive ideas. Fundamental 'particles' are not precisely located but are 'smeared' through space; 'particles' that have interacted continue to be 'entangled', however far apart they are; whole atoms have wave functions which are not just the sum of those of the constituent particles; and the properties of the very small depend on those of the whole universe, and vice versa. It is even suggested by some that the wave-functions of 'particles' collapse only when they are observed by consciousness.

Ecology started out with some similarly radical ideas. The nature of an organism is not fixed but depends on its interaction with its environment; an assemblage of organisms develops a behaviour of its own, which is not simply the sum of the individual behaviours; flows of matter, energy and information join communities of living things together into a dynamic web; life and the inorganic systems of the biosphere influence each other and evolve together.

The affinity between these two revolutions has led some physicists to make creative contributions to ecological thought. Feinberg and Shapiro in 1980 proposed that life be defined as 'the activity of a biosphere', and they envisaged all sorts of possible biospheres without carbon or water.[12] Hoyle and Wickramasinghe, in 1978, proposed that life is the product of the whole Universe.[13] As for this world, one of the best ecological accounts was written by a physicist, Fritjof Capra.[14]

There is no reason why a flow-based view should not be adopted by a society. At the core of the Buddha's teaching is the doctrine of interdependent origination (*pratitya samutpada*): that each thing exists only in relation to everything else, which is virtually the same as the founding principle of ecology. This is bound up with the *anatma* doctrine: that there is no permanent, independent *atman* – 'self' or 'soul'. Many aspects of Buddhist cultures are to be understood in these terms, including diet and agriculture.

The Chinese classic Tao Te Ching is the purest expression of the view that common sense entities have no independent reality. By the powerful use of a few thousand words, this astonishing book undermines the power of words and points to a way that 'one who speaks does not know, one who knows does not speak'. Its starting point is that 'naming is the mother of the myriad things', which in the present context might be translated as 'using nouns is the origin of innumerable entities'. The wise person learns to live in accordance with the nature (*zi-ran*) of things and to avoid forcing them (*wu-wei*). Although not the dominant element in Chinese culture, Taoism contributed to the development of various arts and techniques, most notably the world's oldest tradition of landscape painting.

It looked for a while as though a new understanding of reality would spread from physics and ecology to all areas of western culture; however, when it came to including our own species in ecology, radical thinking failed, and humans and 'the Environment' reverted to being separate entities. The trouble was that the very basis for the separation of sciences and arts was being called into question. Our relations with our environments of soil, water, plants and animals can be seen as part of human biology, but our relations with our social environment of houses, neighbourhoods, markets, transport systems and so on looks like the subject matter of a social science. As for relations with what may be called our cultural environment of knowledge and belief, fashion and tradition, that looks like an arts subject. Faced with the prospect of ceasing to be a 'proper science', ecology shrank back into an increasingly mechanistic biology.

It is no solution to cordon off an area and call it 'human ecology', for the same neo-Cartesian frontiers will prevent that from being unified either within itself or with the rest of ecology. Nor is it enough to say that ecologists can study the physical interaction between people and other species, while sociologists and economists look at socio-economic aspects, and psychologists and historians of ideas search for people's motivation. If the findings of these specialists are to be put together coherently, a framework will be needed. It will have to be one in which mind and matter can both be accommodated. The concept of pattern is potentially the basis for such a unification.

'Matter' is an ancient name for that which can be made into visible or tangible patterns, but it has itself proved to owe its character to pattern; for example, diamond and graphite differ only in the arrangement of their atoms. At a still deeper level, patterns of energy and of matter interweave, and the two are interconvertible. When pursued down to levels of sufficient smallness, the distinction between pattern and thing patterned breaks down; the entities in the strange world of quantum physics can behave either like particles or like waves, and they appear to be best described as patterns of excitation in 'fields'. Pattern seems to be more fundamental than matter. (Or are they two aspects of but one 'parent'? After all, the words derive from Latin *pater* and *mater* – 'father' and 'mother'.)

'Mind' is an ancient word for all the patterns of which we can become conscious, in the form that consciousness gives them – Aristotle's 'form of forms'. The English word is related to 'meaning' and is cognate with Latin *mens* and Sanskrit *manas*, suggesting an origin at least six thousand years old. Its history has intertwined with that of other words such as 'soul' and 'spirit', whose Latin, Greek and Hebrew homologues derived from roots meaning 'breath' or 'breeze' and referred to that which gives life to breathing things. The restriction of mind or soul to humans is a later development; our word 'animal' is Latin, meaning a thing possessed of *anima*, 'soul'.

In Christian popular dualism, and in its tidied-up Cartesian version, English-speakers use the word 'mind' in secular contexts and 'soul' in religious ones, though both are equally vague. French-speakers can use *esprit* for both. A mind/soul is conceived as a sort of 'thing' that a person 'has', and as, at any rate in principle, detachable from the body; it is a bit like a pocket handkerchief – very private and with nasty bits. People take seriously the prospect of one day being able to 'download' the contents of their minds into a computer. When the members of the Heaven's Gate cult killed themselves, it was in the belief that their minds or souls would be 'beamed up' into the space-ship that they believed to be following Comet Hale-Bopp.

For the purpose of exact thinking it might be better to avoid the words 'mind' and 'soul' altogether. The true bipolarity of the world is between consciousness and the patterns of which it is conscious. Consciousness is not a 'thing', with characteristics that can be examined, yet it is that by which all things are known. Nor is it 'something' that an organism either 'has' or does not 'have'. There are degrees of consciousness, from the edge of sleep to the alertness of crisis, from infancy or senility to healthy maturity, and – why not – from animals whose nervous systems can present only simple alternatives, to our own big-brained species.[15]

Since Freud, there has been a widespread belief in 'the Unconscious' – a part of mind of which we are not conscious – but the whole point of Freudian analysis is to bring the contents of this 'Unconscious' into consciousness, which makes them no different in principle from any of the other things of which we are not currently conscious. As for recovered memory, perhaps it takes a therapist to persuade people that 'remembering' something they never knew is the same as remembering something they had forgotten.[16] It will be assumed here that what we cannot become conscious of is not part of mind.

Consciousness will here be accepted simply as a reality known to all, in no need of explanation. In a sense it is the very source of reality, since without it no reality can be known. However, it is not a separate or separable realm but an integral part of the life of bodies with nervous systems of a certain complexity. The Indian concept of *advaita* – non-duality – expresses very well the relationship; consciousness and the patterns of matter and energy in bodies and nervous systems are not one and the same, but neither are they separable.

The patterns perceived by consciousness are not simple copies of patterns of events in the world or in the nervous system. Indeed tastes and smells do not seem like patterns at all; the flavour of a banana is neither like the pattern of firings in the neurons of tongue and nose, nor like the pattern of atoms in the molecules contained in the banana. For an organism to interact with its environment, there is no need for such complexities; nervous systems have evolved to present simple sensations to consciousness. However, even apparently patternless sensations come together with other sensations to form patterns – for example, the smell, taste, shape and colour of banana, perceived together.

The concept that unifies mind and matter, then, is form or pattern. The latter word has been preferred here, because 'form' tends to be thought of as a static pattern in space, and 'pattern' is more easily understood as extending also in time. Indeed, every static form is the product of the pattern of events by which it is produced. All knowledge is consciousness of patterns, and there is no fundamental division between consciousness of simple, concrete patterns, such as the motion of pendulums, and consciousness of complex abstract patterns, such as the movements of symphonies. The mathematics of complexity offers a basis on which to unify our understanding of all pattern.[17]

Seeing the world in terms of pattern is different from seeing it in terms of the enduring entities of western tradition. Patterns ebb and flow, spread and merge, part and fade away. Figure can become ground and ground become figure. Repeating motifs can all be subtly different and yet be reckoned as repetitions. The boundaries of motifs can be variously drawn or can shift backwards and forwards. With a change of scale, what seemed to be a motif becomes a pattern of yet smaller motifs, or what seemed to be a pattern becomes a motif in a larger pattern. This can be illustrated in many ways: by examining a tree from different distances, by studying certain designs by M.C.Escher, by listening to the later work of Jean Sibelius, by playing with the Mandelbrot set, or simply by lying down and watching white clouds drifting in a blue sky. A number of recent books have proposed impressive schemes for unifying what is known of brain and mind.[18]

It is easy enough to produce a series of paired examples to show the seamless transition from the patterns of the non-human to those of human culture: guarding of aphids by ants and of cattle by people; dams constructed by beavers and by people; tools made by chimpanzees and by people; docile wolves living as camp-followers and people hunting with dogs . . . Some philosophers will say – though as the series goes on it gets harder and harder to do so – that in each case the second example shows intentionality; humans know what they are aiming to do and animals do not. However, for the ecological consequences, all that matters is that behaviours that increase survival are promoted, whether the patterns between which selection is made are genotypes or imagined futures.

An ecology that includes conscious humans will not be different in kind from one that includes conscious chimpanzees or wolves or elephants. The position taken here is that human mind differs from animal mind only in three ways: in the number and variety of patterns that human consciousness can view (including imagined futures); in the ability to communicate them; and in the capacity for abstracting patterns of patterns and patterns of patterns of patterns. These are all matters of degree, but they have created the increasingly vast, complex, interconnected and fast-changing own-worlds of human cultures.

Principles of a universal ecology.

The foundation of natural ecology is the notion of adaptation between, on the one hand, the patterns of development and behaviour of organisms and, on the other, the patterns of events in their living and non-living environments. Adaptation is also the foundation of evolution by natural selection, and 'coevolution' is a reasonable alternative term for universal ecology. All that is needed is to add the adaptation between the patterns perceived by consciousness and their physical, biological and cultural environments. This has already been achieved to some extent in theories of gene-culture coevolution and evolutionary psychology.[19]

In going from the organisms of natural ecology to the ideas or mental patterns that enter into a universal ecology, there is no sudden leap. Many ideas belong in a biological 'habitat'; they are part of patterns of human behaviour that interact with patterns of plant growth and animal behaviour. Obvious examples are the thoughts of farmers, hunters or campers about their activities, but it should also be obvious that everybody's ideas about food fall into this category. However lofty our ideals, it is primarily by the use of our teeth and our stomachs that we shape the planet.

There is no difference in kind between this fit of ideas in their biological 'habitat' and their relationship with what has been made by humans. Examples include ideas about clothing, housing, transport, manufacturing and most activities in an urban setting. Through the economic system, these are translated into demands for resources that either come from the biosphere or impinge on it, and into an output of waste products that are returned to planet-wide systems. Again, gene-culture coevolution provides examples.

These two categories contain a large part of what goes on 'in the mind', but they are dwarfed by the volume of thoughts about relations between people. These too have a physical 'habitat' in the patterns of activity that bring people into contact with each other – the human equivalent of the subject matter of ethology. They too have economic consequences that feed through into the rest of the biosphere. In the form of ideas about relations between the sexes, they also have a direct impact on birth rates and population growth rates. Ideas about relations between adults and children affect the fidelity with which a culture is transmitted between generations.

The kinds of ideas so far mentioned are abstractions from experience of patterns in the world. There remains the category of abstractions from abstractions. These are the ones that seem to be evoked by the term coined by Sir Geoffrey Vickers in 1968: 'ecology of ideas'.[20] The concept of Cartesianism, for example, can exist only in 'an ecosystem of ideas' that includes concepts such as matter, mind, reality, logic and so on. The remarkable thing is that some of the highest-order abstractions are the patterns with the greatest power to influence practical action. They provide the criteria according to which a wide range of lower-order patterns are accepted or rejected.

One of the fundamental ideas in ecology is that no one species controls an ecosystem; causation is cyclical. For example, carnivores regulate the numbers of herbivores, which determine the numbers of carnivores . . ., and some little microbe may change the balance by devastating the population of either. Anyone who supposes that humans can control the biosphere or manage 'the Environment' has failed to learn this lesson, but it should be applied also to the ecology of ideas: no idea can securely dominate the planet.

Although no species controls an ecosystem, a few may have a preponderant influence, by virtue of their combined biomass, individual longevity, range of activity or some other factor. A population of trees, for example, creates a forest soil, moderates the local climate and provides the food and the habitat for a great range of creatures. Deforestation radically changes the whole community. One of the tasks for a universal ecology is to identify sets of ideas or behaviours that have an equivalent influence. To do so they need to be sufficiently faithfully transmitted to continue to affect a wide area over a long time period.

A preponderant set of ideas and behaviours seems to require a number of essential components, including the idea that its components belong together, the idea of its own importance and ideas of the rewards for maintaining it and the punishment for deserting it. There is also what may be called a 'demography' of ideas, which includes both the ability to promote the births and delay the deaths of their adherents, the ability to be transmitted faithfully between generations and the ability to spread to new adherents and to hold on to them. Any long-lasting set of ideas, like any enduring life form, must combine enough flexibility to adapt to changing conditions with enough constancy to preserve existing adaptations.[21]

A key feature of successful ideologies is the capacity to give meaning to existence, relating the acts and artefacts in individual lives to the overall pattern of the world. The human desire for meaning can be even stronger than the desire to live. Throughout history there have been examples of martyrdom, in which healthy people in the prime of life have died willingly, even joyfully, making their death the final symbolic act. Conversely, those who fail to fit their life to a larger pattern may live miserably or even kill themselves, seeing their existence as 'meaningless'. A meaningful death can thus be preferred to a meaningless life.

It is this combination of qualities, rather than any concern with a 'supreme being', possession of a priesthood or whatever, that characterizes what European languages call a 'religion'. There is however one of the traditional criteria that applies: possession of a sacred text, which helps to ensure constancy from generation to generation. The Jewish Torah and the Muslim Koran are good examples, containing explicit references to most of the main features of their respective religions.[22] The Christian Bible is a rather poor example, given the doctrine that the Old Testament laws have been superseded, and given the primacy of the Church as the source of authority.

There should be no doubting the ecological significance of religions.[23] Most of them pay considerable attention to sexual behaviour and family responsibility, with consequences for population growth.[24] There has been a process of natural selection, according to which the most numerous societies are those with the most strongly pro-natalist religions, while anti-natalist and weakly pro-natalist religions such as those of the Jains and the Parsees have condemned their followings to dwindle.

Most religions concern themselves also with what their followers eat and how their food is produced, and there may also be animal sacrifice. Christianity is an exception in this respect, having deliberately rejected the food laws of Judaism. Religions may also lay down laws on land ownership and inheritance. All of this influences land use and hence the availability of rural resources for other species and their ecosystems.

A third kind of interaction between religion and the rest of the biosphere comes from ideas about relations between humans and other living things, ranging from the extreme human-centredness of Christianity to the universal non-violence of Jainism. Ideas about the 'souls' of animals and in some cases plants have practical importance for the way they are treated. Many religions attach particular value to particular species, for example the sacred cows of Hinduism and the sacrificial rams of Islam.

Secular ideologies have generally failed to achieve enduring preponderance, partly because they lack any core of ideas of supposedly permanent importance for individuals, partly because they do not propose sufficient rewards or punishments for accepting or rejecting them, partly because they do not make life meaningful. It will be interesting to see how long the current phase of western culture will last. By undervaluing children, who are neither producers nor significant consumers, and by giving production priority over parenthood, it risks bringing about a demographic collapse as dangerously rapid as the previous – and in many places current – expansion.

Propagating a universal ecology.

Other chapters in this book describe what must be called the present ecological crisis. Action is urgently needed, but human actions are shaped by beliefs and feelings. It has been argued above that the neo-Cartesian world view promotes false beliefs and undervalues feelings, with grave ecological consequences, and that it should be replaced by a universal ecology.

The demise of Descartes' ideas has been announced many times,[25] yet they remain preponderant, for the very reason that makes them wrong: mental and physical cannot be separated, and Cartesianism is not just an abstract philosophy but also a way of living and thinking, followed by hundreds of millions of people and embodied in all sorts of artefacts and structures. Universal ecology therefore faces a very ecological problem: how to take over the territory of a highly successful rival.

Newly introduced organisms spread opportunistically, first lodging themselves wherever a patch of suitable habitat appears, then moving outwards to oust rival species. So with ideas, they must first establish themselves in the way of life of individuals and small groups, then expand to those around them. Young people, having the least vested interest in existing ideas, are the most likely to adopt new ones, so parents and teachers – and school-friends – are in a privileged position.

In the Cartesian system of education, knowledge is broken up into subjects taught by specialists, and there is no explicit over-arching framework that includes all branches of knowledge; at the most there is mathematics, which serves all the sciences. The usual way to introduce a new set of ideas is to insert a new subject into the curriculum. If universal ecology is a framework for knowledge rather than a subject, then it will not fit into this scheme at all. Rather than be taught explicitly, it will need to be conveyed through the teaching of whatever is taught.

Young children certainly do not divide their knowledge of the world into compartments. The separation of subjects begins in primary school, but it is very incomplete, if only because one or two teachers cover them all. The separation of mental and physical learning too is only partial. An ecologically minded teacher should have no difficulty in helping pupils to see that each of them is part of everything that happens around them.

It is in secondary schools and institutions of higher education that the fragmentation of knowledge is completed. To set them free from the legacy of Descartes, we should perhaps return to something like the system that preceded him. The medieval university had a 'foundation course' for all students. It was referred to indiscriminately as 'the seven arts' or 'the seven sciences', and consisted of the *Trivium* – grammar, rhetoric and logic – and the *Quadrivium* – arithmetic, geometry, astronomy and music. This was a wonderfully balanced combination of 'arts' and 'sciences', of verbal and physical skills, and of human and universal.

A modern equivalent of the Trivium and Quadrivium could provide the integrating element missing from subject-based teaching; in other words, it should be ecological in the broadest sense. Many schools and some universities already have a space in their timetable for some form of 'general studies'. However, it should be a matter not of teaching vague generalities, but of taking specific topics, looking at them from widely different points of view, and relating them to action that students can take in their everyday lives.

In the longer term, the spread of a universal ecology, if it comes about, will put an end to the division of knowledge into subjects. People will instead start from centres of interest and will seek out all the relevant knowledge in widening and overlapping circles. Natural ecology will become the ecological approach to living things and to the biosphere.

An essential part of knowledge is research – the quest for new knowledge. In the neo-Cartesian scheme, scientific and technological research is virtually sacred, with its promise of ever greater productivity, efficiency and prosperity. Lured by such prospects, governments and firms spend huge sums of taxpayers' and shareholders' money on research and development. In the competition between nation-states, the proportion of national income spent in this way is a key statistic.

Judged by comparison with this norm of 'real' research, much of what goes on in the arts and social sciences is regarded as hardly research at all. The work of the solitary scholar in the library, or of the research assistant administering questionnaires, is seen as personal, subjective and incapable of being verified. It never seems to reach any final answers, and there is always room for new studies of old topics. Fortunately for its survival, it is not usually very expensive, though if it cost more it would no doubt be taken more seriously.

The modern notion that proper research is expensive work for experts has moved the word a long way from its original meaning. It comes via old French from the Low Latin *recircare* 'to go round and round [looking for something]', from *circa* 'around'. It should simply mean to search persistently and intently, and its application should not be limited to teams using costly equipment to solve abstruse problems.

One of the effects of the cult of expensive science has been to destroy people's confidence in personal research. If reliable knowledge has to be handed down by experts, the ordinary person becomes a passive receiver with no control over the advancement of knowledge. If a large part of the fruits of costly research is the confidential property of the businesses or military establishments that financed it, the alienation of the public is made still worse.

Whatever the apparatus and the expense, the ultimate judge in all research is always the person. Whether the topic is the interaction of sub-atomic particles or the effect of a food on one's feelings of well-being, patterns of events have to be detected by the human nervous system so that a judgement can be made as to whether a belief 'feels' right or wrong. The word 'expert' means simply 'experienced', and everyone can hope to become expert in living his or her own life and making judgements about new knowledge relevant to it. It is not even a matter of high intelligence; every bird or mammal that survives is expert in how to live.

The kind of knowledge that has come to be known as scientific differs not in dispensing with feeling but in demanding ruthless honesty in discriminating between those feelings of rightness that are based on the evidence and those that come from a comfortable fit with other beliefs, from social approval and so on. If the result feels right in the first sense, it must be defended even though it contradicts some cherished theory, upsets the director of the laboratory, or clashes with the editorial policy of a learned journal. Some 'science' in fact falls short of this ideal, and much of what is not counted as science lives up to it.

If personal research is to produce valid results, it needs the honesty of science. In the example of choosing healthy food, it is important to discount feeling that it is right because an advertisement says so, or because it is the fashion among your friends, or because you saw an article in the paper saying that scientific research has proved that it is good; ultimately it must feel right to you. It is equally important to resist the temptation to pick the food that others say is wrong just to prove that you are independent. The ability to feel your own feelings confidently is not something trivial that can be taken for granted; on the contrary, it is a skill that needs to be patiently developed and sturdily defended.

Universal ecology has been presented here as a framework for knowledge. Its essential ideas are integration, context and relevance. Whatever expensive knowledge has to be integrated is already being generated by the various disciplines. Research in universal ecology is essentially personal: the individual's attempt to integrate all the knowledge available. This does not mean that it has to be private. As in the natural sciences, one of the safeguards against error and bias is readiness to open results to public inspection and criticism. Indeed, given that no one person can possess more than a fraction of all the knowledge that is to be integrated, collaboration is of the greatest value, provided that at the end each person remains loyal to his or her own judgement.

'Integration' is from Latin *integer*, whole, from the same root as 'intact' – untouched, unspoilt. Another derivative, 'integrity', has come to have the moral sense of being whole, pure or honest (in French *entier* 'entire', also from Latin *integer*). The English 'whole', cognate with Greek *holos*, from which is derived 'holistic', is at the root of the word 'health'. It is appropriate that the integration of knowledge should awaken all these echoes. Its effect should be to enable the knower to relate to life as a whole, to achieve health and integrity, and to give meaning to actions.

It is common experience that wholeness is very difficult to achieve. We muddle through from one thing to another, not so much borne along on a stream of consciousness as groping our way between the irregular flashes of a strobe of consciousness. We waste thought on trivia and casually allow important decisions to take themselves. We are overwhelmed by present circumstances and ignore what is out of sight. Most comically, we thread together scattered beads of memory to make a life story that we fancy holds together. Yet on this precious self, whose unity is so doubtful, we lavish extraordinary generosity, thinking it worthy of far more than its share of the world's riches.

Collectively we muddle along in the same way, creating group selves out of the accidents of kinship, geography and history, persuading ourselves that there is something fundamental and enduring about being British or French, Serb or Croat, 'white' or 'black', helping these group selves to pursue their interests at the expense of the interests of the human and non-human world that surrounds them. At a higher level of abstraction, we fancy that our species has unique claims on the rest of the biosphere, that human life alone is sacred, and that all manner of destruction is justified in the interests of humanity.

If we are to understand our place in the whole, there has to be a fiercely honest critique of the selves that create frontiers across the unity of the world. Where does my self end and the rest of the world begin? What does my self really need from the world? What does the world need from me? Is there any difference in kind between my being and the being of the world? Does the self have any permanent and separable existence at all? These questions are posed not for abstract philosophical speculation, but for careful self-observation in the business of everyday living.

Out of personal research into such questions, we may hope to achieve a heightened consciousness, becoming aware of more of the processes in which we are involved, and of our interactions with the human and non-human world, moving towards greater wholeness in our personal lives. We may hope to adapt and re-adapt ourselves, balancing flexibility and constancy, realizing in personal evolution our membership of the living planet, finding and creating a wealth of pattern, and preparing eventually to let go of self in death. If a universal ecology provides people with a framework in which to fulfil themselves in such ways, it can indeed be said to open the way to a new Renaissance.

Acknowledgements
I am grateful to Christopher Stewart and Ulrich Loening for many helpful comments but they must not be held responsible for any shortcomings.

Chapter 19

Where next?

Duncan Poore

The Author

Duncan Poore is a classicist turned scientist. After a period during the war working with Japanese naval codes and ciphers, he took a degree in natural sciences and a doctorate in plant ecology at Cambridge. Since then he has worked at the interface between environment and economic land use: with Hunting Technical Services in Cyprus and the Middle East; as Professor of Botany in the University of Malaya; as Director of the Nature Conservancy (Great Britain); with IUCN (the World Conservation Union) as Scientific Director and Acting Director-General; and as Professor of Forest Science in Oxford. For the past 15 years he has been closely associated with the International Institute of Environment and Development and with international work in environmental and forest policies, mainly in the tropics.

Chapter 19

Where next?

Duncan Poore

Introduction

These essays speak for themselves. They are so rich in material that it would be impertinent to attempt to summarise. Instead, in this last chapter, I shall provide a personal commentary, and shall build this partly round a simple model.

The various authors highlight many important issues, but three stand out – two broadly social and one environmental, all symptoms of a serious underlying malaise – the inability of human beings *en masse* to come to terms with their place in the world they inhabit.

The first is the increase of human numbers beyond the capacity of the earth to support them all at an acceptable standard of living; indeed vast numbers of people barely survive in conditions of extreme poverty, destitution, starvation and misery. The second is the placing of excessive demands on the environment, demands which reduce its capacity both to deliver goods and services and to absorb wastes. This is demonstrated in many ways: the degradation and pollution of the soil, inland waters and oceans; the increase of greenhouse gases; the depletion of the ozone layer; the increasing scarcity of readily available energy; and the loss of biological diversity. The combined effects of these are unknown, perhaps unknowable; each or any might, unpredictably, flip into catastrophe. The third issue is the appalling proliferation of wars, conflict, displacement, terrorism and violence.

The combination of these problems (and the many others listed in the first chapter) is unprecedented; the causes of each are manifold. While, for practical reasons, there is no alternative to tackling them piecemeal, such an approach is unlikely to succeed if the measures taken are not part of a overall design and inspired by a comprehensive vision – my ideal would be one such as Philip Stewart's 'universal ecology'. Within this framework, I suggest that the problems, and their possible solutions, lie in the interaction of five elements: technology, economic mechanisms, governance, ethics, and – underpinning and reinforcing the others – knowledge and understanding.

Since the appearance of the human species, men and women have changed the face of the earth – and their own condition – by inventions and their practical application, from the simplest of stone tools to the computer, the cello and the internal combustion engine. These inventions have penetrated every element of living. They have affected population numbers in many ways – through more available food, improved drinking water and sanitation, advances in medicine and contraceptive devices; but, at the same time, they have promoted the prodigal use of natural resources. The ways in which we live, the material standard of living and the quality of life have been transformed by technical artefacts and constructions ranging from the sewing needle to Concorde and the Taj Mahal.

Globalisation (the in-word of 1999) has been promoted by phenomenal technical advances in mobility and in communication; people and goods can now be moved almost anywhere within days, and information and money almost instantaneously. Justin Arundale emphasises the revolutionary effects of information technology and how it may influence both cultural convergence and divergence.

Moreover, technology (as James O'Connell argues) has often, indirectly, generated the inequalities which have been a potent cause of wars, violence and terrorism; and, directly, has produced ever more devastating weapons of mass destruction.

In fact, the application of inventions (technology) is, and always has been, the *motive power* of change. But the *engine* of change is provided by economics. Man is, as far as I have been able to ascertain, the only species of animal that barters and trades. This feature distinguishes men and women quite as effectively from other animals as the nature of their consciousness. Barter, trade and economic transactions provide, it seems, the only means by which people can obtain a livelihood; and by which the world-wide human system can operate. However desirable an alternative might be, we have not yet succeeded in finding it.

Meanwhile, the combination of advances in technology and the prevailing economic system – if both are left to their own devices – seems inevitably to favour those who are already advantaged or well endowed. They have led, and continue to lead, to great inequalities between people and nations, inequalities which tend to increase with each new invention. Technology gives power, but a power that can best be exercised by those who are already powerful – both Shridath Ramphal and O'Connell make this point forcibly. But it is possible to conceive a different world, in which the exercise of power is neither harmful nor destructive.

The only *regulator* of the process, which may be able to prevent it running out of control, is provided (as far as it exists at all) – and can only be provided – by a shared code of ethics, mediated by a respect for law and controlled by effective institutions of governance.

One can, of course, seek even deeper for motivations. Behind 'invention', there lie two elementary human characteristics: intellectual curiosity and the wish to see what happens – the urge to experiment. Behind 'economics' lie the urge to trade, the stimulation of gambling, the temptation of riches and power, or simply the desire for security. Behind 'an ethical code', lie altruism and a conviction that there are values that go beyond the material. Behind all these lies the universal human desire to find meaning and purpose in existence.

Accordingly, there are alternative or complementary ways forward: to alter the mechanisms (the application of technology or economic instruments), or to seek to change prevailing human values. Rom Harré argues that we are emerging from postmodernist thinking to have confidence in the validity of science in those fields that are proper to science; Nigel Simmonds that the law cannot be effective if there is no common ethical base; Ramphal that governance can only be fully effective if it is seen to be unbiased; Martin Palmer that everything may be possible, given a system of belief – and hope.

How is change going to occur? By new technology? Through new economic instruments? Through revised and better agents of governance? Through changed values? I believe that it must be by all of these. But all depend upon knowledge, upon understanding and upon the wisdom that comes from full understanding. If people are truly to influence their own destiny, they must be enabled to think, to understand and to have a vision of a better future. The alternative? The dissolution of society?[1]

Technology

Much change will certainly come from new technology. The history of inventions shows that the final, widespread and influential application of any new machine has often been far from the original intention of the inventor or the way in which it was initially used. Accordingly, the careful nurture of curiosity, invention and innovation, for their own sakes, should be an essential part of our educational culture. But there is much room, in addition, for the encouragement of technologies which would promote initiatives such as the 'sustainable livelihoods' of Ashok Khosla or the 'Lean Economy' of David Fleming[2] and generate new devices that would be economical in their use of energy and raw materials – those, in fact, which might lead to a stable, secure and sustainable future.

On the other hand, technology has a momentum of its own. Once a discovery has been made, the pressure to apply it is enormous – especially if the application is likely to be profitable. The world has constantly been taken unawares by the way in which it has been turned upside down by new inventions. This will continue – an unpredictable element in human history. Almost every new device or process has unforeseen consequences. Biotechnology is the most recent to appear on the stage and shows every sign of becoming a powerful actor. So far, human society has not proved very skilful at anticipating and controlling the incidental harm caused by such advances. We should exercise all our critical abilities to extract the maximum good with minimum damage. The precautionary principle is fine – if it does not strangle valuable innovations at birth. Control is fine – if it does not stifle invention and initiative.

Economics

We come now to economics. Among the contributors to this book, there is some disillusionment with the market economy as the only mechanism for a functioning world polity. David Burns satirises it; Ramphal and O'Connell castigate its evil consequences – the burden of debt, especially, inhibits recovery in the Third World; Khosla calls the system bankrupt; Robert Goodland considers it incompatible with 'environmental sustainability' and Fleming believes that it has passed its usefulness and should be replaced by 'a climax economy'. All these seem agreed that the present style of market economics cannot by itself be the engine to get us 'from here to there'.[3] Some analysts also consider that the present global financial system is inherently unstable and does not generate and sustain the confidence upon which it depends and without which it may collapse. What then are the alternatives? It would clearly be disastrous to attempt to dismantle the market economy wholesale – revolutions have never accomplished their long-term objectives – but it might be possible to supplement it, reduce its dominance and even, perhaps, gradually replace it. Meanwhile it should be subject to greater regulation; Ramphal argues convincingly that there should be an Economic Security Council in the United Nations, which would *inter alia* exercise a degree of control and might design a long-term strategic framework for economic activity.

One challenge is, therefore, to design, and thereafter control, an economic system (or system with sub-systems) which will generate enough wealth to provide the basis for 'sustainable livelihoods' (in the sense of Khosla). This is the first prerequisite; but, if we are to reach fulfilment, a mere livelihood is not enough. The economic system must also generate enough wealth for common services such as health, education and environment.[4] It should also be adequate to support the higher manifestations of human culture and creativity, by encouraging the arts and stimulating the pursuit of knowledge. But *more wealth* will not, by itself, solve all problems. There is also the question of how it is shared; the market and unregulated free trade do not, apparently, encourage equitable distribution! Would the Economic Security Council help?

Interactions

The application of any technology and the operation of any economic system raise questions of both ethics and governance; for it only through governance that the values of society are institutionalised and can become accepted practice.

Examples could be chosen from many fields: the development of nuclear technology, the use of the world wide web in matters of copyright, the use of electronic transfer in currency dealing, the feeding of hormones to beef cattle, or the processes which underlie the proliferation of supermarkets.

I shall illustrate the interaction of these five – technology, economics, ethics, governance and understanding – by an example from medicine.[5] The study of the human genome is opening up many new possibilities for the prevention and treatment of genetically transmitted disease. This is where technology comes in. But, whether or not remedies can be applied will depend, whether we like it or not, upon their cost; and any illnesses, which result from a failure to prevent or treat, will also incur costs. These have to be assessed in relation to other calls upon health services and upon government or private budgets. There will also be effects on insurance premiums and possibly also on job prospects. If particular services are unavailable at government expense, those who can afford to do so may be prepared to pay for them or to procure them in another country. Hence the economic dimension will intervene. But powerful ethical issues arise over the possible use of genetic manipulation on human beings. These have to do, for example, with the way in which the technology is applied, the relation of doctor to patient, and the extent to which information should be made widely available. As a result, there are, in the United Kingdom at least, many committees set up to consider and reach some consensus on these ethical questions. Governance is needed to decide what any society ultimately does. The conclusions may be enshrined in national or international law and become subject to the jurisdiction of the courts; an example is *The Convention on Human Rights and Biomedicine* (1996). To complicate matters further, different countries (or even different parts of the same country) may reach different ethical conclusions and have different laws.

Furthermore, as everyone will be affected by the resulting laws and regulations, everyone – in a democratic society – should have the opportunity to influence the form they take. If this influence is to be beneficent, the facts must be presented to them accurately and without bias; and they need an education that equips them to make wise and balanced judgements.

Environmental limits

Two of the three problems that I identified at the beginning of this chapter are inextricably linked: on the one hand, the growth of human numbers and the desperate plight of many of them; on the other, the excessive demands made on the natural environment. Any attempted cure must address both. One thing is clear: *at some stage the environment will set a ceiling.*[6] When and how?

Both Charles Pereira and Virginia Abernethy consider that population has already surpassed the carrying capacity of the earth – which some believe may be as low as 2 to 3 billion people, compared with our present 6 billion. It may be argued that technology can provide the answers, or that we can escape by substituting another kind of capital for natural capital – in fact, that we can move the boundaries that confine us. Anyway, if we are able to choose, do we really want to live right up against the limits? It would certainly be most uncomfortable! We should pause to think who we are, what kind of world we wish to live in and what kind of life we wish to lead. And then: how can we optimise the potential of the human species within these constraints? Or are these stupid questions; do we just blunder on from crisis to crisis in the same old way?

In fact, of course, things are not like this at all; these questions are posed at an unrealistic level of generalisation. Who is meant by 'we'? Each person can only act within his or her particular environment. It is not at all clear why and in what circumstances new movements and systems of ideas succeed or fade away; nor what influence, if any, universal communication will have on the evolutionary success of any particular idea. Stewart touches upon these problems. But, how, then, can the individual hope to exercise a constructive influence? A large part of our problem, as individuals, may be our feeling of impotence in the face of such complexity – an impotence that can perhaps only be countered by 'belief'.

Fleming, too, is pessimistic about on the mismatch between population and resources:

'The developed market economy, world-wide, is on the point of having to fight for its life in the face of environmental degradation and resource depletion, and many of the less developed regions of the market are already having to do so. That is to say, the idea that we should try to develop in a way which does not damage the environment is the inverse of the truth: the agenda that faces us now is to find those forms of economic management which remain open to us when the effects of the environmental damage that has already been inflicted begin to bite.'

Abernethy, also, is far from complacent. She argues that there is a reciprocal relationship between environment and population growth. The accepted wisdom, the 'demographic transition', has comforting overtones – that increases in prosperity and standard of living will lead to reductions in birth rate. The alternative, for which she presents evidence, is that the choice to have children or to refrain from having them is strongly influenced by the parents' perception of their future prospects: if these look good, more children; if poor or uncertain, fewer. It could be important which of these is, or becomes, the dominant process (perhaps both my operate together). The former would presuppose that stability would result from a more prosperous world population but one which, given constantly increasing expectations, would be likely to make even greater demands on resources; the latter, a population which would be squeezed by the constrictions of declining access to resources. The former, perhaps, might be looked upon as a voluntary response; the latter as one imposed by reaching the global carrying capacity, either physically (determined by energy, food , water or disease), or psychologically (influenced by domestic and international tensions). This is no idle scare-mongering. Stewart speculates about the possibility of a 'dangerously rapid' demographic decline if there should be a failure of confidence in western culture. Fleming envisages the conditions in which confidence might indeed fail.

Relieving the pressure – livelihoods

There are, however, many things that can be done *now* to relieve the pressure. They are in many fields: increasing food supply; encouraging the reduction in fertility; optimising the balance between intensive and extensive uses of the land; reducing dependence on non-renewable sources of energy; conserving biological diversity; reducing greenhouse gas emissions; and taking measures to anticipate the effects of climate change on agriculture, forestry, natural ecosystems and human settlements. In principle, it is already known how to do all of these things.[7]

I am convinced, however, that among the most effective, both socially and environmentally, will be those which involve human livelihoods.

Every human being (male or female) has to spend his waking hours in some way or another. What he does may be described in a multitude of ways: useful, pleasurable, tedious, destructive, profitable, playful, inventive *et cetera*. (The story of the Garden of Eden might be considered an allegory of the origins of work from play, or of the transition from a hunter-gatherer existence to the toil of farming.) Moreover, every human being has to find a livelihood, some way of supporting himself and his dependants. The way of the modern world is mainly through paid employment.

Employment, and the lack of it, is everywhere a vexed question. It is a major preoccupation of politicians in all countries; a large body of unemployed spells trouble – discontent, migration and social unrest. In the developing world, the labour force is growing more rapidly than jobs can be created; in the developed world, it seems highly unlikely that the political objective of 'full paid employment' is anything but an illusion, given human numbers, the constraints of resources and ever more mechanisation and automation. The pattern of work is changing so rapidly that there is rarely a balance between money supply, the provision of employment and the available labour force. Recent events in Thailand, Indonesia and Malaysia have shown how fragile the equilibrium can be.

For the poor, work is often unsatisfying and degrading drudgery. Even those who earn more than is needed for necessities organise their lives in a perverse fashion. On the job, they may overwork to breaking point, making an obsession of efficiency, productivity and earning power; in their leisure time, they spend both time and money on things that give them pleasure. Productivity and efficiency rarely feature in this second life; the criteria are often the very opposite! Only the rare few, either of the rich or the poor, spend their working life doing the things that give them real satisfaction and pleasure. If one considers this objectively, it is a bizarre way of spending one's life; but to change it will require a fundamental overhaul of attitudes to work and of the economic system that underlies it.

Nevertheless, moves are being made in this direction. Elements of a practicable course of action are sketched out in three chapters: those of Khosla, Goode and Fleming.

Khosla, on the basis of careful analysis and much practical experience, identifies 'sustainable livelihoods' as a means to alleviate the problems of the poor and the unemployed, and to give self-respect and autonomy to all individuals. He believes that the establishment of 'sustainable livelihoods' offers the only solution that would be efficacious for people and countries in all stages of development. It requires the deployment of the most appropriate – and often 'advanced' – technology, kinds of enterprise, economic instruments and management skills. While it could operate in enterprises of all sizes, it is most needed, and perhaps most effective, at what he terms the level of 'mini-enterprises – enterprises requiring, in India at least, an investment of between US$250 and US$250,000.

Goode argues that the world population will only be able to live within environmental limits if cities become self-sustaining to a much greater degree than hitherto. This is because of the immense influence of cities on what takes place outside them, even in parts of the world remote from them – their 'ecological footprints'. He gives examples of the ways in which, slowly and tentatively, this is beginning to happen.

Fleming develops a similar theme in a recent paper addressing what he terms 'The Lean Economy'.[8] His argument goes thus. Demand in the developed market economies is not increasing at the same rate as sustained advances in labour productivity. Rather than being invested in further goods and services, surplus money is being diverted to security in the form of savings and 'relative advantage' as, for example, better housing, schooling etc. The result is likely to be endemic, growing unemployment. As well as orthodox steps to increase employment, there should be an alternative approach – a move towards what he terms 'the climax economy'. His proposed solution is 'the New Domestication', the adoption at a domestic level of the 'advanced technologies, systems and competence of industry'. At the same time, economies should move from the 'complicated' (in which large centres of production are separated by great distances from centres of distribution and consumption – entailing high transport costs) to the 'complex' (in which all the various elements are near to one another, enriched by other local activities – both more economical and more resilient). To quote:

' . . . the pressing reality of today . . . is the patchwork of unemployment which stands in contempt of hopes of growth on a sufficient scale to solve the problem. The prescription for present action arises out of the opportunity to transform those unemployed resources into the integrated local economies of the future. Local small-scale lean production technologies and skills can be built – now. Energy systems, water conservation, the first steps in urban food production, small scale secondary production and services for local use, local decision making – it is these fundamentally creative initiatives which should be the focus of local development today, in place of the mirage of employment tomorrow. Some local development of this kind is beginning already, but it is half-hearted, blighted by the assumption that the real solution is a return to full employment . . . At present, local economic development lacks context in the form of a story about where it is designed to lead. It needs to be based on an understanding that it is not a mere holding operation for poor areas, but that it represents the starting-point of a transformation to the climax form of economic development – the lean economy.'

There is a considerable convergence between these three views – a convergence which I believe to be significant and which is also in line with much of the recent work and thinking of the International Institute for Environment and Development (IIED).[9] The transformation to the 'climax economy' will require root-and-branch changes in developed economies – an unravelling of the 'complicated'; in developing economies, however, starting from a simpler base, the transition might be less painful.

All of these possible ways forward, by Fleming, Goode and Khosla, stress that there are two essential preconditions: that they need the support of a favourable climate of government policy; and that decision-making and operation should be devolved to the greatest practicable extent. One or the other is not enough; without *both*, there is little chance of success. And both depend upon people who are well informed, involved and capable of wise judgements – and, therefore, *educated*.

Conflict and violence

Both Ramphal and O'Connell devote much space to the multiplication of wars, particularly within states, and their appalling results, in slaughter, destitution and displacement. The desperate inequalities in the contemporary world and the number of the absolute poor cannot help but contribute to the present rash of wars. Differences in race and religion may be partial causes – as may fear, historical grievances and the ambitions of power-hungry autocratic leaders. It is hard to know which cause is most important in any instance; and one may be used as a scapegoat for others – religious differences being invoked, for example, when the more direct cause may be economic. But it seems reasonable to suppose that the generation of sustainable livelihoods, and any measures that would relax the pressure of population on resources, would help to raise the living standards of the poor and to iron out some of the most conspicuous inequalities. This, in turn, should reduce the tensions that lead to internecine strife.

We are now faced with serious problems of governance – at many levels. Although Marxist communism has largely disappeared, there are almost as many systems of government as there are nations, some more suitable than others for particular circumstances and temperaments. Issues at the sub-national and national levels grade into those at the regional and global. One present feature of governance is the tension between the need for ever larger units in the cause of economic efficiency (or power) and the desire for local identity and autonomy. This takes a number of forms: a desire for separatism – as in the Basques, the Kurds or the people of Acheh – rather than being a part of a larger nation; devolved authority *versus* government from the centre; populism *versus* central leadership; 'top-down' *versus* 'bottom-up'. These tensions are also a potent cause of war, violence and terrorism. Simmonds comments on them:

> 'The European Union which set aside age-old rivalries is the prime example of a new and co-operative political structure – that, while [countries] may compete with one another, their interests still converge. Such convergence is much less clear in developing countries, whose economies seldom complement one another, and whose newly formed states, new ethnic proximities within them, and changed relations between groups create no small temptation to conflict.'

As so often in antitheses of this kind, the answer must surely lie between the extremes; but how and in what proportions? I believe that many of our troubles stem from over-generalisation, sticking too strongly to a conviction that there is a best, single way; and that what is right for one country is necessarily right for all. Are we now mature enough to drop our arrogance and pursue the golden mean? Have we sufficient self-confidence and self-discipline?

There is now a compelling need for effective global governance to deal with such issues as the proliferation of weapons of mass destruction, peace-keeping, refugees, poverty, indebtedness, terrorism and crime, drugs, the international money market, global environmental standards, regulation of trade and many others. Nations need to pull together on global problems to an extent they never have before.

Ramphal believes strongly in the United Nations Organisation and its specialised agencies. There is no doubt that we would be much worse off if the UN had not existed. The specialised agencies have accumulated much knowledge and skill that would not otherwise have been available: FAO has made a great contribution to food security; the World Health Organisation to the control of disease; and the other agencies each in their own fields. The monetary institutions have played their part in staving off financial crises. The system as a whole provides a foundation for future building and reform.

But our attitudes today tend to be dominated by cynicism. There is dissatisfaction with the bureaucracy of the system; some of the most influential countries do not live up to their commitments either in financial contributions or to the noble-sounding declarations that they have signed; and, above all, the structures designed for the problems of the post-war years are poorly adapted to the problems of today.

Nevertheless, it is generally better to work with existing institutions, rather than invent new and totally untested alternatives. Accordingly, improvement should lie in the reform of the United Nations. A change in the structure of the Security Council is clearly overdue; and there is now also need for an Economic Security Council. Ramphal argues that many serious problems could be averted by early, firm action. This requires a 'rapid response arm'; but it also requires rapid and firm response! The plaintive cry 'Too little and too late' is far too often justified. Also, if reform is to be effective, decisions must be seen to be reached in a democratic manner and not in the special interests of any one state or group of states; the United States as the single present world super-power has particular responsibilities – and difficulties – in this connection; and there must be the resolution to act quickly and decisively once a decision has been reached. None of these conditions is met at present.

The issue of whether or not to intervene in the internal affairs of sovereign states poses particularly difficult ethical (and legal) questions. Are the present sovereign states always the best unit of government? In what circumstances, if any, is intervention justified? If it is, who should intervene? Who bears the cost? How does one prevent intervention making a bad situation worse? How long should international responsibility last? What are the particular features of a 'crime against humanity'? Is there a time after which we should 'forgive and forget'? These are highly topical questions, which are very far from being resolved.

Attitudes and values

There is little doubt that all the developments outlined above would be greatly assisted by shared values – a common ethical code. Palmer gives an account of the way in which the main religious faiths are converging towards a common understanding of the place of the human species within nature. This is an encouraging beginning; but only a beginning. Ramphal tells us that the Commission on Global Governance has suggested that people of all cultural, political, religious or philosophical backgrounds could unite in upholding certain core values – respect for life, liberty, justice and equity, mutual respect, caring and integrity. It would, indeed, be marvellous if this were to come about; but, unfortunately, even these terms are open to different interpretations and the ideals they represent are often more honoured in the breach than in the observance. Rom Harré concludes his chapter with these words: 'I think it is true to say that the sciences find themselves in less trouble from the rise of relativism than do our moral intuitions. We are surrounded by the products of that science which is about the material world. We look in vain for the signs of a universal morality of decency and respect for persons.'

In the final analysis, it is through national and international law that such shared values become part of the structure of society. In this respect, the outlook is not encouraging; three quotations from Simmond's chapter make the point.

'The modern vision of the ideal polity as radically pluralistic makes law of vital importance, for it is law that must demarcate the bounds of liberty in a world where people lack other shared standards; but this vision simultaneously makes law immensely problematic, for it is hard to see how law could possess the necessary determinacy of meaning once detached from the shared background of understandings upon which it has always relied.'

'The modern political community aspires to be a peaceful and orderly realm of rights founded upon no shared conception of excellence or value. This pluralistic vision necessitates the ascription of a central role to law, for it is law that must demarcate the bounds of otherwise conflicting liberties. Yet, if liberal pluralism makes law central and necessary, it also makes its existence problematic. For how can we sustain a shared set of laws when we share no other moral conceptions?'

'If the law is to demarcate the bounds of liberty for autonomous agents bent upon the advancement of their own projects, it must possess a determinacy that is independent of the will of those same agents. In the absence of some shared set of values or base-lines, however, it would seem that the law can enjoy no such determinacy. Rather than being the framework of rules within which social life is conducted, and which sets the limits to competitive self-seeking between individuals, the law becomes itself a resource over which the contending factions struggle. The unrestricted assertion of rights, the invocation of unspecified standards of justice or equality, and the litigious willingness to exploit the lee-ways of causal attribution, are becoming common features of modern society. What is less common is any willingness to face up to the political landscape that these features present.'

Most people have become not only cynical but fatalistic. Everything is far too complicated; they feel they just have to go with the stream wherever it may lead. But there are two other reactions, neither of them healthy or helpful. One is to retreat into small private worlds, insulated as far as possible from the mainstream of events; and the other is to use the system in order to exploit it, through highly organised fraud and crime using all the resources of modern technology.

I suggest, therefore, that we should look hard and critically at some of our prevailing attitudes – some of them very sacred indeed. While it seems to be axiomatic that profit cannot be avoided as the driving force behind world economic systems, profit need not be the same as greed. There seems to be a confusion between equality and equity. Equality is impossible – not everyone can sing like Maria Callas; the elevation of equality to an imperative only breeds envy. There is an obsession with rights to the virtual exclusion of responsibilities. Every right can be better expressed as a responsibility; animal rights, for example, as the moral responsibility of every person not to cause unnecessary suffering or pain to any other animal (including human animals). The insistence upon right for everyone and everything leads to a *cul de sac*. It has become the fashion, too, both for people and for nations, to blame every mishap upon the other party; and this is accompanied as often as not by litigation. This is perhaps a twist of the issue of rights versus responsibilities. One has a right to make silly mistakes and blame someone else for them, rather than a responsibility to be careful! None of these attitudes is healthy.

If the future is to be stable, it must be founded on co-operation. I suggest, with diffidence, that this requires, above all, some of the cardinal virtues recognised (but seldom widely practised) by most civilisations and religions; above all upon integrity, compassion, moderation and tolerance. Within this framework it should be possible to treat with mutual respect the cultural, philosophical and religious differences which lend such richness and colour to the peoples of the world.

The future

I do not intend to indulge in predictions. History has an awkward habit of dealing jokers from the pack. It may be a revolutionary new technology, a natural disaster, a political upheaval, a new and pernicious virus, or a fundamental disruption of natural processes – and all our forecasts go awry. This, however, is no excuse for doing nothing. History also shows that underlying trends in ideas and social behaviour may continue, perhaps bent but unbroken, through serious disruptions.

We human beings have one great potential advantage as agents of change; our adaptation is primarily cultural not genetic, and therefore largely under our control. The changes we induce are also much more rapid. This places the responsibility fairly and squarely upon us to influence events in the right direction.

When it comes to the broad topic of the evolution of *Homo sapiens*, we become tangled in the twin moral mazes of eugenics and of the nature-nurture argument, where strongly held prejudices tend to override balanced judgement.

We show our humanity in believing firmly that any attempt to manipulate our 'nature' is bad: we condemn utterly such eugenic policies and experiments as were carried out by the Nazis and equally oppose the attempted elimination, the 'genocide', of racial groups such as we have been seeing in the Balkans and Rwanda. We condemn the sterilisation of those with heritable deficiencies or deformities and, quite differently from our simpler ancestors, we do everything to keep alive every human foetus. We are also showing ourselves to be strongly opposed to the idea of cloning human beings. In these respects, our moral stand is, therefore, to retain human genetic diversity. We cannot predict what the effect of these attitudes will be, but we would feel ourselves less than human if we were to do otherwise. We fervently trust that eugenics are no longer an issue! But can we be totally sure?

At the same time, we cannot altogether rule out the possibility of further evolution of physical characteristics in man. Selective pressures may still operate on survival through the reproductive years due to the incidence of new diseases such as AIDS, although we can be sure that the resources of medical science will be mobilised to reduce the possibility; and that such intervention will be supported by the great majority of people.

The fight against disease is of very great significance at the personal level, but is probably much less important for human evolution at the level of the population.

Now that the genetic basis of many organic diseases is known, many new moral problems have arisen: whether or not to have children when there is a significant risk of serious genetic disorder; how to decide what should, or should not, be considered serious in this context; whether to use all available resources of modern medicine to prolong life; to what extent the inheritance of genetic deficiencies should be made known, and many others. The danger of the wider spread of such diseases in the population is, however, now being offset to some degree by increased knowledge and avoiding action. Moreover, the much greater mixing of human populations that have been previously isolated from one another is significantly reducing the chances of both parents having a copy of any damaged gene. The greater this inter-racial or cross-cultural mixing, the lesser will be the danger of children suffering from genetic diseases. Perhaps it will also erode racism.

One geneticist holds the view that selective pressures through the differential survival of children have almost ceased to operate in many parts of the world (although there may be nasty surprises such as AIDS).[10] Instead future physical evolution will be mediated, as it often has been in the past, by patterns of social behaviour: by factors affecting fertility; by cultural mixing and by the differential growth of populations. He considers that the physical evolution of *Homo sapiens* may almost have run its course.

Even if he is wrong, the future mainstream of human evolution is likely to be in the cultural sphere: 'by the progressive extension of the faculties of manipulation (technology), invention, self-awareness, abstract thought, philosophy, religion and the arts; and in increasingly complex social interactions and means of communication with fellow human beings'.[11]

The nature-nurture controversy continues to generate more heat than light. It should surely be self-evident that, if all physical characteristics are under ultimate genetic control, the brain cannot be totally exempt. Moreover, it should be equally evident that the innate mental equipment of any human being can be altered by his or her surroundings during the whole course of his or her life. (There is an open question of how far this influence can go back in the unborn foetus.) The issue that should exercise us is not 'either/or' but 'how much?' – the extent to which environmental influences (which are potentially under our control) can be invoked to bring out the innate potentialities of every individual from the cradle (or earlier) to the grave. The further evolution of humanity in the psychosocial sphere is uniquely the province of education in the widest sense of that word. It will flourish from the full flowering of every single individual but, beyond that, it will depend on the extent to which the wisdom thereby acquired can become part of the wisdom passed down from generation to generation.[12]

Mass communications pose a real danger, that of homogenisation of culture. Huxley, rather optimistically to my view, believed that his 'single inter-thinking group' could come into being without the sacrifice of human diversity. I have already said that I am disinclined to believe in the desirability of the single inter-thinking group, even if the world becomes potentially a single inter-breeding group. But there is little doubt that the ease of movement (both of bodies and of ideas) poses a real threat to the retention of cultural diversity.

Now that we are reaching a phase in our history when there is almost universal access to information and the autonomy of the individual has become axiomatic, the ways in which we behave must depend upon cultural transmission.

Information, knowledge, experience and wisdom should be cumulative, transmitted from person to person and from generation to generation by education – a process that continues from the day one is born until the day one dies.

I make no excuse for entering this subject at length, for it is only through a wide and liberal education that people will be able to find their way wisely through the maze of personal and democratic choices with which they will be faced, to adapt to change and to promote intelligent, holistic and humane approaches to the future.

Hermann Bondi describes what we are trying to achieve through education in these terms:

> 'Surely our purpose is to help the individual to develop and to become a useful and appreciated member of society, that is somebody who can strongly and purposefully interact with others, whose talents and interests are continuously developing, somebody who contributes to the well-being of the whole of society through useful work . . . .'

This is, or should be, what is meant by education – to provide the tools to enable a woman or man to become a complete and integrated person, intellectually, socially and culturally. Moreover, Bondi strongly draws the important distinction between education and training; education should equip for life, training for a job. Yet, although a future of hope can only lie with education, many nations and cultures appear to have adopted, or failed to grow out of, a self-defeating segmentation of knowledge and a narrow and materialistic concept of the purpose of education; standard education tends to kill the enthusiasm for learning. Stewart believes his concept of a 'universal ecology' should provide a comprehensive and integrative approach to education.

What might be the essential elements of such an education? Stewart argues for a modern equivalent of the mediaeval *Trivium* and *Quadrivium*, which contained between them grammar, rhetoric, logic, arithmetic, geometry, astronomy and music – 'a wonderfully balanced combination of arts and sciences, of verbal and physical skills, and of human and universal'. My choice of essential elements would include the following – in no particular order of importance: the fostering of intellectual curiosity and an enthusiasm to learn; an ability to speak, understand, read and write one's own language effectively; an ability to manipulate figures; a critical faculty, including an ability to argue logically and to distinguish truth from falsehood; a love of beauty embodied in nature and symbolised in the visual arts, music and literature; an urge to seek quality in all things; an appreciation of the excitement of science and an understanding of the nature and fallibility of scientific hypotheses; an appreciation of the confidence or lack of confidence to be placed on statistics; an appreciation of probability and 'risk'; the ability to relate to other people and get on with them; a sense of identity derived from learning about the history and culture of one's own country; and the knowledge of another language, both as an introduction to the structure of language and to the understanding of another culture.[13]

This, I believe, is the kind of intellectual and cultural equipment that is required for individual fulfilment; intelligent, co-operative and constructive living; and an adaptive and flexible approach to a changing world. However, these elements, by themselves, do not provide the *integration* advocated by Stewart – an objective which I applaud. One way of tackling this would be to teach comprehensively, from many different perspectives, about a particular topic, such as one period in the history of one locality. This should include the geographical, ecological and biospheric setting; international connections; religion; politics; trade and industry; social changes; ethics; advances in science, technology and philosophy; and literature, drama, music and the arts. This would teach, above all, how everything is connected to everything else.

Any complete system of education should recognise, also, that different individuals have different needs and talents; some, for example, learn better through their hands. So, practice should be added to theory to make the complete person – the opportunity for individual and group participation in activities such as debating, drama, music, art, sports, carpentry, engineering, electronics *et cetera*.

I have left to the last the most important and most elusive element of all: a belief that there is a purpose in existence. Without this, all may be in vain. The future lies with the young in every continent – Max Nicholson's New Renaissance.

At a recent prize-giving at my grand-daughter's school, the headmaster, at the end of a thought-provoking speech about the power of 'mind', asked the two most senior children (a girl and a boy) to give some thought to the question: 'Do you believe that you will be able to use your education to try to make the world a better place?' When they spoke at the end of the ceremony, each, without hesitation and full of confidence, said: 'Yes. I do'.

Notes and References

Chapter 1 – Setting the Scene

1. The terms 'mankind' and 'man' are used in many places in this book. These are proper English usage for the human species, corresponding to the Latin *homo* and the Greek ἄνθρωπος; they are not intended to have any connotations or overtones of sex (gender). Where 'man' is to be distinguished from 'woman' (Latin *vir* and Greek ἀνήρ), this should be evident from the context.

2. Huxley, Julian (1946) *UNESCO: Its Purpose and its Philosophy*. Preparatory Commission of the United Nations Educational Scientific and Cultural Organisation, Paris.

3. Huxley's views and those of a number of his contemporaries were published in 1961 in Huxley, Julian (Ed.) *The Humanist Frame*. George Allen and Unwin, London.

4. Huxley coined the term 'psychosocial' for this phase of evolution; C. D. Waddington preferred the term 'sociogenetic'.

5. Huxley believed that there was still a place for religions, but as 'organs of psychosocial man concerned with human destiny, and with experiences of sacredness and transcendence', emphatically not as agents for divine intervention in human affairs. Religions, too, were subject to evolution.

6. The idea of the 'Noösphere' first appears in Teilhard de Chardin, Pierre (1959) *The Phenomenon of Man*. Harper and Brothers, New York. Perhaps it is now being partially realised in an unexpected manner by the internet.

7. For example, he suggests that the question of human numbers should be closely linked with the question of 'what are people for?'; and that the approach to education must abandon 'the democratic myth of equality'.

8. Similarity in brain patterns in all human beings and potentially universal communication, with English becoming the common language, are moving us closer to the possibility of a 'single inter-communicating group', however widely we may be separated in 'systems of thought'.

Chapter 2 - The New Renaissance

1. Exceptions are the Earthwatch field research project at S'Albufeira on the Bay of Alcudia in Mallorca and the neighbouring Tour du Valat Station in the French Camargue.

2. The work done by the Nuffield Unit of Tropical Animal Ecology in the Queen Elizabeth National Park in Uganda, and more recent research at Seronera in the Serengeti National Park in Tanzania are further exceptions.

3. An ambitious new Encylopaedia of Life Support Systems (EOLSS) is being prepared, with the encouragement and support of UNESCO. It is to be hoped that it may partly meet this need.

Chapter 3 - Soil, Water and People

1. Serageldin, Ismael (1996) *The CGIAR at 25: into the Future*. World Bank, Washington, DC.

2. ISNAR (1997) Research Report **13**. The Hague, Netherlands.

3. FAO (1996) *African Agriculture: the next 25 years*. FAO, Rome.

4. Farugee, R. and R. Gulhati (1983) *Population Growth and Policies in Sub-Saharan Africa*. World Bank No. **559**. World Bank, Washington, DC.

5. MacNamara, Robert S. (1986) Sir John Crawford Memorial Lecture. World Bank, Washington, DC.

6. Pereira, H. C., A. Aboukhalid, A. Felleke, D. Hillel and A.A. Moursi (1979) *Opportunities for Increase of World Food Production from the Irrigated Lands of Developing Countries.* IRDC, Ottawa.

7. Pereira H.C. (1989) *Policy and Practice in the Management of Tropical Watersheds.* Westview Press, Boulder, Colorado.

8. World Resources Institute, UNEP, UNDP, World Bank (1998) *World Resources*, p. 152. Oxford University Press.

9. Brown L. R. *et al.* (1996) *State of the World* p. 79. World Watch Institute, Washington.

10. Oldeman L.R., V.W.P. van Engelen and J.H.M. Pulles (1991) *The Extent of Human-Induced Soil Degradation.* International Soil Reference & Information Centre, Wageningen, Netherlands.

11. World Resources Institute etc. (1998) *Op. cit.* p. 157 [See Note 8]

12. UN Population Division (1997) *World Population Prospects 1950-2050 (1996 Revision).*

13. Tickell, Sir Crispin (1996) *Journal of the Royal Agricultural Society of England*, **157.**

14. Prance, Sir Ghillean (1998) Royal Agricultural Society of England, Eighth Annual Lecture.

15. Royal Society (1998) 'Genetically modifed plants for food use' 16pp. Statement **2/98.** The Royal Society, London.

16. Royal Agricultural Society of England (1998) *Old Crops in New Bottles.* Royal Agricultural Society of England, Stoneleigh.

17. Martin, C.J. (1965) *World Population Kaleidoscope 2000AD.* Finance & Development, IMF, Washington, DC.

18. Kendall, H.W. *et al.* (1995) *Meeting the Challenges of Population, Environment and Resources : the Costs of Inaction.* World Bank,Washington, DC.

Chapter 4 - Population and Environment

1. Bongaarts, John (1998) Demographic Consequences of Declining Fertility. *Science* **282**, pp. 419-420 (Oct. 16).

2. Ventura, Stephanie J., Joyce A. Martin, T.J. Mathews and Sally C. Clarke (1996) *Advance Report of Final Natality Statistics, 1994.* Supplement **44** (11), Tables 10 & 11 (pp. 41-42). Division of Vital Statistics, National Center for Health Statistics, Hyattsville, MD.

3. Umpleby, Stuart (1990) The Scientific Revolution in Demography. *Population and Environment* **11**(3), pp. 159-174.

4. Bongaarts, John and Griffith Feeney (1998) On the Quantum and Tempo of Fertility. *Population and Development Review* **24** (2), pp. 271-291.

5. United Nations Secretariat (1998) *World Population Projections to 2150.* Population Division, Department of Economic and Social Affairs, UN. (February).

6. World Wide, North Korea's Population has Shrunk. *Wall Steet Journal* p.1, Feb. 18, 1999.

7. Ahlburg, Dennis A. (1993) The Census Bureau's New Projections of the US Population. *Population and Development Review* **19** (1), pp. 159-174.

8. Camarota, Steven A. (1999) *Immigration in the United States – 1998.* Center for Immigration Studies, Washington, DC. (Jan 8).

9. Ahlburg, D.A. and J.W. Vaupel (1990) Alternative Projections of the U.S. Population. *Demography* **27** (4), pp. 639-647.

10. Pflaumer, Peter (1992) Forecasting US Population Growth with the Box-Jenkins Approach. In Ahlburg, D.A. & Kenneth C. Land (Eds) *International Journal of Forecasting*, Special Issue **8**(3), pp. 329-338.

11. Ahlburg and Vaupel (1990) *Op. cit.* [See Note 9].

12. Campbell, Colin J. and J.H. Laherrere (1998) The End of Cheap Oil. *Scientific American* (March) pp. 78-83; Younquist, Walter (1997) *GeoDestinies.* National Book Company,

Portland, Oregon; Youngquist, Walter (1998) Spending Our Great Inheritance – Then What? *Geotimes*. Eugene, Oregon (July); Duncan, Richard C. (1997) *The World Petroleum Life-cycle: Encircling the production peak*. Space Studies Institute, Princeton University (May 9); Ivanhoe, L. F. (1995) Future World Oil Supplies: There is a Finite Limit. *World Oil* **77** pp. 80-88 (October).

13. Watt, Kenneth (Ed.) (2000) *Human Ecology: The Encyclopaedic Legacy of H. G. Wells*. Rutgers University, New Jersey.

14. For further discussion, see Campbell, Colin and J.H.Laherre (1998), based in Ireland and France, respectively; and in the United States, Ivanhoe L.F. (1995); Duncan, Richard (1997); Youngquist, Walter (1997 & 1998); and Watt, Kenneth (2000). [See Notes 12 and 13].

15. See Watt (2000) *Op. cit.* [See Note 13 and Chapter 11].

16. Population Reference Bureau. (1971) Man's Population Predicament. *Population Bulletin* **27**(2), pp. 39, p. 19.

17. Ohlin, G. (1961) Mortality, Marriage and Growth in Pre-industrial Populations. *Population Studies* **14** (3) pp. 190-197; Davis, Kingsley (1963) The Theory of Change and Response in the Modern Demographic History. *Population Index* **29**(4) pp. 345-365; Demeny, Paul (1968) Early Fertility Decline in Austria-Hungary: A Lesson in Demographic Transition. *Daedalus* **97**(2) pp. 502-522; Demeny, Paul (1988) Social Science and Population Policy. *Population and Development Review* **14**(3) pp. 451-480; Abernethy,V. (1979) *Population Pressure and Cultural Adjustment*. Human Sciences Press, New York; Abernethy, Virginia Deane (1993) *Population Politics: The Choices that Shape Our Future*. Plenum Press, New York; Abernethy, Virginia Deane (1994) Optimism and Overpopulation. *The Atlantic Monthly* pp.84-91 (December); Abernethy, Virginia Deane (1997) Allowing Fertility Decline: 200 Years After Malthus's Essay on Population. *Environmental Law* **27**(4) pp. 1097-1110.

18. Abernethy (1993, 1994 & 1997) *Op. cit.* [See Note 17]; Abernethy, Virginia (1998) Why Asian Population Growth is Winding Down. *Chronicles* pp.46-47 (October); Abernethy, Virginia Deane (1999) United States-Supported Population Policies in the Third World. *World Development: Aid and Foreign Direct Investment 1999/2000*. World Business Council and Kensington (Eng.) Investments.

19. Walt, Vivienne (1995) Women's work. *Mother Jones* pp. 34-51, p. 48 (October).

20. Moffet, George D. (1994) *The Global Population Challenge: Critical Masses*. Viking Press, New York p.137.

21. Moffet (1994) *Op. cit.* p. 229. [See Note 20].

22. Walt (1995) *Op. cit.* p. 38. [See Note 19].

23. Abernethy (1998) *Op. cit.* [See Note 18].

24. Courbage, Youssef (1995) Fertility Transition in the Mashriq and Maghrib. In: Obermeyer, Carla M. (Ed.) *Family, Gender and Population in the Middle East*. The American University in Cairo Press, Cairo.

25. Abernethy (1997) *Op.cit.* [See Note 17].

26. Abernethy, Virginia Deane (1999) Population Dynamics Revisited: Lessons for Foreign Aid and US Immigration Policy. In: Baudot, Barbara S and William R. Moomaw (Eds) *People and Their Planet* pp. 143-156. St. Martin's Press, New York.

27. Abernethy (1997 & 1998) *Op. cit.* [See Notes 17 and 18].

28. Caldwell, John C., I.O. Orubuloye and P. Caldwell, P. (1992) Fertility Decline in Africa: A New Type of Transition? *Population and Development Review* **18**(2) pp. 211-243.

29. Caldwell *et al.* (1992) *Op. cit.* pp. 236-237. [See Note 28].

30. Caldwell *et al.* (1992) *Op. cit.* p. 226. [See Note 28].

31. Caldwell *et al.* (1992) *Op. cit.* p. 229. [See Note 28].

32. Nabarro, David N. and Elizabeth M. Tayler (1998) The 'Roll Back Malaria' Campaign. *Science* **280** pp. 2067-2068 (June 26).

33. Phoolcharoen, Wiput (1998) HIV/AIDS Prevention in Thailand: Success and Challenges. *Science* **280** pp.1873-1874 (June 19).

34. Anon. (1998) HIV Incidence: 'More Serious Than We Imagined.' *Science* **280** p. 1864 (June 19).

35. Way, Peter O. and Karen Stanecki (1993) How Bad Will It Be? Modeling the AIDS Epidemic in Eastern Africa. *Population and Environment* **14**(3) pp. 265-278.

36. Anon. (1998) *Op. cit.* p. 1864 [See Note 34].

37. Anon. (1998) The Science of AIDS: A Tale of Two Worlds. *Science* **280** pp.1843-1844 (June 19); Spector, Michael (1998) Doctors Powerless as AIDS Rakes Africa. *New York Times*, p. A1 (August 6).

38. Altman, Lawrence K. (1998) Parts of Africa Showing H.I.V. in 1 in 4 Adults. *New York Times* pp. A1, A8 (June 24).

39. Anon. (1998) Hutu Rebels Butcher 73 People in Burundi *USA Today* p.13A (April 23); McKinley, James C. Jr. (1998) AIDS Brings Another Scourge to War-Devastated Rwanda. *New York Times* pp. A1-6 (May 28); Mydans, Seth (1998) Indonesia in Drama's Grip, Warily Awaits Finale. *New York Times* p. A4 (June 2); French, Howard W. (1998) Congo Catches Rwanda's Disease. *New York Times* p. 3 (August 9) .

40. Homer-Dixon, Thomas F., Jeffrey H. Boutwell and George W. Rathjens (1992) Environmental Change and Violent Conflict. *Scientific American* pp. 38-45 (February).

41. Esman, Milton J. (1994) *Ethnic Politics*. Cornell University Press, Ithaca, New York.

42. Smil, Vaclav (1997) Global Population and the Nitrogen Cycle. *Scientific American* pp. 76-81 (July).

43. Gever, J., R. Kaufmann, D. Skole, and C. Vorosmarty (1986) *Beyond Oil: The Threat to Food and Fuel in Coming Decades*. Ballinger, Cambridge.

44. Koshland, Daniel E. (1991) War and Science. *Science* **251**(4993) p. 497.

45. Holdren, John (1991) Population and the Energy Problem. *Population and Environment* **12**(3) pp. 231-256.

46. Bartlett, Albert A. (Unpublished, 1998). *How Population Growth Drives Greenhouse Gas Emissions in the United States*. University of Colorado, Boulder, Colorado.

47. Pimentel, David and Marcia Pimentel (Eds) (1997) *Food, Energy and the Environment*. University Press of Colorado, Niwot, Colorado.

48. Younquist, Walter (1997) *GeoDestinies*. National Book Company, Portland, Oregon; Bartlett, Albert A. (Unpublished, 1998) *Op. cit.* [See Note 46].

49. Pimentel, David and Marcia Pimentel (Eds) (1996) *Food, Energy and Society*. University Press of Colorado, Niwot, Colorado; Pimentel, David and Marcia Pimentel (1997) *Op. cit.* [See Note 47]; Pimentel, David, C. Harvey, P. Resosudarmo, K. Sinclair, D. Kurz, M. McNair, S. Crist, L.Shpritz, L. Fitton, R. Saffouri and R. Blair (1995) Environmental and economic costs of soil erosion and conservation benefits. *Science* **267** pp.1117-1123 (Feb. 24); Pimentel, David, M. Giampietro and Sandra G.F. Bukkens (1998) An Optimum Population for North and Latin America. *Population and Environment* **20**(2) pp. 125-149.

50. Ahlburg, D.A. & J.W. Vaupel (1990) *Op. cit.* p. 646. [See Note 9].

The following publications are also relevant:

Arrow, Kenneth, B. Bolin, R. Constanza, P. Dasqupta, C. Folke, C.S. Holling, Bengt-O. Jansson, S. Levin, K.-G. Maler, C. Perrings and D. Pimentel (1995) Economic Growth, Carrying Capacity, and the Environment. *Science* **268** pp. 520-521 (April 28).

Atiyah, M. and F. Press (1992) *Population Growth, Resource Consumption, and a Sustainable World*. Royal Society of London and National Academy of Sciences, London and Washington, DC.

Cohen, Joel E. (1995) *How Many People Can the Earth Support?* W.W. Norton, New York.

Demeny, Paul. Early Fertility Decline in Austria-Hungary: A Lesson in Demographic Transition. *Daedalus* **97**(2) pp. 502-522.

Anon. (1995) Disease Fights Back. *The Economist* pp.15-16 (May 20).

Duncan, Richard C. (1993) The Life-Expectancy of Industrial Civilization: The Decline to Global Equilibrium. *Population and Environment* **14**(4) pp. 325-358.

Gellert, J.A. (1993) International Migration and Control of Communicable Diseases. *Soc. Science Med.* **37** pp.1489-1499.

Umpleby, Stuart (1988) Will The Optimists please Stand Up? *Population and Environment* **10**(2) pp. 122-132.

Williams, G.C. (1975). *Sex and Evolution*. Princeton University Press, Princeton.

Wilson, E.O. (1975). *Sociobiology*. Harvard University Press, Boston.

Chapter 5 - *Diversity: Insurance for Life*

1. The term 'biodiversity' (shorthand for 'biological diversity') has become fashionable since the ratification of the international *Convention on Biological Diversity*, negotiated at Rio de Janeiro in 1992. 'Biological diversity' is defined in the Convention as: 'The variability among living organisms from all sources including, *inter alia*, terrestrial, marine and other aquatic ecosystems and the ecological complexes of which they are part; this includes diversity within species, between species and of ecosystems.'

2. Three terms are commonly used by biologists in this context. (a)'Community', which refers to the various organisms which live together in one area. Different communities usually contain different combinations of species. (b) 'Habitat', which refers to the environment in which an organism or community lives — the climate which surrounds it, the soil in which it grows etc. (c) 'Ecosystem', which encompasses the community, the habitat and the processes linking species with one another and with the habitat. To some extent, these terms are interchangeable and are often used loosely as equivalent to one another.

   Since each ecosystem contains a collection of species, the maintenance of species diversity (and thus of the genetic diversity within species) depends upon the conservation of the full variety of ecosystems, with their many internal interactions. It is often difficult, however, to reach a precise definition or classification of any community or ecosystem. The line between different units is hard to draw and they may be classified at many different levels and in many different ways. For example, ecosystems are sometimes defined by characteristics of their habitat, such as the soil type they occupy or the climate in which they occur. Rainforest is found in those parts of the warm tropical belt in which rainfall is not too seasonal; tundra occurs in cold northern latitudes; and desert where there is little or no rainfall. Different communities occur on acid and on alkaline soils, on sandy and clay soils. At the macro level, communities are most frequently defined by the type of vegetation and the component plant species that occur. However, much smaller units are also communities, such as the moths and algae that inhabit the fur of an Amazonian sloth or the colonies of bacteria that inhabit a human body.

   An example of the finer division of plant communities is found in igapó vegetation on the banks of the Rio Negro in Brazil where a student of mine, Shirley Keel, studied a transect from low-flood level to the forest behind the levée [Keel, S.H.K. and G.T. Prance (1979) Studies of the vegetation of a white-sand black water igapó (Rio Negro, Brazil) *Acta Amazonica* **9** pp. 645-655]. The transect showed that the species composition changed every 10m. along the line because of the corresponding change in length of flood time. At least four distinct communities were apparent. At the low water mark two species of Myrtaceae, *Myrciaria dubia* and *Eugenia inundata*, dominate. Further up the beach *Pithecollobium adiantifolium*, *Schisostemon macrophyllum* and *Eugenia chrysobalanoides* dominate in a slightly more species rich community. At the top of the transect, in a lightly flooded area with less than two months underwater, there is a species-diverse community containing about 40 species and dominated by the palm, *Leopoldinia pul-*

chra. Behind the levée is a further community of tall igapó forest also with many species. In this small distance of 60m. along a transect there are at least four distinct plant communities. This demonstrates the importance of understanding the community aspect of biodiversity in order to conserve plants.

3. May, R.M. (1990) How many species? *Phil. Trans. Roy. Soc. London* B **330** pp.171-182.

4. Hawksworth, D.L. (1991) The fungal dimensions of biodiversity: magnitude, significance and conservation. *Mycological Research* **95** pp. 441-56.

5. Sibley, C.G. and B.L. Monroe (1990) *Distribution and Taxonomy of Birds of the World.* Yale University Press, New Haven.

6. McNeely, J.A., K.R. Miller, W.V. Reid, R.A. Mittermeier and T.B. Werner (1988) *Conserving the World's Biological Diversity.* IUCN, Gland, Switzerland.

7. Valencia, R., H. Balslev and G. Paz y Mino (1994) High tree alpha diversity in Amazonian Ecuador. *Biodiversity and Conservation* **3** pp. 21-28.

8. Gentry, A.H. (1988) Tree species richness of upper Amazonian forests. *Proceedings of the US National Academy of Sciences.* **85** pp.156-159.

9. Dodson, C. and A.H. Gentry (1978) Flora of the Rio Palenque Science Centre. *Selbyana* **4** pp. 16/28 pp.

10. Whitmore, T.C., R. Peralta and K. Brown (1985) Total species count in a Costa Rican tropical rain forest. *Journal of Tropical Ecology* **1** pp. 375-378.

11. Ashton, P. (1964) Ecological studies in the mixed dipterocarp forests of Brunei State. *Oxford Forestry Memoirs* **25**. 110 pp.

12. Valencia, R., H. Balslev and G. Paz y Mino (1994) High tree alpha diversity in Amazonian Ecuador. *Biodiversity and Conservation* **3** pp. 21-28.

13. Gentry, A.H. (1988) *Op. cit.* [See Note 8].

14. Gentry, A.H. (1988) *Op. cit* [See Note 8].

15. Faber-Langendoen, D. and A.H. Gentry (1991) The structure and diversity of rainforests at Bajo Calima, Chocó Region, western Colombia. *Biotropica* **23** pp. 2-11.

16. Balslev, H., J. Luteyn, B. Øllgaard and L.B. Holm-Nielsen (1987) Composition and structure of adjacent unflooded and floodplain forest in Amazonian Ecuador. *Opera Botanica* **92** pp. 37-57.

17. Wright, D.D., J. Heinrich Jessen, P. Burke, and H.G. da Silva Garza (1997) Tree and liana enumeration and diversity on a one-hectare plot in Papua New Guinea. *Biotropica* **29** pp. 250-260.

18. Proctor, J., J.M. Anderson, P. Chat and H.W. Vallack (1983) Ecological studies in four contrasting lowland rainforests in Gunung Mulu National Park, Sarawak. *Journal of Ecology* **71** pp. 237-260.

19. Kochummen, K.M., J.V. Lafrankie Jr, and N. Manokoran (1990) Floristic composition of Pasoh Forest Reserve, a lowland rainforest in Peninsular Malaysia. *Journal of Tropical Forest Science* **3** pp. 1-13.

20. Gentry, A.H. and J. Terborgh (1990) Composition and dynamics of the Cocha Cashu 'mature' floodplain forest (pp. 141-157) In: A.H. Gentry (Ed.) *Four Neotropical Rainforests.* Yale Univ. Press, New Haven.

21. Prance, G.T., W.A. Rodrigues and M.F. da Silva (1976) Inventário florestal de um hectare de mata de terra firme km 30 da estrada Manaus-Itacoatiara. *Acta Amazonica* **6**(1) pp. 9-35.

22. Gentry, A.H. (1988) *Op. cit.* [See Note 8].

23. Campbell, D.G., D.C. Daly, G.T. Prance and U.N. Maciel (1986) Quantitative ecological inventory of terra firme and várzea tropical forest on the Rio Xingu, Brazilian Amazon. *Brittonia* **38** pp. 369-393.

24. Pires, J.M., T.H. Dobzhansky and G.A. Black (1953) An estimate of species of trees in an Amazonian forest community. *Botanical Gazette* **114** pp. 467-477.

25. Reitsma, J.M (1988) Vegetation forestière du Gabon. *Technical Series, Tropenbos* **1** pp. 1-142.

26. Phillips, O.L., P. Hall, A.H. Gentry, S.A. Sawyer, S.A. and R.Vasquez (1994) Dynamics and species richness of tropical rainforests. *Proceedings of the National Academy of Sciences of the USA* **91** pp. 2805-2809.

27. Boom, B.M. (1987) Ethnobotany of the Chacobo Indians, Beni, Bolivia. *Advances in Economic Botany* **4** p. 168.

28. Adis, J.J., Y.D. Lubin and G.G. Montgomery (1984) Arthropods from the canopy of inundated and terra firme forests near Manaus, Brazil, with critical considerations of the Pyrethrum-fogging techniques. *Studies on Neotropical Fauna and the Environment* **19** pp. 223-236; Erwin, T.L. (1990) Canopy arthropod biodiversity: a chronology of sampling techniques and results. *Revista Peruana de Entomologia* **32** pp. 71-77; Stork, N.E. (1991) The composition of the arthropod fauna of Bornean lowland rain forest trees. *Journal of Tropical Ecology* **7** pp. 161-180.

29. Erwin, T.L. (1982) Tropical forests: Their richness in *Coleoptera* and other arthropod species. *Coleopterists Bulletin* **36** pp. 7-75; Erwin, T.L. (1983) Tropical forest canopies: The last biotic frontier. *Bulletin of the Entomological Society of America* **30** pp. 1-19; Erwin, E.O. (1988) The tropical forest canopy: The heart of biotic diversity In: Wilson, E.O. and F. M. Peter (Eds) *Biodiversity* pp.123-129. National Academy Press, Washington.

30. IUCN, UNEP, WWF (1991) *Caring for the Earth: A Strategy for Sustainable Living.* 228 pp. Gland, Switzerland.

31. O'Brien, S.J., M.E. Roelke, L. Marker, A. Newman, C.A. Winkks, D. Mectzer, L. Colly, J.F. Evermann, M. Bush and D.E. Wilot (1985) Genetic basis for species vulnerability in the cheetah. *Science* **227** pp. 1428-1434.

32. Personal communication: Peggy Olwell and Michael Maunder.

33. Rick, C.M. (1976) Genetic and biosystematic studies on two new sibling species of *Lycopersicon* from interandean Peru. *Theor. Appl. Genetics* **47** pp. 55-68.

34. Iltis, H.H. (1988) Serendipity in the exploration of biodiversity: What good are weedy tomatoes? In: Wilson, E.O, and F.M. Peter (Eds) (1988) *Op. cit.* pp. 98-105 [See Note 29].

35. Prescott-Allen, C. & R. (1986) *The first resource: Wild species in the North American Economy.* Yale Univ. Press, New Haven.

36. A 'clade' is a group of biological taxa (e.g. species, genera etc.) which are descended from a common ancestor. A 'cladogram' is a family tree of clades.

37. Vane-Wright, R.I., C.J. Humphries and P.H. Williams (1991) What to protect? Systematics and the agony of choice. *Biological Conservation* **55** pp. 235-254; Williams, P.H., G.T. Prance, C.J. Humphries and K. Edwards (1996) Priority-areas analysis and the Manaus Workshop — 90 areas for conserving diversity of neotropical plants (families Proteaceae, Dichapetalaceae, Lecythidaceae, Caryocaraceae and Chrysobalanaceae). *Biological Journal of the Linnaean Society* **58** pp. 125-137.

38. The term is synonymous with 'biological diversity'.

39. Heywood, V.H., G.M. Mace, R.M. May & S.N. Stuart (1994) Uncertainties in extinction rates. *Nature* **368** p.105.

40. Smith, F.D.M., R.M. May, R. Pellew, T.H. Johnson and K.R. Walter (1993) How much do we know about the current extinction rate? *Tree* **8** pp. 375-378.

41. Other aspects of the growth of population are discussed in Chapters 3 and 4 of this book.

42. Myers, N. (1988) Threatened biotas: 'Hot spots' in Tropical Forests. *The Environmentalist* **8** pp. 187-208; Myers, N. (1990) The Biodiversity Challenge: Expanded hot-spot analysis. *The Environmentalist* **10** pp. 243-256

43. WWF (1998) *Living Planet Report.* WWF International, Gland, Switzerland 38pp.

44. Vitt, L.J., J.P. Caldwell, D.C. Smith and H.M. Wilbur (1990) Amphibians as harbingers of decay. *BioScience* **40** p. 418; Wake, D.B. (1991) Declining amphibian populations. *Science* **253** p. 860; Phillips, K. (1994) *Tracking the vanishing frogs.* St Martins Press, New York.

45. Pounds, J.A., M.P.L. Fogden and J.H. Campbell (1999) Biological response to climate change on a tropical mountain. *Nature* **398** pp. 611-614.

46. World Commission on Environment and Development (1987) *Our Common Future.* Oxford University Press, Oxford. 400 pp.

Chapter 6 - *The Urgency of Environmental Sustainability*

1. Goodland, R. and H.E. Daly (1996) Environmental sustainability: universal and non-negotiable. *Ecological Applications* **6**(4) pp.1002-1017.

2. As human and social sustainabilities are best dealt with by interdisciplinary teams led by social scientists and sociologists, they are not amplified here.

3. Hueting, R. (1974) (English edn 1980) *New Scarcity and Economic Growth: more welfare through less production?* North-Holland Publishing Co., Amsterdam, New York, Oxford. 269 pp.; Goodland, R. (1999) An appreciation of Dr Roefie Hueting's Ecological Work (Ch. 9: pp. 1-19); van Ierland, E.C. *et al.* (Eds) *Valuation of Nature and Environment.* Royal Netherlands Academy of Science, Amsterdam; Daly, H.E. (1972) In defence of a steady-state economy. *American Journal of Agricultural Economics* **54**(4) pp. 945-954; Daly, H.E. (Ed.) (1973) *Toward a Steady-State Economy.* Freeman, San Francisco. 332 pp.; Daly, H.E. (1974) The economics of steady state. *American Economic Review* (May) pp. 15-21; Daly, H.E. (1999) *Ecological economics and the ecology of economics.* E. Elgar, Cheltenham. 192pp.

4. Constanza, R., J. Cumberland, H. Daly, R. Goodland and R. Norgaard (1997) *An Introduction to Ecological Economics.* St. Lucie Press (CRC Press), Boca Raton, Florida. 275 pp.

5. Hueting, R. and L. Reijnders (1997) *Sustainability is an Objective Concept.* Den Haag, Netherlands Bureau of Statistics. MS 10 pp.

6. Mill, J.S. (1848) Revised edn 1900 *Principles of Political Economy.* Collier, New York. 2 vols.

7. Malthus, T.R. (1798 [1970]) *An Essay on the Principle of Population.* Penguin, Harmandsworth. 291 pp.

8. Ehrlich, P.R. and A. Ehrlich, A. (1989a). Too many rich folks. *Populi* **16**(3) pp. 3-29; Ehrlich, P.R. and A. Ehrlich (1989b) How the rich can save the poor and themselves. *Pacific and Asian Journal of Energy* **3** pp. 53-63; Hardin, G. (1968) The tragedy of the commons. *Science* **162** pp.1243-1248; Ehrlich, P.R. (1997) *The Stork and the Plow: The equity answer to the human dilemma.* Yale University Press. 384 pp.; Ehrlich, P.R. (1998) *Betrayal of Science and Reason: How anti-environment rhetoric threatens our future.* Island Press, Covelo, California. 335pp.; Hardin, G. (1993) *Living within Limits: Ecology, Economics and Population Taboos.* Oxford University Press, New York, 339 pp. The author (Robert Goodland) also calls himself a biocentric Neo-Malthusian.

9. Daly, H.E. (1972) In defense of a steady-state economy. *American Journal of Agricultural Economics* **54**(4) pp. 945-954; Daly, H.E. (Ed.) (1973) *Toward a Steady State Economy.* Freeman, San Francisco 332 pp.; Daly, H.E. (1974) The economics of the steady state. *American Economic Review* (May) pp. 15-21.

10. Daly, H.E. (1977) *Steady-state Economics.* (2nd edn) Island Press, Washington, DC. 302 pp.

11. Ricardo (1817[1973]) *Principles of Political Economy.* Dent, London. 300 pp.

12. World Bank (1992) *World Development Report 1992: Development and the Environment.* Oxford University Press, New York. 308 pp.; World Bank (1997) *World Development Report 1997: The State in a Changing World.* The World Bank, Washington DC. 262 pp.

13. Boulding, K.E. (1966) The economics of the coming spaceship earth, pp. 3-14, in Jarret, H. (Ed.) *Environmental Quality in a Growing Economy.* John Hopkins, Baltimore. 173 pp.; Boulding, K.E. (1968) *Beyond Economics.* University of Michigan, Ann Arbor. 302

pp.; Boulding, K.E. (1973) The economics of the coming spaceship earth, pp. 121-132, in Daly (Ed.) (1977) *Op. cit.* [See Note 10]; Boulding, K.E. (1992) *Towards a New Economics: Ecology and Distribution.* Edward Elgar, Aldershot. 344 pp.; Mishan, E.J. (1967) *The Costs of Economic Growth.* Staples Press, London. 190 pp.; Mishan, E.J. (1977) *The Economic Growth Debate: An Assessment.* Allen and Unwin, London. 277 pp.; Daly, H.E. *Op. cit.* (1972, 1973, 1974, 1999) [See Notes 4 & 10].

14. Toman, M.A. (1994) Economics and sustainability: Balancing trade-offs and imperatives. *Land Economics* **70**(4) pp. 399-413.

15. Beckerman, W. (1974) *In Defense of Economic Growth.* Jonathan Cape, London. 287 pp.; Beckerman, W. (1992). Economic growth: Whose growth? Whose environment? *World Development* **20** pp. 481-492; Beckerman, W. (1994) Sustainable development: Is it a useful concept? *Environmental Values* **3** pp. 191-209; Beckerman, W. (1995) *Small is Stupid: Blowing the Whistle on the Greens.* Duckworth, London. 202 pp.; Daly, H.E. (1995) On Wilfred Beckerman's critique of sustainable development. *Environmental Values* **4** pp. 49-55; El Serafy, S. (1996) In defense of weak sustainability: a response to Beckerman. *Environmental Values* **5** pp. 75 -81.

16. Meadows, D., D. Meadows, J. Randers, and W. Behrens (1972). *The Limits to Growth.* Universe Books, New York 205 pp.; Meadows D. H., D.L. Meadows and J. Randers (1992) *Confronting Global Collapse: Envisioning a Sustainable Future.* Chelsea Green Publ., Post Mills VT. 300 pp.

17. Barney, G.O. (Ed.) (1980) *The Global 2000 Report to the President of the USA.* Penguin, Harmondsworth. 2 vols.

18. Simon, J. and H. Kahn (Eds) (1984) *The Resourceful Earth: a Response to Global 2000.* Blackwell, Oxford. 585 pp.

19. Panayotou, T. (1993) *Green Markets: The Economics of Sustainable Development.* International Center for Economic Growth, San Francisco. 169 pp.; Summers, L. (1992) Summers on sustainable growth. *The Economist,* 30 May, p.191; Fritsch, B., S. Schmidheiny and W. Seifritz (1994) *Towards an Ecologically Sustainable Growth Society: Physical Foundations, Economic Transitions and Political Constraints.* Springer, Berlin. 198 pp.

20. IUCN (1980) *The World Conservation Strategy* IUCN, WWF, UNEP, FAO, UNESCO, Gland,Switzerland. 7 and 54 pp.

21. Clark, W.C. and R.E. Munn (Eds) (1987) *Sustainable Development of the Biosphere.* Cambridge University Press, Cambridge. 491 pp.

22. Daly, H.E. and J. Cobb (1989) *For the Common Good.* Beacon Press, Boston. 492 pp.

23. The growth debate and sustainability are usefully synthesized in Korten, D.C. (1991) Sustainable development *World Policy Journal* (Winter) pp.156-190, and Max-Neef, M. (1995) Economic growth and the quality of life. *Ecological Economics* **15** pp.115-118.

24. World Commission on Conservation and Development (1987) *Our Common Future.* Oxford University Press, Oxford and New York. 383 pp.

25. Hueting, R. (1990) The Brundtland report: A matter of conflicting goals. *Ecological Economics* **2** pp. 109-117.

26. World Commission on Environment and Development (WCED) (1992) *Our Common Future Reconvened.* Centre for Our Common Future, Geneva. 32 pp.

27. Haavelmo, T. and S. Hansen (1992) On the strategy of trying to reduce economic inequality by expanding the scale of human activity (pp. 38-51), and Tinbergen, J. and R. Hueting (1992). GNP and market prices: Wrong signals for sustainable economic success that mask environmental destruction (pp. 52-62) in Goodland, R., H.E. Daly and S. El Serafy (Eds) *Population, Technology, Lifestyle: The Transition to Sustainability.* Island Press, Washington DC. 154 pp.

28. Compare his writings in 1974 with those in 1993. Solow, R. (1974) The economics of resources or the resources of economics. *American Economic Review* **15** pp. 1-14; Solow, R. (1993a) An almost practical step toward sustainability. *Resources Policy* **19** pp. 162-172; Solow, R. (1993b) Sustainability: An economist's perspective (Ch. 10 pp. 179-187)

in Dorfman, R. and N.S. Dorfman (Eds) *Selected Readings in Environmental Economics* Norton, New York. 517 pp. c.f. Daly, H.E. (1999) *Op. cit.* [See Note 3].

29. See, for example, Ahmad, Y., S. El Serafy and E. Lutz, E. (Eds) (1989) *Environmental Accounting for Sustainable Development.* The World Bank, Washington DC. 100 pp.; Serageldin, I., H.E. Daly and R. Goodland (1995) The concept of sustainability in Van Dieren, W. (Ed.). *Taking Nature into Account.* pp. 99-123. Springer Verlag , New York. 332 pp.

30. Goodland, R., H.E. Daly and S. El Serafy (1992) *Op. cit.* [See Note 28].

31. Daly, H.E. (1992) Allocation, distribution and scale: Towards an economics that is efficient, just and sustainable. *Ecological Economics* **6**(3) pp.185-193; and Daly, H.E. (1996) *Beyond Growth: The Economics of Sustainable Development.* Beacon Press, Boston 253 pp.

32. Since the late 1980s there has been a substantial corpus of literature on 'ecological economics'. See Costanza, R. (Ed.) 1991. *Ecological Economics: The Science and Management of Sustainability.* Columbia University Press, New York 525 pp. and Constanza *et al. Op. cit.* (1997). [See Note 4]. These have largely espoused stronger types of sustainability. See: Hueting (1974, 1980) *Op. cit.* [See Note 3]; Collard, D., D.W. Pearce and D. Ulph (Eds) (1987) *Economics, Growth and Sustainable Environments.* St. Martin's Press, New York. 205 pp.; Archibugi, F. and P. Nijkamp (Eds) (1989) *Economy and Ecology: Towards Sustainable Development.* Kluwer, Dordrecht. 348 pp.; Tisdell, C. (1992) *Environmental Economics: Policies for Environmental Management and Sustainable Development.* Edward Elgar, Aldershot. 259 pp.; IIASA (1992) *Science and Sustainability.* International Institute of Applied Systems Analysis, Laxenburg, Vienna. 317 pp.; Barbier, E. (Ed.) (1993) *Economics and Ecology: New Frontiers and Sustainable Development.* Chapman and Hall, London. 205 pp.; Turner, R. (Ed.). (1993) *Sustainable Environmental Economics and Management; Principles and Practice.* Belhaven, London. 389 pp.; Netherlands (1994) *The Environment: Towards a Sustainable Future.* Kluwer Academic, Dordrecht 608 pp.; Jansson, A., M. Hammer, C. Folke and R. Costanza (1994) *Investing in Natural Capital: The Ecological Economics Approach to Sustainability.* Island Press, Washington DC. 504 pp.; Prugh, T., R. Constanza, J. Cumberland, H. Daly, Robert Goodland and R. Norgaard (1999) *Natural Capital and Human Economic Survival.* Lewis Publications, Boca Raton. 180 pp.

33. Ierland, van (1999) *Op. cit.* [See Note 3].

34. Daly and Cobb *Op. cit.* (1989) [See Note 22]; Daly (1999) *Op. cit.* [See Note 3].

35. Goodland and Daly (1996) *Op. cit.* [See Note 1].

36. World Commission on Conservation and Development (1987) *Op. cit.* [See Note 24].

37. Goodland, R. and S. El Serafy (1998) The urgent need to internalize CO_2 emission costs. *Ecological Economics* **27** pp. 79-90.

38. For example, the controversial log exports to Japan from north-western America show the limits of importing and exporting. Goodland, R. and H. E. Daly (1996) If tropical log exports are so perverse, why are there so many? *Ecological Economics* **18**, pp. 189-196.

39. El Serafy (1996) *Op. cit.* [See Note 15]

40. El Serafy (1996) *Op. cit.* [See Note 15]

41. Daly (1999) *Op. cit.* [See Note 3].

42. See, for example El Serafy (1981) Absorptive capacity, the demand for revenue and the supply of petroleum. *Journal of Energy and Development* **7**(1)

43. As many as one third of all species may be tropical beetles, yet there are less than 200 taxonomists in the world who can identify them. Biodiversity sceptics question the utility of beetles. But, because it was not known that beetles pollinated oil palm (it was thought they were wind pollinated), the oil palm industry of Asia lost billions of dollars of profits foregone, until the pollinating beetles (*Eleaedobius camerunikus*) was introduced from Cameroon nearly 100 years after the Asian palm industry began.

Chapter 7 - Economic Fiction

The notes in this chapter are given as footnotes to the text.

Chapter 8 - Information, Culture and Technology

1. T.S. Eliot (1934) from *Choruses from 'The Rock'*.
2. The term was invented by the science fiction writer William Gibson in his novel *Neuromancer* (1984).
3. This is true in principle. However, the legality of many activities on the internet is increasingly being challenged, and the law may eventually prohibit what technology makes so easy.
4. Network Wizards Internet Domain Survey, January 1999. http://www.nw.com/
5. *The Economist*, 17 April 1999, p. 102.
6. *OECD Communications Outlook 1999*.
7. A crucial document in this process was the report of the 'McBride Commission', set up by UNESCO to investigate these problems. The report was called *Many Voices One World* (UNESCO 1980).
8. These issues are interestingly explored in B. D. Loader (Ed.) *The governance of cyberspace*. Routledge, 1997.
9. Judgement in: Sayre v Becket, 1785.
10. *Report of the Committee to consider the law of copyright and design* (1997). Cmnd 6732. HMSO.
11. The distinction between expression and idea is perhaps clearer in the USA than in the UK.
12. *Computers and privacy* (1975). Cmnd 6353. HMSO.

Chapter 9 – Cities as a Key to Sustainability

1. Satterthwaite, David (1999) *The Earthscan Reader in Sustainable Cities*. Earthscan, London.
2. Ward, B. and R. Dubos (1972) *Only One Earth: the care and maintenance of a small planet*. Penguin, London.
3. IUCN (1980) *The World Conservation Strategy*. IUCN, WWF, UNEP, FAO, UNESCO, Gland,Switzerland. 7 and 54 pp.
4. World Commission on Environment and Development (1987) *Our Common Future*. Oxford University Press.
5. Celecia, John F. (1997) *Redefining concepts, challenges and practices of urban sustainability*. European Foundation for the Improvement of Living and Working Conditions, Dublin.
6. Rees, William E. (1992) Ecological footprints and appropriated carrying capacity: what urban economics leaves out. *Environment and Urbanisation* 4(2) pp. 121-130.
7. Rees, William E. (1995) Achieving sustainability: reform or transformation? *Journal of Planning Literature* 9(4) pp. 343-361.
8. Girardet, Herbert (1999) Creating Sustainable Cities. *Schumacher Briefings* 2 Green Books, Dartington.
9. Celecia (1997) *Op. cit.* [See Note 5].
10. Girardet (1999) *Op. cit.* [See Note 8].
11. City of New York (1999) *High performance building guidelines*. Department of Design and Construction, City of New York.

12. London Waste Action (1999) *Towards a Waste Reduction Plan for London*. Report for London Waste Action by M.E.L. Research Limited, Birmingham.

13. Rabinovitch, Jonas (1992) Curitiba: towards sustainable urban development. *Environment and Urbanisation* 4(2).

14. Rees (1995) *Op. cit.* [See Note 7].

15. Newman, Peter (1996) Transport: Reducing automobile dependence. *Environment and Urbanisation* 8(1) pp. 67-92.

16. Rogers, Richard (1997). *Cities for a Small Planet*. Faber and Faber, London.

17. Rogers (1997) *Op. cit.* [See Note 16].

18. Newman (1996) *Op. cit.* [See Note 15].

19. D.E.T.R (1999) *Towards an Urban Renaissance*. Final report of the Urban Task Force, D.E.T.R., London.

20. Elkin, Tim, D. McLaren and M. Hillman (1991) *Reviving the City: towards sustainable urban development*. Friends of the Earth, London.

21. Goode, David (1990) *A Green Renaissance* in D. Gordon (Ed.) *Green Cities: Ecologically sound approaches to urban space*. Black Rose, Montreal.

22. Barker. G. (1987) European approaches to urban wildlife programmes. In: *Integrating man and nature in the metropolitan environment*. Ed. Adams, L.W. and D.L. Leedy. *Proceedings National Symposium on Urban Wildlife*, pp. 183-190. National Institute of Urban Wildlife, Columbia, Maryland; Goode, D.A. (1989) Urban nature conservation in Britain. *Journal of Applied Ecology* 26 pp. 859-873.

23. Goode, David (1998) Integration of nature in urban development. In: *Urban Ecology* Ed. J. Breuste, H. Feldmann and O. Uhlmann. Springer Verlag, Berlin.

24. Goode, David A. and Niall Machin (in press) Biodiversity Action Plans in Santiago, Chile and London, UK. *Proceedings of the Fourth International Symposium on Urban Wildlife Conservation*, University of Tucson.

25. ICLEI (1995) *Towards Sustainable Cities and Towns*. Report of the European Conference on Sustainable Cities and Towns, Aalborg, Denmark, May 1994. International Council for Local Environmental Initiatives, Freiburg.

26. Roseland, Mark (1998) *Toward sustainable communities*. New Society, Gabriola Island.

27. Spirn, Anne W. (1984) *The Granite Garden: Urban nature and human design*. Basic Books, New York.

Chapter 10 - Sustainable Livelihoods

1. The position of minerals is different from that of hydrocarbons. The latter are unrenewable in the fullest sense. Minerals are rather different. They are refined and concentrated for use and generally dispersed after use. In theory, they can be re-concentrated and re-used. Whether this is practical is a matter of available technology and cost.

2. The figure of one million dollars is for capital investment in industrial (manufacturing) jobs. It seems to vary from country to country. Sample estimates (derived from a variety of business sources) are the following: DM2.3 million for Germany; about US$1 million in the US; and US$1.9 million for Japan (perhaps because of the heavy automation in that country). The Treuhand Agency, in charge of privatising East German industry, worked with a figure of DM 1 million direct investment per workplace — a figure which does not include the capital investments in infrastructure needed to support an industrial plant. [See also Chapter 4, p. 61]

Chapter 11 - The Limits to Sustainable Development

1. The phrase was used by Robert Allen (1980) in *How to Save the World*, Kogan Page, London, which summarised *The World Conservation Strategy* [IUCN (1980) *World Conservation Strategy*. IUCN, WWF, UNEP, FAO, UNESCO, Gland, Switzerland]. David Pearce, Anil Markandya and Edward Barbier (1989) in *Blueprint for a Green Economy*, Earthscan, London, claim to be the first to use it in the context of a strategy which integrates economics and the environment. The project of looking for a single strategy, or formula, which would reconcile the potentially conflicting agendas of the environment and economics, developed in part as a reaction against the warnings of resource depletion and pollution by Donella Meadows, Dennis Meadows, Jørgen Randers and William Behrens (1972) in *The Limits to Growth*, Earth Island, London.

2. Popper, Karl (1945) *The Open Society and its Enemies* Vol. II, Chapter 11. Routledge, London.

3. Pearce, *et al.* (1989) *Op. cit.* [see Note 1].

4. This is from the report of the World Commission on Environment and Development (1987) *Our Common Future* (The Bruntland Report), London. The exact words used are, 'Humanity has the ability to make development suitable — to ensure that it meets the needs of the present without compromising the ability of future generations to meet their own needs'.

5. UK Government (1999) *A Better Quality of Life: A Strategy for Sustainable Development for the United Kingdom*. Cmnd 4345. HMSO, London.

6. The claim that utterances that do something, such as casting a spell, are statements despite their lack of a main verb, is justified, if not fully borne out, in J.L. Austin (1962) *How to Do Things with Words*. Oxford.

7. Elkington, John (1997) *Cannibals with Forks: The Triple Bottom Line of 21st Century Business*. Capstone, Oxford.

8. Elkington (1997) *Op. cit.* [See Note 7]

9. See, for example, Ivanhoe, L.F. (1996) Updated Hubbert Curves Analyze World Oil Supply. *World Oil*, November, pp. 91-94; Hooshang Amirahmadi (1996) Oil at the Turn of the Twenty First Century. *Futures* **28**(5) pp. 433-452; Colin Campbell (1997) *The Coming Oil Crisis*. Multi Science, Brentwood, UK; R.W. Bentley (1998) *UK Energy: The Next 5-10 Years*. (Report submitted to the UK Department of Trade & Industry, UK). Department of Cybernetics, The University of Reading. [See also Chapter 4].

10. International Energy Agency (1998) *World Energy Outlook*. OECD, Paris.

11. According to one detailed study, it would require fifty years to build renewable energy systems up to a level equivalent to just 35 percent of present-day energy demand. LTI-Research Group (editor) (1998) *Long-Term Integration of Renewable Energy Sources into the European Energy System*. Physica Verlag, Heidelberg.

12. Estevan, Antonio and Alfonso Sanz, 'Hacia La Reconversión Ecologica del Transporte in España', Centro de Investigación para la Paz, Madrid, quoted (n.d.) in Richard Douthwaite (1996) *Short Circuit*. Green Books, Hartland.

13. Naisbitt, John (1995) *Global Paradox*. Nicholas Brealey, London.

14. von Weizsäcker, Ernst, Amory B. Lovins and L. Hunter Lovins (1997) *Factor Four: Doubling Wealth — Halving Resource Use*. Earthscan, London

15. Keepin, Bill and Gregory Kats (1988) 'Greenhouse Warming: Comparative Analysis of Nuclear Efficiency and Abatement Strategies' *Energy Policy* **16**(6) pp. 538-561.

16. Davidow, William and Michael Malone (1992) *The Virtual Corporation*. Harper Collins, London p. 128.

17. Fleming, David (1997) Tradable Quotas: Using Information Technology to Cap National Carbon Emissions. *European Environment* **7**(5) Sept.-Oct. pp. 139-148; David Fleming (Ed.) (1998) *Proceedings: Domestic Tradable Quotas Workshop, Brussels* 1-2 July 1998. (EUR 18451).

18. Fleming, David (forthcoming, 2000) *The Lean Economy.* (C.f. Richard Norgaard (1998) *Development Betrayed.* Routledge, London; and the contributors to Majid Rahnema with Victoria Bawtree Eds (1997) *The Post-Development Reader.* Zed Books, London).

19. Meadows *et al. Op . cit.* [See Note 1]

20. The International Energy Agency's projections of gas supplies on the basis of business as usual (BAU) suggest that gas production from all OECD sources will peak in 2015, and that the OECD's share of the market will fall from around 45 percent at present to 30 percent, and falling, in 2020. The major sources of supply will be the Transition Economies (the former Soviet Union), the Middle East and South & East Asia, which will, collectively, have a 55 percent share of the market, while China will have only a 2 percent share of the market. This concentration of supply, the declining production from the OECD and the political tensions arising from gas-rich economies bordering a gas-poor China suggest the probability of both price- and geopolitical-instability in the gas market. Moreover, the rise in oil prices means that there will be both a higher demand and higher prices for gas than the BAU projections indicate. This suggests that the peak in OECD gas output should be brought forward to around 2010. After that gas prices will be much higher, and a global peak in production, with severe price volatilities, can be expected to develop during the second decade of the century. See Laherrère, J.H. , A. Perrodon and C.J. Campbell (1996) *The World's Gas Potential,* Geneva: Petroconsultants S.A., and IEA (1998) *Op. cit.* [See Note 11].

21. As suggested by Michio Kaku (1998) *Visions.* Oxford University Press.

Chapter 12 - Contemporary Order, Peace and Conflict

1. Fukuyama, Francis (1997) *The End of Order.* pp. 28-32. London.

2. By analogy with Marx' 'lumpen proletariat' — a 'lower level and emerging élite'.

3. One should not underestimate the malaise that marks the European consciousness in the wake of failures to intervene in a timely way in the earlier stages of the post-Yugoslav debacle and blood bath. The initial Croatian moves against its Serbian citizens that prompted a coalition of frightened Serbs, Greater Serbia protagonists and the command of the Yugoslav federal army led to years of cruelty and bloodshed, dislocation and destruction. All the inhabitants of the former Yugoslavia suffered loss through the wars but the greatest losers were Serbs displaced from Croatia and Bosnian Muslims. In the aftermath Kosovans have undergone enormous displacement and tribulations but, even here, Serbs seem likely to be the eventual greatest losers.

4. One has only to remember the devastation of the 30 Years War (1618-1648) whose psychological desolation is conveyed so powerfully by Brecht in *Mother Courage.*

5. Obviously Germany still worries countries whose historic memories remain painful in the aftermath of conflicts since the 1860s as well as those countries, especially Poland and the Czech Republic, that gained territory from Germany after 1945.

6. In a different way France has found itself in the same position as the smaller states, having realised that its age-long power tussle with Germany has been distracting and ruinous, and whose main aim for a long time has been not to be left out of central decision-making in the Community/Union, and Europe.

7. The United Kingdom, in its sense of a special relationship with the United States, has consistently weakened attempts to create a military unity in Europe. The most recent example came in the bombing of Iraq during 1998, in which the United Kingdom took part but which was opposed by other European countries. However, recent proposals on European defence, put forward by the British, suggest an incipient will to promote greater European defence integration. Also, though the style of the British intervention in the Kosovo war upset several European countries, in the longer run, they may well be glad to have an unusual display of British decisiveness.

8. In the wake of the Soviet Union's break up, ethnic Russians are the biggest minority to have been left in various countries and regions without proper security protection.

9. The Baltic countries and Slovakia face substantial minority problems: Latvia's population is almost half Russian, while Estonia's is about a quarter; Slovakia's population is made up of more than a sixth Hungarian and other groups; and Lithuania's is more than a sixth composed of Russian, Polish and other groups.

10. The military defeat of Iraq, which was a former Western ally against Iran and whose defeat was not accompanied by the overthrow of its regime nor by a resolution of other problems (including an attack on poverty in the Arab world and the Middle East), has been seen in the West as scarcely a definitive victory. Rather, it has been seen as an indication that trouble may come to the West from other unforeseen sources and that military preparedness should be maintained to deal with future uncertainty and trouble.

11. One of the most elementary reforms required is to change the permanent membership of the Security Council, finding places for Brazil, India, and Japan and commuting British and French seats for European Community representation. Moreover, the present veto system needs to be replaced by a less negative system and by a device (as in the EU) for majority voting decisions.

Chapter 13 - *Improving the World's Governance*

1. Maynes, Charles William (1998) The Perils of (and for) an Imperial America. *Foreign Policy* (Summer), p.37.

2. Maynes (1998) *Op. cit.* p. 44. [See Note 1].

3. Halloran, Richard (1996) The Rising East. *Foreign Policy* (Spring).

4. Anon (1997) *Preventing Deadly Conflict.* Report of the Carnegie Commission, New York, p. 6.

5. Cited in *Our Global Neighbourhood* (1995) p. 130. Report of the Commission on Global Governance. Oxford University Press.

6. *New York Times.* 8 June 1998.

7. *The Economist.* 13 June 1998.

8. *The Economist.* (1998) *Op cit.* [See Note 7]..

9. UNDP (1997) *Human Development Report.* New York.

10. UNDP (1997) *Op. cit.* [See Note 9]

11. *The Economist* (1998) *Op. cit.* [See Note 7]

12. UNDP (1997) *Op. cit.* [See Note 9]

13. UNDP (1997) *Op. cit.* [See Note 9]

14. Sachs, Jeffrey (1998) International Economics: Unlocking the Mysteries of Globalization. *Foreign Policy* (Spring), p. 103.

15. *International Herald Tribune.* 26 September 1998.

16. UNDP (1997) *Op. cit.* [See Note 9].

17. Birdsall, Nancy (1998) Life is Unfair: Inequality in the World. *Foreign Policy* (Summer), p. 76.

18. Dale, Reginald (1998) The One-Time G-22 Looks Useful. *International Herald Tribune.* 28 April 1998.

19. The Commission on Global Governance. *Op. cit.* pp. 153-62. [see Note 5].

20. *International Herald Tribune.* 14 July 1998.

21. The Commission on Global Governance. *Op. cit.* [see Note 5].

Chapter 14 - The Dissolution of Law?

1. The nature of autonomous choice is taken to be an unproblematic matter of fact. In reality, its nature is very far from unproblematic. When I choose goods in a supermarket, are my choices not determined by a multitude of cultural and contextual influences of which I am scarcely aware? Does that not render my action heteronomous rather than autonomous? Must an autonomous action then be arbitrary and irrational? Or must it (as Kant would have held) be an action determined not by my subjective desires but by the objective requirements of a moral law? 'Autonomy as obedience' is a notion unlikely to appeal to enthusiasts for modern individualism.

2. Franz Wieacker *History of Private Law in Europe,* translated by Tony Weir (1995) p.282. Clarendon Press, Oxford.

3. Simmonds, N.E. (1998) 'Rights at the Cutting Edge'. In: M.H. Kramer, N.E. Simmonds and H. Steiner, *A Debate Over Rights* pp. 168-171. Oxford: Clarendon Press.

4. Kant, Immanuel, *The Metaphysics of Morals,* translated by Mary Gregor (1991) pp. 555-6. Cambridge University Press, Cambridge.

5. [1901] AC 495.

6. Lord Lindley was familiar with German Kantian jurisprudence. See Simmonds, *Op. cit.* [See Note 3], at p. 171. Kantian jurisprudence exerted a powerful influence upon Victorian treatise writers such as Sir Frederick Pollock, and through them upon the common law.

7. Hohfeld, W.N. (1923) *Fundamental Legal Conceptions as Applied in Judicial Reasoning.* Yale University Press, New Haven.

8. Hegel, G.W.F. *The Philosophy of Right* (s.135), translated by T.M. Knox (1952), Oxford University Press, Oxford and H.B. Nisbet (1991), Cambridge University Press, Cambridge; Schopenhauer, Arthur. *The World as Will and as Representation* (3rd edn 1859), translated by E.F. Payne (1966). Vol.1, p. 528. Dover, New York..

9. Marx and Hegel contrast 'civil society' with 'the state'. Thus, by 'civil society', Marx refers to the institutions of economic production, the market and the family.

10. Marx, Karl, 'On the Jewish Question'. In: *Karl Marx: Early Writings (1975)* p. 234. Penguin, London.

11. *The Ethics of Aristotle* translated by J.A.K. Thomson (1955) p. 145. Penguin, London.

12. Hart, H.L.A. (1961) *The Concept of Law.* Chap. 8. Clarendon Press, Oxford.

13. Karl Marx (1974) Critique of the Gotha Programme. In: *The First International and After* p.347. Penguin, London.

14. Marx (1974) *Op.cit.* p. 344 [Note 13 above].

15. For example, when people seek to demonstrate the lack of moral consensus in modern society, they tend to focus upon certain well-worn examples, such as the debate over abortion. What they fail to see is that such disagreements are only made possible by the existence of extensive background agreement. At a minimum, we need to agree that it is wrong to kill innocent human beings for reasons of personal convenience before we will find much sense or significance in the abortion debate.

16. A point admirably made by Tom Campbell (1996), *The Legal Theory of Ethical Positivism* p. 173 Dartmouth, Aldershot.

17. E.g. The European Convention on Human Rights.

18. Some interesting, albeit misleading, reflections upon the distinction between private and public law are to be found in Weinrib, Ernest (1995) *The Idea of Private Law.* Harvard University Press.

19. Simpson, A.W.B. (1987) *Legal Theory and Legal History.* p. 178. Hambledon Press, London.

20. The law of tort (or 'delict' as it is called in the Roman legal tradition) deals with civil (that is, non-criminal) injuries other than breaches of contract or trust: it thus includes injuries caused by strangers who may be quite remote from the plaintiff.

21. Hart, H.L.A. and Tony Honoré (1985) *Causation in the Law.* 2nd edn Oxford.

Chapter 15 - The Practice of Conservation by Religions

1. Edwards, Jo and Martin Palmer (Eds) (1997) *Holy Ground - the guide to faith and ecology* (Buddhism Section) p. 59. Pilkington Press, Yelverton.
2. Edwards and Palmer (1997) *Op. cit.* pp. 47-48 Jain Statement of 1991. [See Note 1].
3. Anon. (1990) *Orthodoxy and the Ecological Crisis.* The Ecumenical Patriarchate and WWF International.
4. Batchelor, Martine and Kerry Brown (Eds) (1992) *Buddhism and Ecology* p. 31. Cassells, London.
5. Batchelor *et al.* (1992) *Op. cit.* p. 38. [See Note 4].

Chapter 16 - And After All That

1. Hacker, P. M. S. (1996) *Wittgenstein and Analytical Philosophy* p. 123. Blackwell, Oxford.
2. Wittgenstein, L. (1953) *Philosophical Investigations* (translated by G. E.M. Anscombe) §90 p.185. Blackwell, Oxford.
3. Cahoone, L. F. (1996) *From Modernism to Post-modernism.* Blackwell, Oxford.
4. Cahoone (1996) *Op. cit.* pp. 14-15 [See Note 3].
5. Cahoone (1996) *Op. cit.* p. 16 [See Note 3].
6. Austin, J. L. (1962) *Sense and Sensabilia.* Clarendon Press, Oxford.
7. Billig, M. (1987) *Arguing and Thinking.* Cambridge University Press, Cambridge.
8. Gergen, K. J. (1991) *The Saturated Self.* Basic Books, New York.
9. Sokal, A. & J. Bricmont (1998) *Intellectual Impostures.* Profile, London.
10. Hare, R. M. (1972) *The Language of Morals.* Clarendon Press, Oxford.
11. Nowell-Smith, P. H. (1954) *Ethics.* Penguin, Harmondsworth.
12. Hume, D (1740 [1962]) *A Treatise of Human Nature* p. 203. Fontana/Collins, Glasgow.
13. Griffin, J. (1986) *Well-being.* Clarendon Press, Oxford.
14. Boas, F. (1940) *Race, Language and Culture.* Free Press, New York p. 272.
15. Boas (1940) *Op. cit.* p. 273. [See Note 14].
16. Boas (1940) *Op. cit.* p. 275. [See Note 14].
17. Boas (1940) *Op. cit.* p. 277. [See Note 14].
18. Tiger, L. and J. Shepher (1975) *Women in the Kibbutz.* Harcourt, Brace, Jovanovich , New York.
19. Fodor, J. R. (1981) *Representations.* Harvester, Brighton.
20. Geertz, C. (1989) Anti Anti-relativism. In: M. Krausz (Ed.) *Relativism: Interpretation and Confrontation.* Notre Dame University Press, Notre Dame.
21. Wittgenstein, L. (1967) Remarks on Frazer's "Golden Bough". *Synthese* **17** pp. 233-253.
22. Geertz (1989) *Op. cit.* p. 31. [See Note 20].
23. Shusterman, R. (1990) Postmodernist ethics of taste. In: G. Shapiro (Ed.) *After the Future.* SUNY Press, Albany.
24. Rorty, R. (1989) Solidarity or objectivity? in M. Krausz (Ed.) *Relativism: Interpretation and Confrontation.* Notre Dame University Press, Notre Dame.
25. Rorty (1989) *Op. cit.* p. 35. [See Note 24].
26. Rorty (1989) *Op. cit.* p. 37. [See Note 24].
27. Rorty (1989) *Op. cit.* p. 43. [See Note 24].
28. Geertz (1989) *Op. cit.* [See Note 20].
29. Rorty (1989) *Op. cit.* p. 43 [See Note 24].

30. Habermas, J. (1984-1987) *Theory of Communicative Action* (translated by T. McCarthy. Heinemann, London.

31. Holiday, A. (1988) *Moral Powers.* Harvester, Hassocks.

32. Aronson, J. L., R. Harré, & E.C. Way (1997) *Realism Rescued.* Duckworth, London; J. Lehman, quoted in Maddox, J. (1997) *What Remains to be Discovered.* Macmillan, London.

33. Appleyard, B. (1993) *Understanding the Present: Science and the Soul of Man.* Picador, London.

34. Dawkins, R. (1998) *Unweaving the Rainbow: Science, Delusion and the Appetite for Wonder.* Allen Lane, The Penguin Press, London.

35. Wilson, E.O. (1998) *Consilience: the Unity of Knowledge.* Knopf, New York.

36. Wilson (1998) *Op. cit.* [See Note 35].

37. Poincaré, H. (1905) *Science and Hypothesis.* Macmillan, London.

38. Goodman, N. (1989) "Just the facts, Ma'm". In: M. Krausz (Ed.) *Relativism: Interpretation and Confrontation.* Notre Dame University Press, Notre Dame.

39. Wittgenstein (1953) *Op. cit.* §241. [See Note 2].

40. Lehman, J. quoted in Murphy, N. (1997) *Anglo-American Postmodernity.* Routledge & Kegan Paul, London.

41. Gooding D.G. (1990) *Experiment and the Making of Meaning.* Kluwer, Dordrecht.

Chapter 17 - Some Thoughts on Education

No notes

Chapter 18 - Eddies in the Flow: towards a universal ecology

1. Haeckel, Ernst (1866) *Die generelle Morphologie der Organismen.* Reiner, Berlin.

2. Tansley, A. G. (1914) Presidential address to the British Ecological Society. *Journal of Ecology* 2 p. 196.

3. Odum, Eugene P. (1953) *Fundamentals of Ecology.* Saunders, Philadelphia.

4. Quinn, James A. (1950) *Human Ecology.* Prentice Hall, New York.

5. Bateson, Gregory (1972) *Steps to an Ecology of Mind.* Chandler, San Francisco.

6. Huxley, Aldous (1978) *The Human Situation.* Chatto and Windus.

7. Huxley, Aldous (1970) The politics of ecology [1963] *Environmental Quality Magazine* 1(1).

8. Carson, Rachel (1962) *Silent Spring.* Houghton Mifflin, Cambridge Mass.

9. Uexküll, Jakob von (1909) *Umwelt und Innenwelt der Tiere.* Springer, Berlin.

10. Damasio, Anthony R. (1994) *Descartes' error: emotion. reason and the human brain.* Putnam, New York.

11. See, for example, Lakoff, George (1987) *Women, fire and dangerous things: what categories reveal about the mind.* Chicago University Press.

12. Feinberg, Gerald and Shapiro, R. (1980) *Life beyond earth.* Morrow, New York.

13. Hoyle, F. and N. C. Wickramasinghe (1978) *Lifecloud: the origin of life in the universe.* Dent, London.

14. Capra, Fritjof (1982) *The turning point.* Simon and Schuster, New York

15. See, for example: Dawkins, Marion (1993) *Through our eyes only?* Freeman Oxford, Durham; and Walker, Stephen (1983) *Animal thought*. Routledge & Kegan Paul, London.

16. Pendergrast, Mark (1995) *Victims of memory: incest accusations and shattered lives.* Upper Access Inc., Hinesberg, Vermont.

17. See, for example, Kauffman, Stuart A. (1993) *The origins of order: self-organization and selection in evolution.* Oxford University Press.

18. See particularly: Clark, Andy (1997) *Being there: putting brain, body and world together again.* MIT Press; Damasio, *Op. cit.* [Note 10]; Edelman, Gerald (1994) *Bright air, brilliant fire.* Allen Lane, Harmondsworth; and Varela, Francisco J., Evan Thompson, and Eleanor Rosch (1991) *The embodied mind.* MIT Press.

19. See, for example: Durham, William (1991) *Coevolution: genes, culture and human diversity.* Stamford University Press; and Barkow, Jerome H., L. Cosmides and J. Tooby (1992) *The adapted mind: evolutionary psychology and the generation of culture.* Oxford University Press.

20. Vickers, Sir Geoffrey (1968) *Value systems and social process.* Basic Books, New York.

21. See, for example: Stewart, P. J. in: Jones, Eric and Vernon Reynolds (Eds) (1995) *Survival and religion.* Wiley, Chichester

22. See Stewart, P.J. (1994) *Unfolding Islam.* Garnet, Reading.

23. Callicott, J. Baird (1994) *Earth's Insights.* University of California Press.

24. Reynolds, Vernon and Ralph Tanner (1995) *The social ecology of religion.* Oxford University Press.

25. See, for example: Lovejoy, Arthur O. (1930) *The revolt against dualism.* Open Court, La Salle, Illinois; and, more recently, Capra, Fritjof (1982) *The Turning Point.* Simon and Shuster, New York.

Chapter 19 - Where Next?

1. Many 'gloom and doom' commentators talk about the possible extinction of the human race. I think this extremely unlikely. More possible is a period of great turmoil and discomfort from which the most adaptable emerge into a transformed world (with carrying capacity reduced for centuries and most big mammals extinct).

2. Ideas leading toward such proposals are developed in Fleming, David (1998) After Growth — Climax: Rising Unemployment as the Cue for Evolution to The Lean Economy. *European Environment* **8**(2). John Wiley & Sons. [See also Chapter 11, Note 18].

3. There are, of course, other economists who take a more optimistic view.

4. Although it has become accepted 'talk' that the environmental costs of any activity should be internalised under 'the polluter plays principle', this is seldom applied in practice. It seems illogical that the same principle is apparently not considered, even in theory, to be applicable to social costs, such as those of throwing people out of work in the interests of greater efficiency. If both of these were to be comprehensively applied, the economic scene would look very different.

5. This is partially based upon material in: British Medical Association (1998) *Human Genetics: Choice and responsibility.* Oxford University Press.

6. The useful term 'outer limits' was first brought into circulation by the United Nations Environment Programme (UNEP). It embraces climate, energy, space (the area of land and water), soil, water, minerals, and biological diversity.

7. The problems posed by environmental limits, serious enough in their own right, are likely to be gravely exacerbated by changes in climate; indeed these changes are considered by an influential body of scientists to be the most serious challenge presently facing mankind. World climate has always fluctuated 'naturally' and will continue to do so, but there now seems no doubt that human activities are superimposing significant man-induced changes upon the natural pattern and that these will have very different

effects in different parts of the globe.

Plants and animals, and their ecosystems, respond to fluctuations of climate by attempting to migrate, keeping always within the range of habitat conditions suitable for them. The result is sometimes a shuffling of the cards to form new ecosystems; sometimes, when no suitable migration route is on offer, to extinction. Human beings have also responded by migration.

But the effect of any present or future climate change will be much more drastic than that of past changes. The migration of 'natural' ecosystems — many already mpoverished by loss of species — will be difficult or impossible, because they are now broken into fragments, separated by cities, roads, fields and plantations. Moreover, human societies and their artefacts will also be under stress: crops and tree plantations being overtaken by conditions for which they are not fitted; cities built for climates to which they are ill-adapted; and low-lying lands threatened by rising sea level. This is a complex of menaces which can only be countered by a combination of the best scientific knowledge, an accurate and unbiased presentation of the facts to the public, and firm, far-sighted political action.

8. Fleming, David (1998 & 2000) *Op. cit.* [See Note 2].

9. IIED and IUCN published in 1998 a series of policy studies (*Policy that works for forests and people*) covering Pakistan, Papua New Guinea, Ghana, Zimbabwe and Costa Rica. A study of India will complete the series.

10. Jones, Steve (1993) *The Language of the Genes.* Harper Collins.

11. Huxley, Julian (Ed.) (1961) *The Humanist Frame.* George Allen and Unwin, London.

12. An important question here is the extent to which this might be through the family (extended parental education being one of the more important advances of the human species over its predecessors) or through institutions of the state — which might be interpreted as an advanced development of the extended family!

13. One technique, at least, should be added to this list to meet the tasks of the present-day world — the ability to use a keyboard!

Acronyms

AIDS	Acquired Immuno-Deficiency Syndrome
ARC	Alliance of Religions and Conservation
ARPANET	(Defence) Advanced Research Projects Agency Network (USA)
BRAC	Bangladesh Rural Action Committee
CAP	Common Agricultural Policy [of the European Union]
CERN	Conseil Européen pour Recherches Nucleaires
CFC	Chlorofluorocarbon
CGIAR	Consultative Group on International Agricultural Research
CIAT	Centro Internacional de Agricultura Tropical (CGIAR), Colombia
CIMMYT	Centro Internacional de Mejoramiento de Maíz y Trigo (CGIAR), Mexico
ECOSOC	Economic and Social Council of the United Nations
EOLSS	Encylopaedia of Life Support Systems
EXPO 2000	The World Exposition, Hanover, 2000
ES	Environmental sustainability
FAO	Food and Agriculture Organization (of the United Nations)
FUFU	Fully Financed Family Units
G7	Group of 7 (wealthy industrialised countries)
GDP	Gross Domestic Product
GM	Genetically Modified
HIPC	Heavily Indebted Poor Countries
HIV	Human Immuno-Deficiency Virus
HYV	High Yielding Varieties
ICSU	International Council of Scientific Unions
IEA	International Energy Agency
IFAD	International Fund for Agricultural Development
IIED	International Institute for Environment and Development
IIMI	International Irrigation Management Institute
IISD	International Institute for Sustainable Development
IITA	International Institute for Tropical Agriculture
INF	Intermediate range Nuclear Forces
IRDC	International Rural Development Centre
IRRI	International Rice Research Organization
ISNAR	International Service for National Agricultural Research (CGIAR)
IUCN	International Union for Conservation of Nature and Natural Resources (World Conservation Union)
IWMI	International Irrigation Management Institute
JANET	Joint Academic Network (UK)
MAB	Man and the Biosphere Programme (of UNESCO)
NARS	National Agricultural Research Service
NPT	Non-Proliferation Treaty
NWICO	New World Information and Communication Order
OECD	Organization for Economic Cooperation and Development
OEM	Original Equipment Manufacturer
OPEC	Organisation of Petroleum Exporting Countries

PC	Political Correctness
PEP	Political and Economic Planning
SEI	Stockholm Environment Institute
SEWA	Self Employed Women's Association
START	Strategic Arms Limitation Talks
TAC	Technical Advisory Committee
TARA	Technology and Action for Rural Advancement
TCP/IP	Transmission Control Protocol/Internet Protocol
TFR	Total Fertility Rate
UFU	Unfinanced Family Units
UNCHE	United Nations Conference on the Human Environment
UNDP	United Nations Development Programme
UNEP	United Nations Environment Programme
UNESCO	United Nations Education, Scientific and Cultural Organisation
UNFPA	United Nations Fund for Population Activities; name changed to United Nations Population Fund
USAID	United States Agency for International Development
WAPDA	Water and Power Development Authority (of Pakistan)
WWF	World Wide Fund for Nature
WWF	Working Women's Forum (India)

Index